Foundations of Electronics, 3rd Edition: Laboratory Projects

by Russell L. Meade

Delmar Publishers

an International Thomson Publishing company I(T)P®

Albany • Bonn • Boston • Cincinnati • Detroit • London • Madrid
Melbourne • Mexico City • New York • Pacific Grove • Paris • San Francisco
Singapore • Tokyo • Toronto • Washington

NOTICE TO THE READER

Publisher does not warrant or guarantee any of the products described herein or perform any independent analysis in connection with any of the product information contained herein. Publisher does not assume, and expressly disclaims, any obligation to obtain and include information other than that provided to it by the manufacturer.

The reader is expressly warned to consider and adopt all safety precautions that might be indicated by the activities herein and to avoid all potential hazards. By following the instructions contained herein, the reader willingly assumes all risks in connection with such instructions.

The Publisher makes no representation or warranties of any kind, including but not limited to, the warranties of fitness for particular purpose or merchantability, nor are any such representations implied with respect to the material set forth herein, and the publisher takes no responsibility with respect to such material. The publisher shall not be liable for any special, consequential, or exemplary damages resulting, in whole or part, from the readers' use of, or reliance upon, this material.

Delmar Staff:
Publisher: Michael McDermott
Acquisitions Editor: Gregory Clayton
Developmental Editor: Michelle Ruelos Cannistraci
Senior Project Editor: Christopher Chien
Production Manager: Larry Main
Art and Design Coordinator: Nicole Reamer

Printed in the United States of America
2 3 4 5 6 7 8 9 10 XXX 03 02 01 00 99

For more information, contact Delmar, 3 Columbia Circle, PO Box 15015, Albany, NY 12212-0515; or find us on the World Wide Web at http://www.delmar.com

International Division List

Japan:
Thomson Learning
Palaceside Building 5F
1-1-1 Hitotsubashi, Chiyoda-ku
Tokyo 100 0003 Japan
Tel: 813 5218 6544
Fax: 813 5218 6551

Australia/New Zealand
Nelson/Thomson Learning
102 Dodds Street
South Melbourne, Victoria 3205
Australia
Tel: 61 39 685 4111
Fax: 61 39 685 4199

UK/Europe/Middle East:
Thomson Learning
Berkshire House
168-173 High Holborn
London
WC1V 7AA United Kingdom
Tel: 44 171 497 1422
Fax: 44 171 497 1426

Latin America:
Thomson Learning
Seneca, 53
Colonia Polanco
11560 Mexico D.F. Mexico
Tel: 525-281-2906
Fax: 525-281-2656

Canada:
Nelson/Thomson Learning
1120 Birchmount Road
Scarborough, Ontario
Canada M1K 5G4
Tel: 416-752-9100
Fax: 416-752-8102

Asia:
Thomson Learning
60 Albert Street, #15-01
Albert Complex
Singapore 189969
Tel: 65 336 6411
Fax: 65 336 7411

Spain:
Thomson Learning
Calle Magallanes, 25
28015-MADRID
ESPANA
Tel: 34 91 446 33 50
Fax: 34 91 445 62 18

Library of Congress Cataloging-in-Publication Data: 98-41278
ISBN 0-7668-0430-5

CONTENTS

PREFACE

The Purposes of this Laboratory Manual

- To confirm and reinforce theory concepts without repetitious busy work.
- To provide hands-on experience in connecting circuits from schematics, making measurements, and analyzing observations.
- To improve critical thinking skills.

Features of this Manual's Design

- Short, stand-alone projects allow flexible scheduling options. You can use this manual in individualized, self-paced programs, or in traditional, group-delivery systems.
- Projects are directly correlated with the companion textbooks, *Foundations of Electronics, 3rd edition*, and *Foundations of Electronics: Circuits and Devices 3rd edition*, and can be used to supplement any introductory dc/ac text.
- Schematic and/or pictorial diagrams *always* appear directly across from the procedural steps that they illustrate.
- The unique, three-column format of this manual reinforces key concepts as the student performs each project. Active feedback during the procedure replaces "cookbook" steps, and ensures that students will not wait until the end of the project to analyze their results.

To Gain Maximum Rewards from Using this Manual

- Read "How to Use This Lab Manual" in the Introduction that follows this Preface.
- Do not be concerned if you don't know what exact word is called for in any of the "Conclusion Column" blanks. The important thing is that you understand the concept being highlighted and can give a word or words that convey your understanding.
- Your rewards will be directly related to your effort.

We wish you great success in your experiences and trust that this guide will be both interesting and meaningful as you prepare to enter the world of electronic technology.

INTRODUCTION

General Information

This student laboratory manual is designed as a companion training guide to be used with ANY electronics fundamentals text. To aid those who will use this manual as an adjunct to *Foundations of Electronics, 3rd Edition*, or *Foundations of Electronics: Circuits and Devices*, you will find a convenient tool called "Project/Topic Correlation Information" at the beginning of each group of projects. The correlation charts relate projects in the manual to topics in the main text.

How to Use this Lab Manual

General Information

The lab manual is laid out in three related columns: Activity, Observation, and Conclusion. To get the maximum benefit from each project, you should use this design feature properly. The following example will help you see the significance of each column's information and understand the interrelationships between these three columns.

Example

ACTIVITY	OBSERVATION	CONCLUSION
1. The purpose of the Activity column is to convey procedural instructions such as: connect; apply; measure; calculate; and so on. After performing the procedural instructions for a given step, you record any results in the Observation column. →	The purpose of the Observation column is to record observations, measurements, and/or results of calculations for each step. After recording this data for the related procedural step(s), you move to the Conclusion column. →	The purpose of the Conclusion column is to trigger your thinking and to cause you to analyze key concepts that should be evident as they relate to what you have just done and observed for the step, or related steps involved. After finishing the Conclusion column up to a horizontal step-divider line, you should move on to the next Activity step.
2. Next Activity step. ←		

Sample from Project in Manual

Note the following things about the sample taken from Project 22 in this manual:

- The general flow through each project is to move from left to right, from Activity to Observation to Conclusion, until you reach a horizontal step-divider line. Then you move on to the next Activity step and continue in the same left-to-right fashion until the next divider line is reached, and so on. For example:
- *Step 1* has a connect-the-circuit activity, but requires nothing to be entered in the Observation or Conclusion columns.
- *Step 2* has procedural activities of apply, measure, and calculate. The results of the measuring and calculating are then entered into the Observation column blanks, as appropriate. Based upon the activities and observations for step 2, you are then asked to do some analysis and fill in some blanks in the Conclusion column, related to step 2.
- The horizontal step(s) divider line indicates the point at which you are to move on to the next Activity step, in this case step 3. You then proceed through the related observations and conclusions for step 3, as you did in the preceding step(s).
- This same process of working from Activity to Observation to Conclusion column, until you reach a horizontal divider line, is continued throughout each project. When you reach a divider line in the Conclusion column, you move on to the next step(s) in the Activity column, and so on.

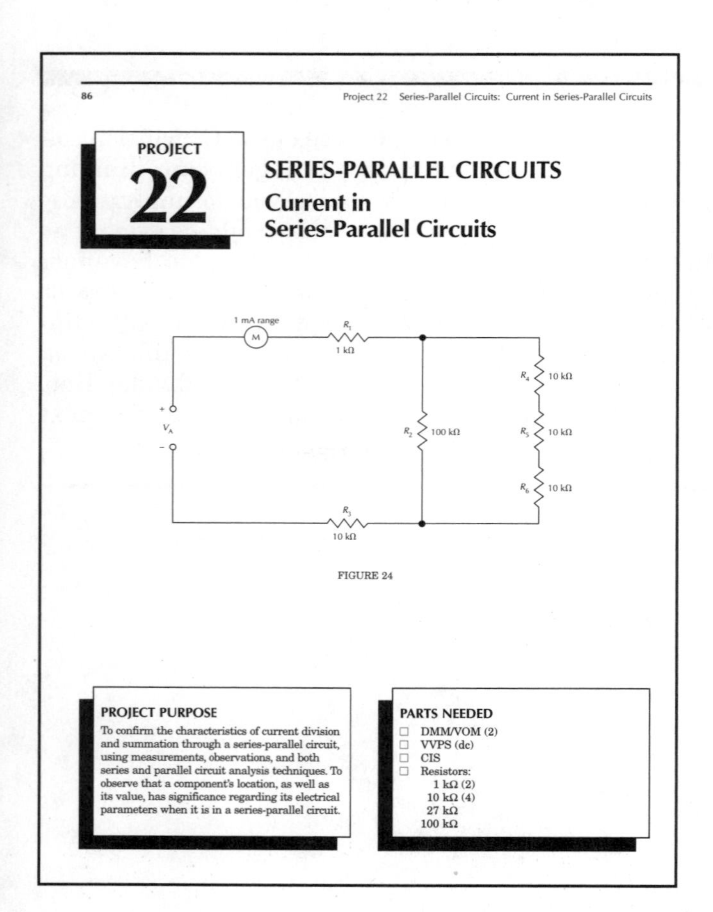

86 Project 22 Series-Parallel Circuits: Current in Series-Parallel Circuits

PROJECT 22

SERIES-PARALLEL CIRCUITS
Current in Series-Parallel Circuits

FIGURE 24

PROJECT PURPOSE
To confirm the characteristics of current division and summation through a series-parallel circuit, using measurements, observations, and both series and parallel circuit analysis techniques. To observe that a component's location, as well as its value, has significance regarding its electrical parameters when it is in a series-parallel circuit.

PARTS NEEDED
- ☐ DMM/VOM (2)
- ☐ VVPS (dc)
- ☐ CIS
- ☐ Resistors:
  - 1 kΩ (2)
  - 10 kΩ (4)
  - 27 kΩ
  - 100 kΩ

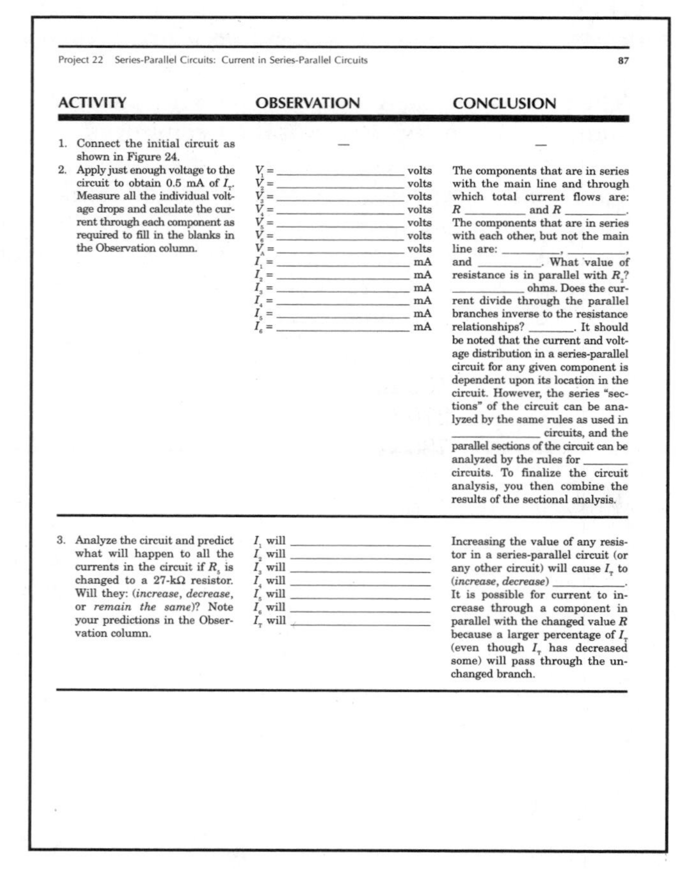

Project 22 Series-Parallel Circuits: Current in Series-Parallel Circuits 87

ACTIVITY	OBSERVATION	CONCLUSION
1. Connect the initial circuit as shown in Figure 24.	—	—
2. Apply just enough voltage to the circuit to obtain 0.5 mA of I_T. Measure all the individual voltage drops and calculate the current through each component as required to fill in the blanks in the Observation column.	V_1 = ________ volts V_2 = ________ volts V_3 = ________ volts V_4 = ________ volts V_5 = ________ volts V_6 = ________ volts V_A = ________ volts I_1 = ________ mA I_2 = ________ mA I_3 = ________ mA I_4 = ________ mA I_5 = ________ mA I_6 = ________ mA	The components that are in series with the main line and through which total current flows are: R ________ and R ________. The components that are in series with each other, but not the main line are: ________, ________, and ________. What value of resistance is in parallel with R_2? ________ ohms. Does the current divide through the parallel branches inverse to the resistance relationships? ________. It should be noted that the current and voltage distribution in a series-parallel circuit for any given component is dependent upon its location in the circuit. However, the series "sections" of the circuit can be analyzed by the same rules as used in ________ circuits, and the parallel sections of the circuit can be analyzed by the rules for ________ circuits. To finalize the circuit analysis, you then combine the results of the sectional analysis.
3. Analyze the circuit and predict what will happen to all the currents in the circuit if R_5 is changed to a 27-kΩ resistor. Will they: (*increase*, *decrease*, or *remain the same*)? Note your predictions in the Observation column.	I_1 will ________ I_2 will ________ I_3 will ________ I_4 will ________ I_5 will ________ I_6 will ________ I_T will ________	Increasing the value of any resistor in a series-parallel circuit (or any other circuit) will cause I_T to (*increase*, *decrease*) ________. It is possible for current to increase through a component in parallel with the changed value R because a larger percentage of I_T (even though I_T has decreased some) will pass through the unchanged branch.

Specific Notations for the User

Many people who use this manual will be using digital multimeters (DMMs) to make measurements. Others will use volt-ohm-milliammeters (VOMs). Either type instrument may be used; however, users should be aware that there will be slight differences in readings, on certain projects. Measurement results are affected by both the component values and characteristics of the circuit being tested and the type of instrument used to make the measurements. DMMs typically have input resistances of 10 megohms or more. VOMs generally have input resistances that are much lower than this. This means that the VOM will cause more meter loading effects on the circuit(s) being tested. For example, when meter loading effects are a factor, the VOM will typically read lower voltage values than the DMM. Instructors should take these minor differences into account when checking student measurements throughout the manual.

Equipment

The general types of equipment required to perform the project experiments include:

1. Component interconnection system (CIS) (proto boards or similar connection matrices)
2. Variable voltage dc power supply (VVPS)
3. DMMs and/or VOMs (two per trainee position)
4. Function generator
5. Oscilloscope (dual-trace, if possible)
6. Digital frequency counter (optional)

Component Parts

The component parts required in order to perform all the projects in this manual include:

For Projects 1–64:

Resistors

Resistor decade box
Kit of 10 various color-coded values
1-watt resistors:
- 100 Ω (2)
- 270 Ω
- 780 Ω
- 820 Ω
- 1 kΩ (3)
- 3.3 kΩ
- 4.7 kΩ
- 5.6 kΩ
- 10 kΩ (4)
- 12 kΩ
- 18 kΩ
- 27 kΩ
- 47 kΩ
- 100 kΩ (2)
- 1 MΩ
- 10 MΩ

Potentiometer:
- 100 kΩ

Inductors

One inductor with very high *L/R* ratio
1.5 H, 95 Ω (dc) (2) or appropriate equivalents
2.5 mH (RF choke)

Capacitors

0.1 μF
1.0 μF (2)

Transformers

12.6 V (w/center-tapped secondary)
Standard power transformer
Audio output transformer
Isolation transformer

Miscellaneous Items

Dry cell, 1.5 V
Speaker, 4 Ω
Meter:
- 0–1 mA (dc) panel meter

For (Devices) Projects 65–89:

Resistors

47 Ω
100 Ω
220 Ω
270 Ω
330 Ω
470 Ω
1 kΩ (2)
2.7 kΩ
4.7 kΩ
10 kΩ (4)
22 kΩ
27 kΩ
47 kΩ
100 kΩ (2)
120 kΩ
270 kΩ
470 kΩ
1 MΩ
1 kΩ linear potentiometer

Diodes

1N4002
Silicon, 1 Amp rating (4)
Zener: 5.1 V, 1 W 1N4733, or equiv.
LED, gal/arsen., red, 20 mA rating

Capacitors

0.01 μF (2)
0.1 μF (2)
1.0 μF
10 μF (2)

Transistors/Semiconductors

NPN, silicon, 2N3904 or equiv.
N-channel JFET, 2N5458 or equiv.
OP-AMP, 741
555 Timer IC
SCR, 2N5060, or equiv.

Miscellaneous Items

1.5-V cell
6-V incandescent lamp (#47)
±15 Vdc power supply
6.3-Vac source

BREADBOARDING CIRCUITS INFORMATION

To "breadboard" a circuit requires properly interconnecting components, jumpers, and electrical devices in such a way as to conform to the connections of a specified circuit. This "breadboarded" circuit then allows for practical measurements and analysis of how the circuit specified in a schematic or other diagram actually operates.

Sample Circuit Board Matrix

Although there are a number of types of circuit connection devices available, a typical type is shown below. The board is made of plastic with a matrix of holes, into which wires and component leads are pushed to make appropriate connections. Refer to the figure below as you study the following points regarding this type circuit board.

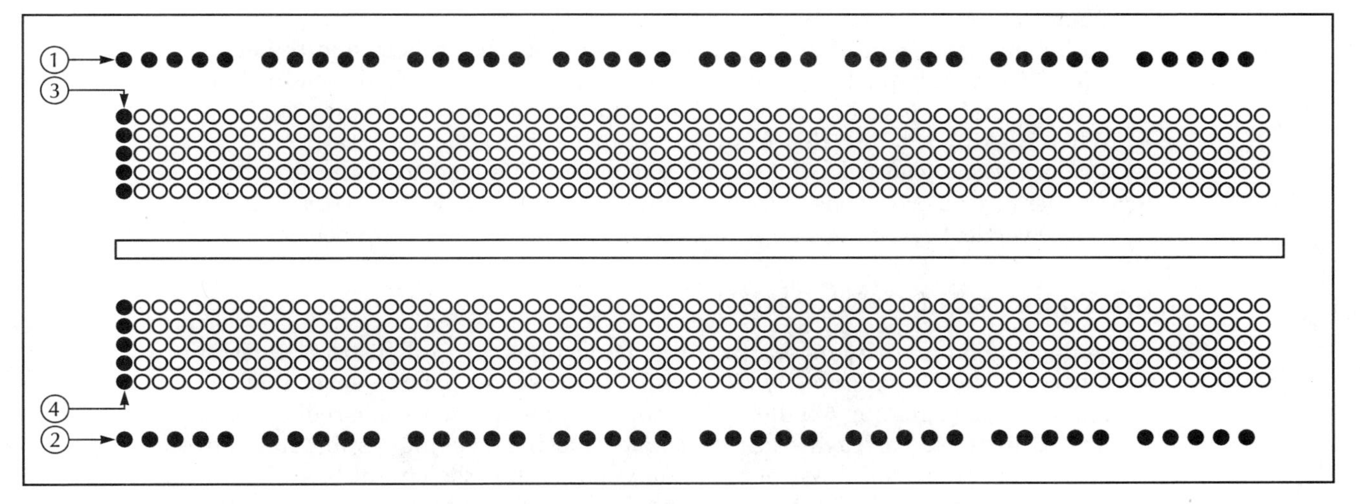

- Each "hole" on the board contains a metallic spring contact. This means that an electrical wire, or component lead pushed down into the hole is making electrical contact with that hole's spring contact.
- The circuit board provides automatic "interconnection" between select holes on the circuit board via metallic "bus" connections made underneath the holes. The matrix of holes is internally interconnected so that:

1. The eight horizontal "groups of five holes" along the top are connected in common.
2. The eight horizontal "groups of five holes" along the bottom are connected in common.
3. Each vertical column of five holes (above the center separator groove) are interconnected; however, each of these sets of five holes is isolated from the horizontally adjacent set of five holes, as well as being isolated from all the holes below the board's center separator groove.
4. Each vertical column of five holes (below the center separator groove) are interconnected; however, each of these five holes is isolated from the horizontally adjacent set of five holes, as well as being isolated from all the holes above the board's center separator groove.

General Comments

Because of the typical circuit board layout and "built-in" interconnections, the following techniques are quite commonly used:

- When connecting power supply or signal source leads to the circuit board matrix:
  - Connect one source lead to one of the top horizontal row (common bus) holes.
  - Connect the other source lead to a hole in the bottom horizontal row (common bus).
  - Insert a jumper lead from each source connection row to the appropriate points in the experimental circuit constructed on the vertical columns portion of the circuit board.

  NOTE: If using alligator clip leads from the source; connections can be made directly to the experimental circuit's component leads, as appropriate, without having to connect the source's leads to the outside rows, then jumpers to connect to the experimental circuit source input points.
- When connecting component leads:
  - Plug one lead from a component into a vertical column hole, and the other lead from that particular component into another vertical column hole (horizontally spaced as convenient for the size of the component).
- When making a connection from one component to the next in a circuit:
  - Connect one lead from the second component to a lead from the first component by inserting one of its leads into an adjacent hole in the *same* (group of 5) vertical column group as the first component's lead is connected.

Sample Circuit Setups

The following sample circuit setups (interconnections) are taken from illustrations used in the *Foundations of Electronics* text used to portray several of the "Chapter Troubleshooting Challenge" circuits. Be aware that the circuit boards illustrated in these portrayals use the same "bussing" techniques for interconnections we have been discussing. (You may notice that the circuit boards used in these illustrations have two horizontal rows of holes at the top and the bottom, rather than just one. The interconnection concepts, however, are the same as we have been discussing.)

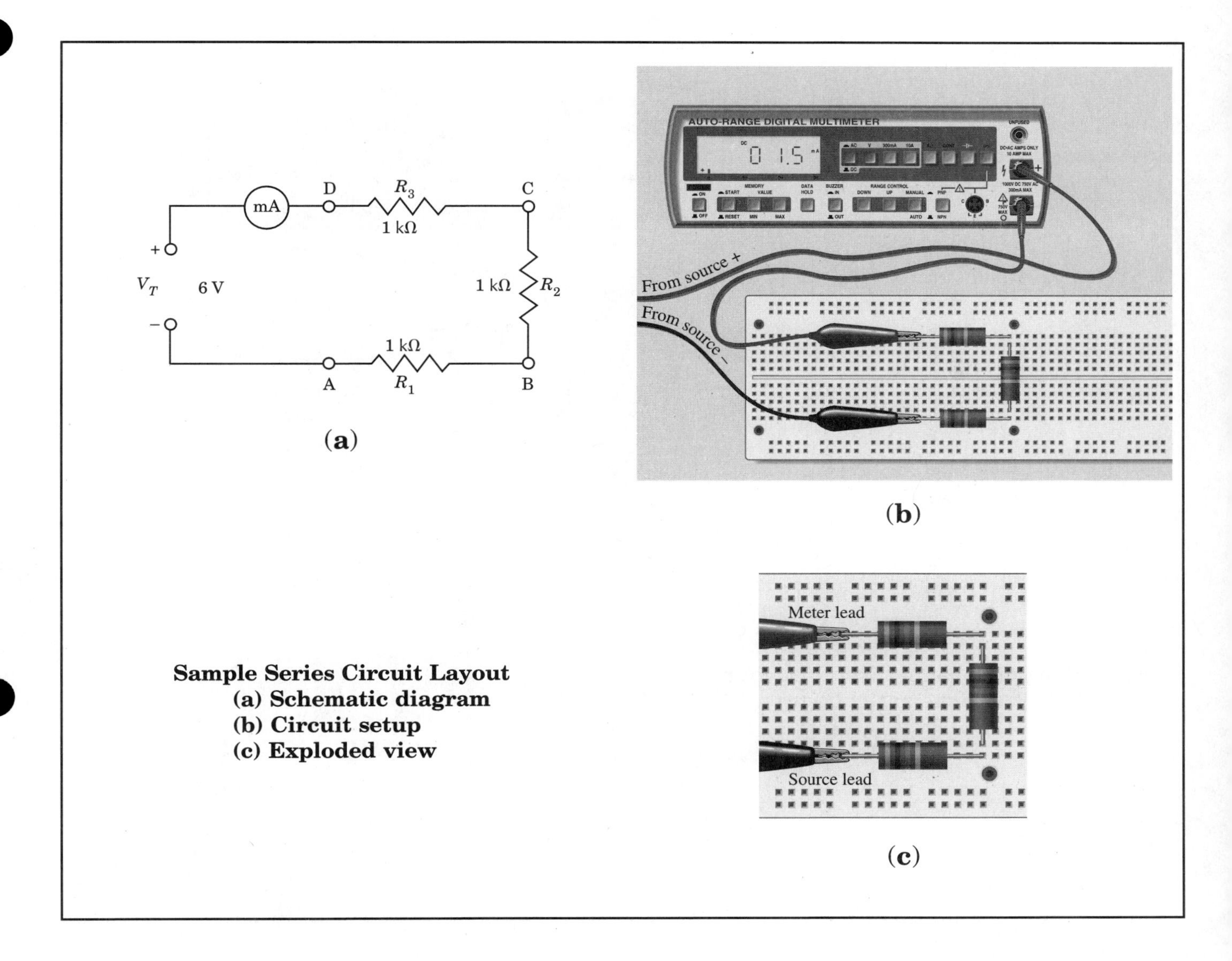

Sample Series Circuit Layout
(a) Schematic diagram
(b) Circuit setup
(c) Exploded view

NOTE: When using the alligator clip leads from the source and meter, as used in the illustration here, we are not connecting the source and meter leads to the "outside" horizontal bus rows, then using jumpers from those rows to connect to the circuit points. If alligator clip leads were not used, it would be typical to connect the source leads to the outside horizontal bus points, then, use a jumper wire from there to the experimental circuit points, as appropriate.

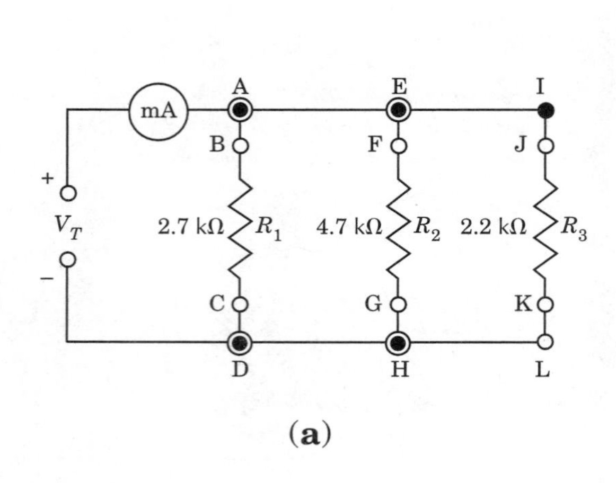

(a)

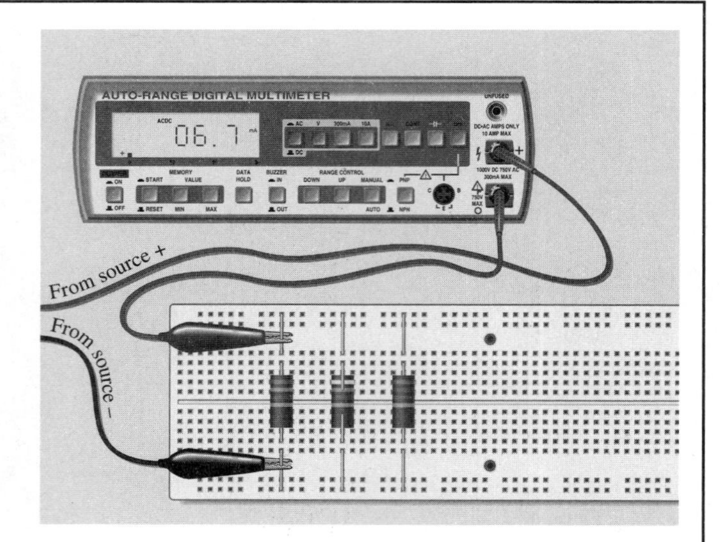

(b)

(c)

Sample Parallel Circuit Layout
(a) Schematic diagram
(b) Circuit setup
(c) Exploded view

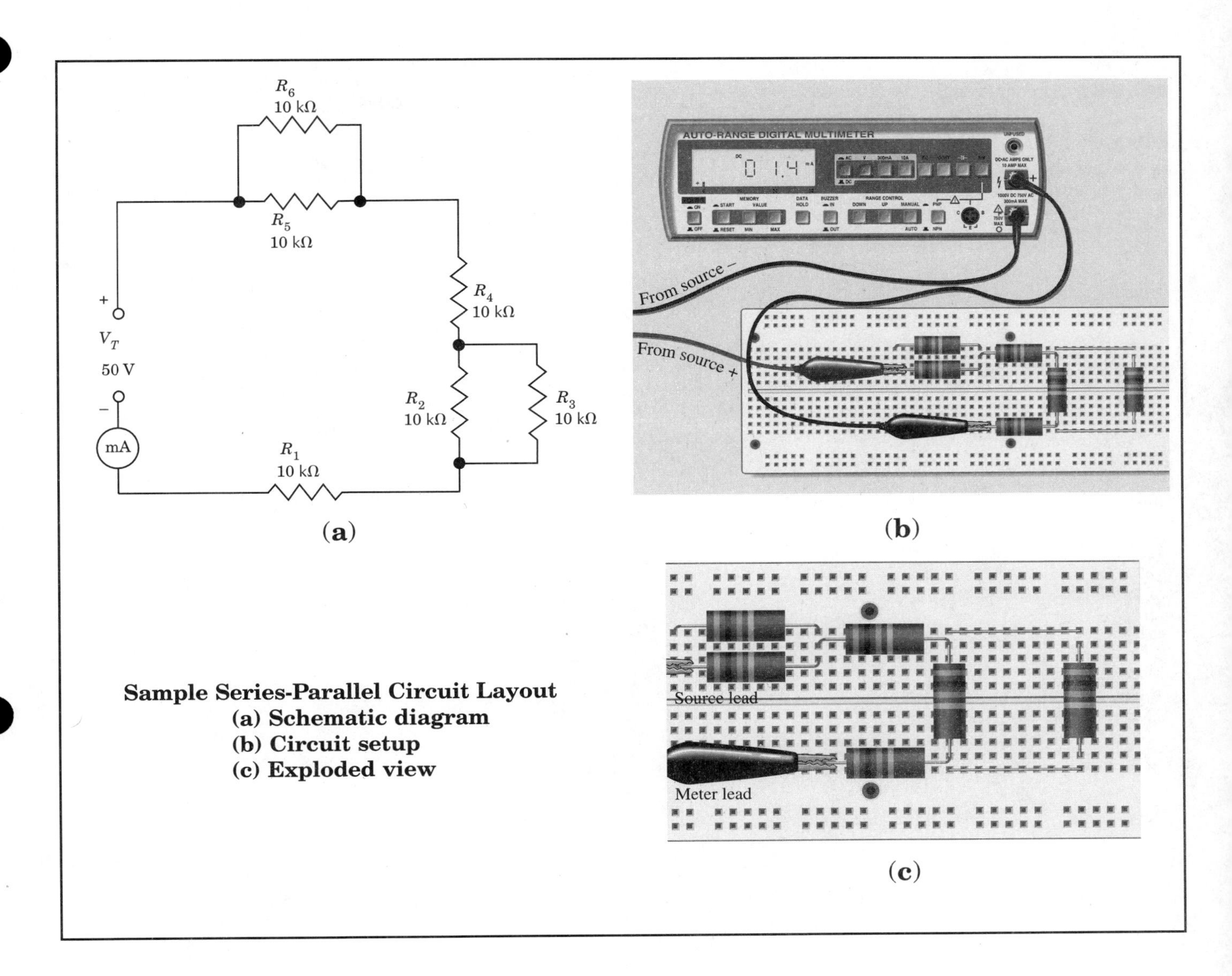

Sample Series-Parallel Circuit Layout
(a) Schematic diagram
(b) Circuit setup
(c) Exploded view

NOTE: Once again, use the alligator clip leads to save using as many jumper wires.

USE AND CARE OF METERS

Objectives

You will become familiar with the basic operation of meters in measuring dc voltage and current. You will also study the basic use of an ohmmeter and learn the normal operating procedures used to protect meter circuits when making measurements.

In completing these projects, you will observe meters, make measurements, draw conclusions, and be able to answer questions about the following items related to the use and care of meters.

- Reading various meter scales
- Use of mode/function selector switch(es) (as appropriate)
- Use of range switch(es)
- Use of test leads
- Proper polarity
- Proper connection to make voltage measurements
- Proper connection to make current measurements
- "Zeroing" an ohmmeter
- Checking continuity with an ohmmeter
- Measuring resistance with an ohmmeter

NOTE: Although you will probably be using digital multimeters in most of your laboratory projects, the projects in this section concentrate on familiarizing you with the analog meters for several reasons:

1. To teach you to read analog scales.
2. To help you become aware of polarity in dc circuits (some DMMs have automatic polarity protection, analog VOMs don't).
3. To help you become aware of the need for selecting proper ranges (some DMMs have "autoranging," analog VOMs don't).

PROJECT/TOPIC CORRELATION INFORMATION

PROJECT	TEXT CHAPTER	SECTION	RELATED TEXT TOPIC(S)
1 Voltmeters	2	2-6	Making a Voltage Measurement with a Voltmeter
2 Ammeters	2	2-6	Making a Current Measurement with an Ammeter
3 Ohmmeters	2	2-6	Making a Resistance Measurement with an Ohmmeter

PROJECT

1

USE AND CARE OF METERS
Voltmeters: Part A

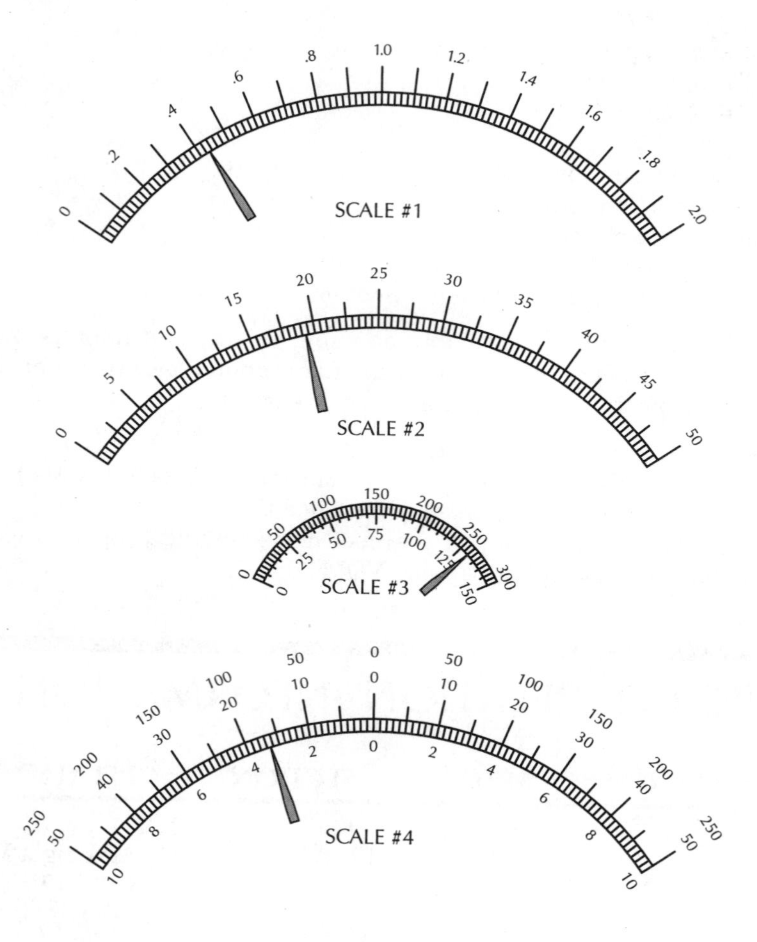

FIGURE 1

PROJECT PURPOSE

To learn how to read analog meter scales and practice reading them.

PARTS NEEDED

- ☐ DMM/VOM
- ☐ Dry Cell 1.5 V
- ☐ VVPS (dc)

SAFETY HINTS

Make sure power is off when connecting meters. Use proper range. Use proper polarity.

Although this project will primarily address the use and care of a dc voltmeter, you will initially be learning how to read the type of scales that are characteristic of nondigital (sometimes called analog) meters. The basic method to interpret the significance of each division on such scales will be useful when using any type of meter with this style of scale.

ACTIVITY	OBSERVATION	CONCLUSION
1. Count the total number of divisions shown on Scale #1 in Figure 1. Include both major and minor divisions.	Total number of divisions = ________________	—
2. If the scale represents voltage, determine the full-scale voltage reading.	Reading at full-scale would represent ________ volts.	—
3. Using the full-scale value and knowing the number of divisions, determine the value of voltage represented by each division.	Each division on the scale represents ________ volt(s).	A simple equation developed from the preceding information is that the value of each scale division = full-scale value divided by the number of ________________. Stated as a formula: each scale division value = f.s. value ÷ ________________
4. Refer to Scales #2, #3, and #4 in Figure 1. Record the voltage value represented by each reading. **NOTE:** For scales #3 and #4, read the *"top" scale* values.	Reading on: Scale #2 = ________ V. Scale #3 = ________ V. Scale #4 = ________ V.	When the pointer is "right on" a scale division, the value can be read directly. When it is "in-between" scale markings, it is necessary to determine the percent of distance it is above the preceding division and below the next ____________ to find the value.

PROJECT

1

USE AND CARE OF METERS
Voltmeters: Part B

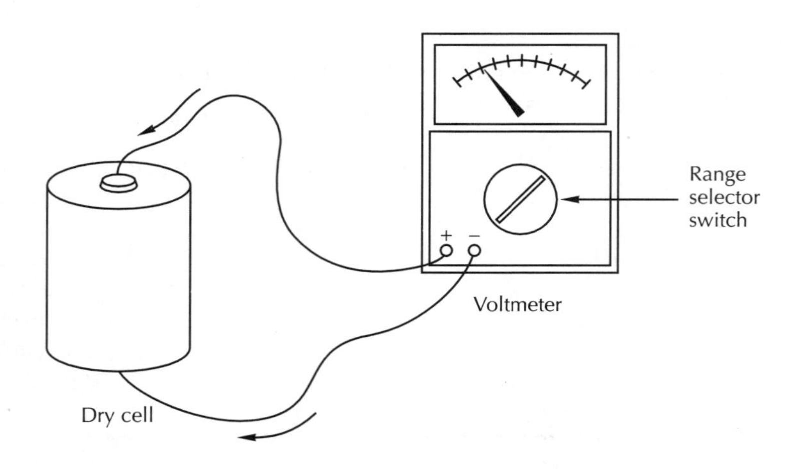

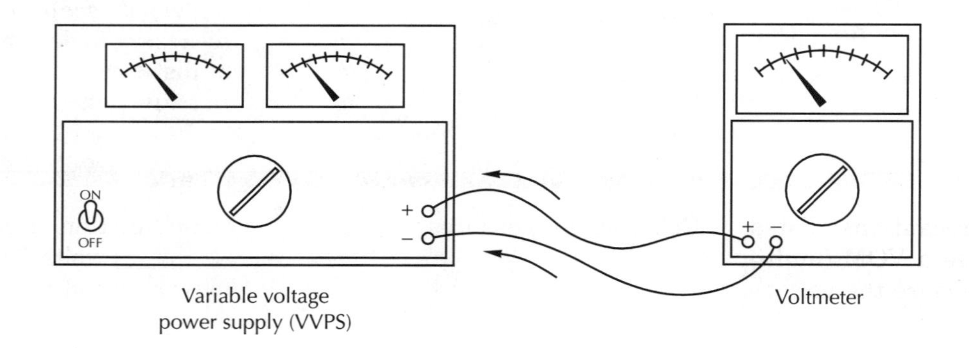

FIGURE 2

PROJECT PURPOSE

To learn how to properly and safely measure dc voltages and to acquire practice in connecting simple circuits from diagrams and in adjusting variable dc source voltage(s) for a desired level of output.

SAFETY HINTS

Be sure power is off on the VVPS when connecting the meter.

Now that you have studied the concept of reading scales, you will begin learning some of the standard procedures for safely using meters to make measurements. To reinforce your thinking, we will take this opportunity to stress key points of meter care in brief form. Think of and properly apply these key factors *every time you use a meter!*

SAFETY HINTS

1. Use the proper METER MODE/FUNCTION! (dc or ac, volts, amperes/mA, or ohms).
2. Be sure the RANGE is high enough for what you will measure.
   (**NOTE:** If not sure, START with the HIGHEST range switch position and work down until the reading causes the pointer to be in the upper two-thirds of the scale, if possible.)
3. Be sure to OBSERVE POLARITY when measuring dc.
4. When MEASURING VOLTAGE, be sure meter is connected IN PARALLEL with the two points having the potential difference to be measured.
5. Use PERSONAL SAFETY cautions! (Power off when connecting test leads or holding only one lead with other hand in pocket, and so on.)

ACTIVITY	OBSERVATION	CONCLUSION
1. Use the precautions listed above and use a VOM (multimeter) to measure the voltage of a dry cell (e.g., flashlight, battery, and so on).	Cell voltage measures ______ V.	Mode/function used was (*dc, ac*) __________ volts. The red test lead was connected to the (+, –) _______ terminal of the cell. The black test lead was connected to the (+, –) ____________ terminal of the cell. The range selector switch was in the ______________ voltage range position.

PROJECT
1
CONTINUED

USE AND CARE OF METERS
Voltmeters: Part B *(Continued)*

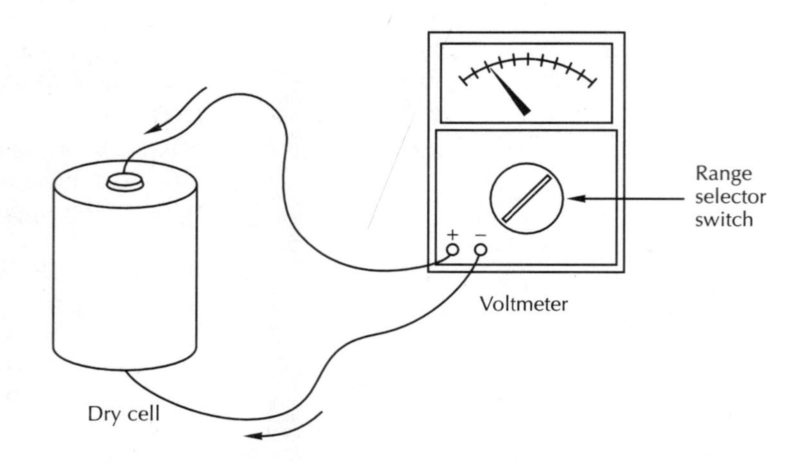

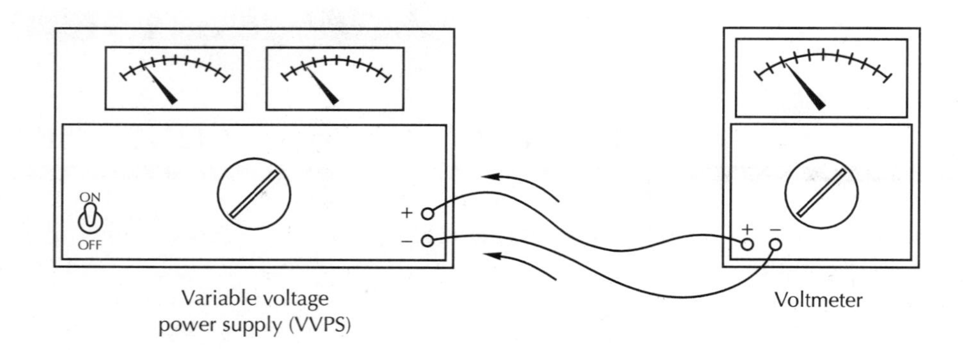

FIGURE 2

PROJECT PURPOSE

To learn how to properly and safely measure dc voltages and to acquire practice in connecting simple circuits from diagrams and in adjusting variable dc source voltage(s) for a desired level of output.

SAFETY HINTS

Be sure power is off on the VVPS when connecting the meter.

ACTIVITY	OBSERVATION	CONCLUSION
2. If a variable-voltage power supply is available, use the voltmeter to monitor the power supply's output voltage terminals and carefully adjust the power supply to 5 V, 10 V, and 15 V output settings, in that order. Have the instructor check your setting each time.	5 V setting OK. (Instructor initial ____________) 10 V setting OK. (Instructor initial ____________) 15 V setting OK. (Instructor initial ____________)	What mode/function was the meter set in? ____________________. The red test lead was connected to the (+, –) __________ output terminal of the power supply. The black test lead was connected to the (+, –) ______ output terminal of the power supply. What meter voltage range setting was used? ____________________.

NOTE: If a digital multimeter (DMM) is available, you may repeat both of the above steps using the DMM. Some DMMs have autoranging so you do not have to set a range switch. Others require setting of a range switch. Use appropriate procedures and make the measurements, as indicated.

PROJECT 1

USE AND CARE OF METERS
Voltmeters: Part C

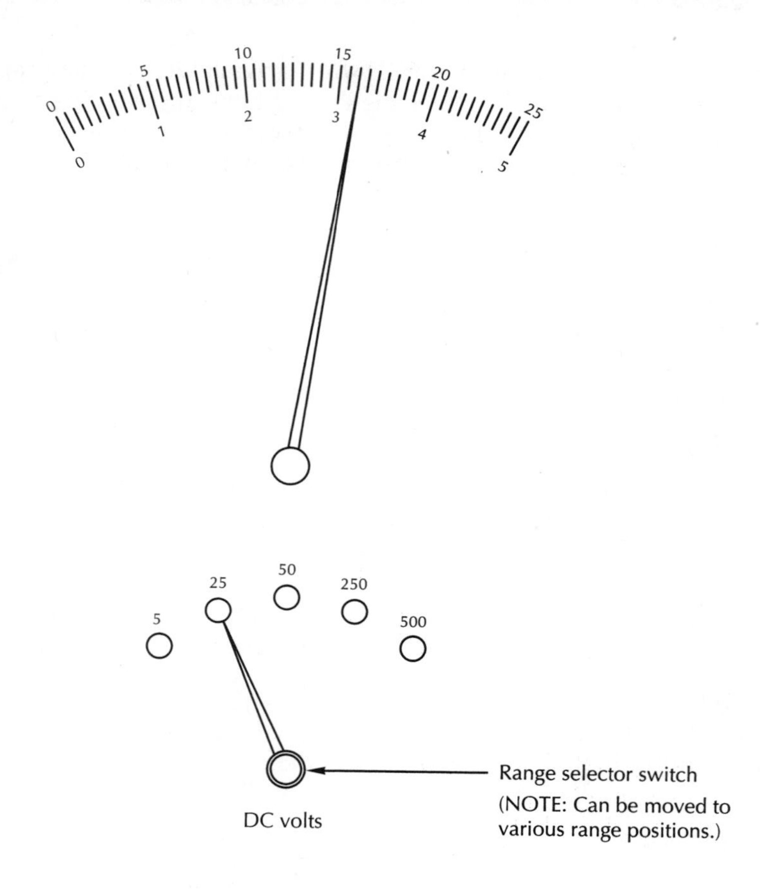

FIGURE 3

PROJECT PURPOSE

To learn how to interpret voltmeter scales versus range switch settings and to obtain practice in interpreting them.

For additional practice interpreting scale readings on analog meters, fill in the chart below, referring to the drawing of a dual-scale meter face and dc volts range selector switch shown in Figure 3.

DC VOLTMETER READING

POINTER POSITION	READING WHEN THE RANGE SELECTOR SWITCH IS SET AT THE RANGE POSITIONS SHOWN				
	5 V	25 V	50 V	250 V	500 V
1 division past the number 3					
3 divisions past the number 10					
4 divisions past the number 2					
1/2 division past the number 15					
2 divisions past the number 20					
1/2 division past the number 5					

ACTIVITY	OBSERVATION	CONCLUSION
1. Observe and list the scales on the face of the VOM or multimeter you are using.	Scales on the meter face are as follows: ______________ ______________	Can some scales be used for more than one voltage range? ________.
2. Observe and list the voltage ranges selectable by the range selector switch.	Selectable ranges are as follows: ______________ ______________	Ranges using the same scale on the meter are: ______________ ______________
3. Observe the voltage ranges selectable on the DMM you are using (if not an autoranging type).	Selectable ranges are as follows: ______________ ______________	The DMM (*does, does not*) ________ ______________ have scales related to the range-switch position.

PROJECT 2

USE AND CARE OF METERS
Ammeters: Part A

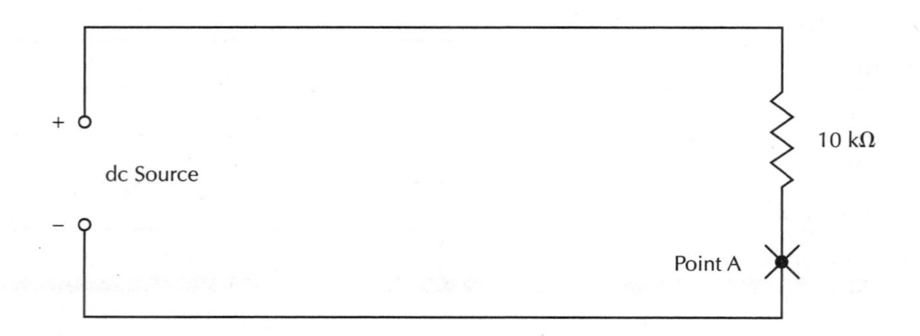

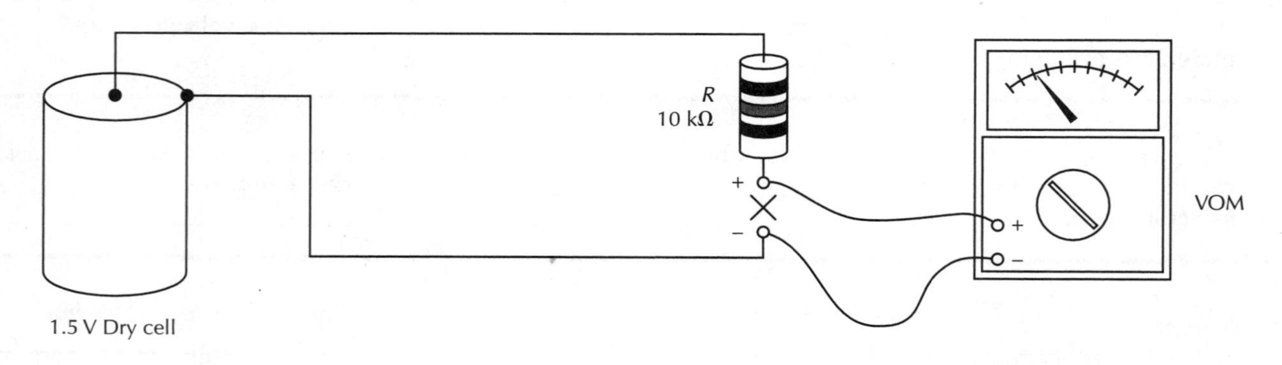

FIGURE 4
Standard schematic diagram (top) and pictorial diagram (bottom)

PROJECT PURPOSE

To learn how to properly and safely measure dc circuit current by connecting simple circuits from schematic or pictorial diagrams and making current measurements.

PARTS NEEDED

- ☐ VOM
- ☐ 1.5 V Dry Cell
- ☐ Component Interconnection System (CIS) (e.g., Proto-Board, etc.)
- ☐ Resistors:
  10 kΩ

SAFETY HINTS

Use proper function, range, and polarity to protect meter.

In this project, you should be aware of the following precautions and procedures related to connecting meters and making current measurements.

SAFETY HINTS

1. Turn POWER OFF circuit into which ammeter will be connected.
2. BREAK THE CIRCUIT in the appropriate place and INSERT THE METER IN SERIES, observing the following rules.
3. Use the proper METER MODE/FUNCTION! (dc current mode).
4. Be sure the RANGE is high enough for the current to be measured. (Start with the HIGHEST range and work down, as appropriate.)
5. Be sure to OBSERVE PROPER POLARITY when measuring dc.
6. After meter is connected properly, then TURN POWER ON AND TAKE READING.
7. Observe appropriate PERSONAL SAFETY rules.

ACTIVITY	OBSERVATION	CONCLUSION
1. Connect the circuit shown in Figure 4.	—	—
2. Set up the VOM selector switches so that the 0–1 mA dc current range is readied to measure dc current.	Mode/Function switch = ________ Range switch = ________________	—
3. Follow the safety rules outlined above, break the circuit at Point A and insert the meter.	Check: (*Yes, No*) Connected in series? ___________ Proper polarity observed? ______ Proper range? ________________	Current will enter the meter through the (*black, red*) ________ lead and will exit the meter through the (*black, red*) ____________ lead.
4. Connect the 1.5-volt cell to the circuit as shown and read the ammeter.	Measured current is: ______ mA.	What change(s) would need to be made in the meter setup if *V* were to be greatly increased or *R* to be greatly decreased? ____________ ________________________________ ________________________________

PROJECT 2

USE AND CARE OF METERS
Ammeters: Part B

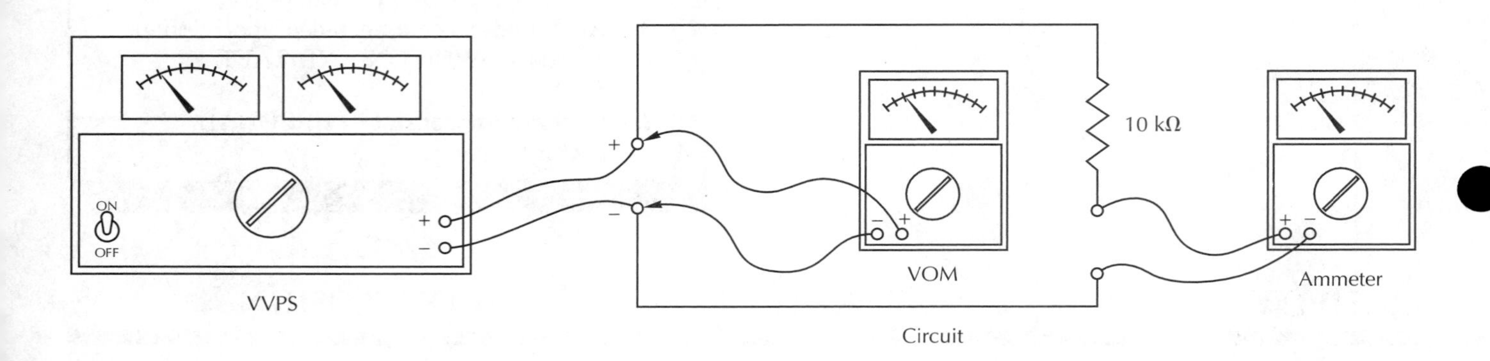

FIGURE 5
Pictorial Diagram

PROJECT PURPOSE

To provide practice in measuring dc circuit current and adjusting source voltage to obtain desired circuit current.

PARTS NEEDED

- ☐ VOM (2) (or 1 DMM, 1 VOM)
- ☐ VVPS (dc)
- ☐ CIS
- ☐ Resistors: 10 kΩ

SAFETY HINTS

Be sure power is off when making circuit changes and/or when connecting meters.

ACTIVITY	OBSERVATION	CONCLUSION
1. Disconnect the 1.5-V cell from the previous circuit in Figure 4. Set up the VOM to read current on the 0 to 10 mA range. Have your instructor check your setup.	Setup OK. (Instructor initial ____________)	—
2. With power OFF, connect a variable-voltage power supply (VVPS) to the circuit so the direction of current through the circuit will be the same as it was when the 1.5-V cell was connected, Figure 5.	Setup OK. (Instructor initial ____________)	—
3. Set the voltage control on the VVPS to the zero output setting. Then, turn power supply on and SLOWLY adjust output until you measure 5 mA through the circuit.	Procedure OK. (Instructor initial ____________)	On the 0–10 mA range of the VOM, 5 mA current caused the meter pointer to rest at (*25, 50, 75*) ____________% of full-scale.
4. If a second DMM/VOM is available, measure the voltage applied to the circuit when there is 2.5 mA of current flow. (*CAUTION!* Remember the safety rules.)	Voltage measures: ________________ volts.	—
5. Turn off power supply and disconnect it from the circuit. Replace the VOM with a 0–1 mA panel meter. When this is done, connect the 1.5-V cell to the circuit and measure current using the panel meter, (if one is available).	Current reading with the panel meter is: ____________ mA.	Is the current reading with the panel meter the same as the VOM reading in step 4 from page 11? ________________

PROJECT 3

USE AND CARE OF METERS
Ohmmeters

OHMMETER SCALES SKETCH
(To be drawn by student)

PROJECT PURPOSE

To familiarize you with a type of ohmmeter scale found on many VOMs and to provide practice in measuring resistances on various meter R ranges, and in checking circuit continuity using an ohmmeter.

SAFETY HINTS

Never have power on in a circuit you are testing with an ohmmeter.

In this project, you should be aware of the following precautions and procedures about using the ohmmeter to make resistance measurements.

SAFETY HINTS

1. Turn POWER OFF and/or DISCONNECT circuit from power source.
2. ISOLATE COMPONENT being measured from the rest of the circuit, whenever possible, to prevent "sneak" paths.
3. Use the proper METER MODE/FUNCTION (dc and Ohms).
4. Be sure the RANGE is appropriate for the range of resistance anticipated, (i.e., $R \times 1$ range, $R \times 100$ range, and so forth).
5. Be sure you have "ZEROED" the meter with the zero adjust control, (i.e., with meter test leads shorted together).
6. Connect test probes across component or circuit to be tested and make the measurement.
7. BE SURE TO TURN SELECTOR SWITCH OFF THE OHMS RANGES to a high voltage range position (e.g., 1,000 V) or to "OFF" position, if your meter has it, when through using the ohmmeter.

ACTIVITY	OBSERVATION	CONCLUSION
1. Refer to the ohmmeter scale on the VOM you are using. Draw a sketch of the ohmmeter scales in the space provided on the opposite page. Only show the major divisions of the scale.	The ohmmeter scale is (*linear*, *nonlinear*) ____________. Zero ohms is indicated on the (*left*, *right*) ____________ end of the scale. Infinite ohms is indicated on the (*left*, *right*) ____________ end of the scale.	The higher the resistance value being measured, the (*higher*, *lower*) ____________ the meter circuit current value will be and the (*greater*, *smaller*) ____________ the pointer deflection will be on the meter scale. The higher the resistance being measured, the closer the pointer will be to the (*left*, *right*) ________ end of the meter scale.

PROJECT 3 CONTINUED

USE AND CARE OF METERS
Ohmmeters *(Continued)*

OHMMETER SCALES SKETCH
(To be drawn by student)

PROJECT PURPOSE

To familiarize you with a type of ohmmeter scale found on many VOMs and to provide practice in measuring resistances on various meter R ranges, and in checking circuit continuity using an ohmmeter.

SAFETY HINTS

Never have power on in a circuit you are testing with an ohmmeter.

ACTIVITY	OBSERVATION	CONCLUSION
2. List the ohmmeter scale multiplying factors available on the meter you are using.	Scales available: $R \times$ __________ $R \times$ __________ $R \times$ __________ $R \times$ __________	—
3. Obtain a 1,000 Ω and a 10,000 Ω resistor. Use proper procedures; measure and record the resistance value of each resistor using the ohmmeter.	The 1,000-ohm resistor measures __________ Ω. The 10,000-ohm resistor measures __________ Ω.	Did the two resistors measure exactly 1,000 Ω and 10,000 Ω, respectively? __________. What could have caused any differences? __________ __________ __________ __________.
4. Obtain a kit of five unknown resistor values from your instructor. Use the ohmmeter and measure and record their values as numbered.	Measured values are: R #1 = __________ Ω R #2 = __________ Ω R #3 = __________ Ω R #4 = __________ Ω R #5 = __________ Ω (Instructor initial __________)	When the value of R to be measured is unknown, it is generally best to start by trying to measure on the (*highest*, *lowest*) __________ R range and then work (*up*, *down*) __________ through the other ranges until the R value can be read in the less crowded portion of the meter scale.
5. If available, obtain several circuits from your instructor, some of which have continuity, some of which do not. Use the ohmmeter and identify the circuits having continuity and those that do not have continuity.	Continuity check: If there is continuity, fill in the blank with "yes," if not, fill in the blank with "no." Ckt #1 __________ Ckt #2 __________ Ckt #3 __________ Ckt #4 __________ Ckt #5 __________	When there was continuity, the ohmmeter reading indicated a (*high*, *low*) __________ resistance reading. When there was NO continuity, the ohmmeter indicated __________ ohms of resistance.

USE AND CARE OF METERS

Complete the following review questions, indicating the appropriate response by placing a check in the box next to the correct answer.

1. To find the value of each division on an analog meter scale, the formula to use is:

   ☐ number of divisions ÷ 2
   ☐ number of divisions × 2
   ☐ number of divisions ÷ full-scale value
   ☐ full-scale value ÷ number of divisions

2. To find the value when a meter pointer is between markings on the scale:

   ☐ Choose the value of the marking the pointer is closest to.
   ☐ Choose the next highest value marking as the value.
   ☐ Interpolate by noting pointer location between markings.
   ☐ Change range switch to try and get reading on a marking.

3. When preparing to measure an unknown dc voltage, the mode/function and range switches should be set at:

   ☐ dc mode, lowest V range
   ☐ ac mode, lowest V range
   ☐ dc mode, highest V range
   ☐ ac mode, highest V range

4. The red test lead on a multimeter should be connected to the meter:

   ☐ positive input jack
   ☐ negative input jack
   ☐ neither of these

5. When measuring dc voltage, the meter's black test lead is normally connected to the more ________________ point of the component or circuit being measured.

   ☐ negative
   ☐ positive

6. Can a single meter scale on the meter face ever be used to read more than one selectable meter measurement "range?"

   ☐ Yes
   ☐ No

7. Precautions for preparing to measure current include:

   - ☐ Circuit off; correct polarity; correct range; connect meter in parallel.
   - ☐ Circuit off; correct polarity; correct range; connect meter in series.
   - ☐ Circuit on; correct polarity; correct range; connect meter in parallel.
   - ☐ Circuit on; correct polarity; correct range; connect meter in series.

8. Precautions when using an ohmmeter to measure the resistance value of a resistor include:

   - ☐ Power off; zero meter; correct range; correct polarity; turn off ohms mode when through measuring R.
   - ☐ Power off; zero meter; correct range; turn off ohms mode when through measuring R.
   - ☐ Power on; zero meter; correct range; correct polarity; turn off ohms mode when through measuring R.

9. If the pointer on an ohmmeter is pointing at 5 on the meter scale and the range switch is on the $R \times 10{,}000$ range, the R value is:

   - ☐ 5 Ω
   - ☐ 500 Ω
   - ☐ 5 kΩ
   - ☐ 50 kΩ
   - ☐ 500,000 Ω

10. Turning the meter off the ohms mode when through may prevent:

    - ☐ Possibility of battery drainage; possibility of meter damage.
    - ☐ Using wrong range; meter damaging a circuit.

OHM'S LAW

Objectives

You will connect several simple dc resistive circuits and make measurements and observations regarding how Ohm's law is applied to practical electrical circuits.

In completing these projects, you will use the color code, connect circuits, make measurements, perform calculations, draw conclusions, and be able to answer questions about the following items relating to the resistor color code and Ohm's law:

- Use of the resistor color code
- The relationship of current to voltage in simple resistive dc circuits
- The relationship of current to resistance in simple resistive dc circuits
- The relationship of power to voltage in simple resistive dc circuits
- The relationship of power to current in simple resistive dc circuits

PROJECT/TOPIC CORRELATION INFORMATION

PROJECT		TEXT CHAPTER	SECTION	RELATED TEXT TOPIC(S)
4	Resistor Color Code Review and Practice	2	2-5	Resistor Color Code
5	Relationship of I and V with R Constant	3	3-1	The Relationship of Current to Voltage with Resistance Constant
6	Relationship of I and R with V Constant	3	3-1	The Relationship of Current to Resistance with Voltage Constant
7	Relationship of Power to V with R Constant	3	3-8	The Basic Power Formula
8	Relationship of Power to I with R Constant	3	3-8	The Basic Power Formula

PROJECT 4

OHM'S LAW
Resistor Color Code Review and Practice

PROJECT PURPOSE

To provide review and hands-on practice in using the resistor color code.

CHART 1

1ST COLOR	2ND COLOR	3RD COLOR	4TH COLOR	OHMS VALUE	TOLERANCE PERCENT
Red	Violet	Yellow	Gold		
Brown	Black	Green	None		
Orange	White	Black	Gold		
Yellow	Violet	Orange	Silver		
Gray	Red	Brown	Gold		
Green	Brown	Black	Gold		
Blue	Red	Brown	Gold		
Green	Blue	Green	Silver		

CHART 2

OHMS VALUE	COLORS		
80 Ω			
3.0 Ω			
1.0 Ω			
13 kΩ			
10 MΩ			
91 Ω			

NOTE: These charts are for practice only and therefore may have some values called for which are not standard available resistor values.

ACTIVITY	OBSERVATION	CONCLUSION
1. List the 10 colors used in the resistor color code to represent 0,1,2,3,4,5,6,7,8,9.	Colors used in the resistor color code are as follows: 0 = ________ 1 = ________ 2 = ________ 3 = ________ 4 = ________ 5 = ________ 6 = ________ 7 = ________ 8 = ________ 9 = ________	—
2. List the other colors in the color code generally used to indicate resistor tolerance or used as special multipliers.	Special colors are: ________ and ________.	The color used to indicate 5% tolerance is ________. The color used to indicate 10% tolerance is ________. The color used to indicate a 0.1 multiplier is ________. The color used to indicate a 0.01 multiplier is ________.
3. Fill in the resistance and tolerance values on Chart 1 (opposite page) as appropriate.	—	—
4. Fill in the colors that can be used to indicate the values in Chart 2 (opposite page) as appropriate.	—	—
5. Obtain a set of 10 resistors having assorted values and tolerances and use the color code to determine their values and tolerances, as appropriate.	Values and tolerances of resistors in the kit are as follows: *R* #1 = ______ Ω ______% tolerance *R* #2 = ______ Ω ______% tolerance *R* #3 = ______ Ω ______% tolerance *R* #4 = ______ Ω ______% tolerance *R* #5 = ______ Ω ______% tolerance *R* #6 = ______ Ω ______% tolerance *R* #7 = ______ Ω ______% tolerance *R* #8 = ______ Ω ______% tolerance *R* #9 = ______ Ω ______% tolerance *R*#10 = ______ Ω ______% tolerance (Instructor initial ________)	—

PROJECT
4
CONTINUED

OHM'S LAW
Resistor Color Code Review and Practice *(Continued)*

PROJECT PURPOSE
To provide review and hands-on practice in using the resistor color code.

CHART 3

1ST COLOR	2ND COLOR	3RD COLOR	4TH COLOR	5TH COLOR	RESISTANCE VALUE	TOLERANCE PERCENT
Brown	Brown	Black	Red	Brown		
Orange	Blue	Black	Red	Red		
Brown	Brown	Black	Orange	Brown		
White	Brown	Black	Orange	Red		
Red	Yellow	Black	Gold	Green		
Brown	Brown	Black	Silver	Blue		

NOTE: These charts are for practice only and therefore may have some values called for which are not standard available resistor values.

ACTIVITY	OBSERVATION	CONCLUSION
6. For precision resistors, a five-band color-coding system is frequently used. List the meaning of each band in the observation column.	Band #1 = ______ Band #2 = ______ Band #3 = ______ Band #4 = ______ Band #5 = ______	—
7. For the fifth band on these precision resistors, list the meaning of each color listed in the Observation column, as appropriate.	Brown = ______% Red = ______% Green = ______% Blue = ______% Violet = ______%	—
8. Fill in the resistance and tolerance values for the five-band precision resistors listed in Chart 3.	—	—

PROJECT 5

OHM'S LAW
Relationship of *I* and *V* with *R* Constant

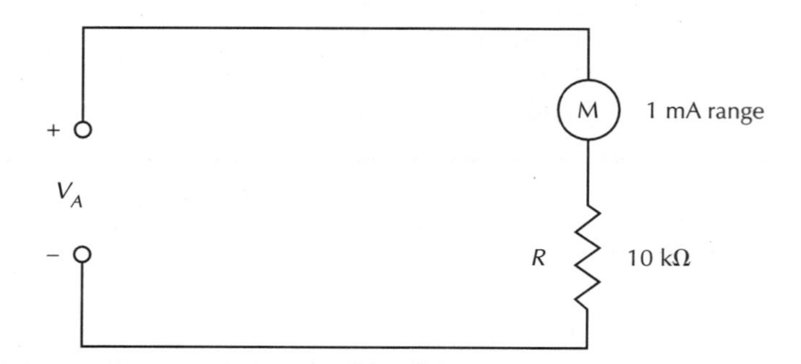

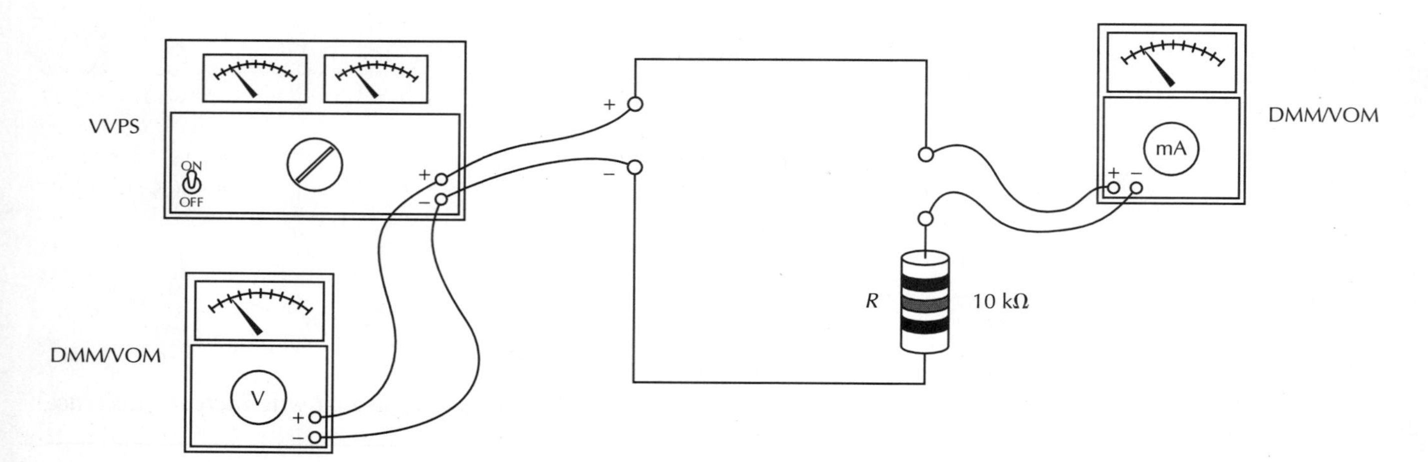

FIGURE 6
Standard schematic diagram (top) and pictorial diagram (bottom)

PROJECT PURPOSE

To demonstrate Ohm's law and the direct relationship of current to voltage, for a given resistance. To provide further practice in connecting circuits and making electrical measurements.

PARTS NEEDED

- ☐ DMM/VOM (2)
- ☐ VVPS (dc)
- ☐ CIS
- ☐ 10-kΩ Resistor

SAFETY HINTS

Be sure power is off when connecting meters.

As you perform this project, remember that Ohm's law states that $I = \frac{V}{R}$.

ACTIVITY	OBSERVATION	CONCLUSION
1. Connect the initial circuit as shown in Figure 6.	—	—
2. Adjust V_A to obtain 1/2 scale deflection on the 1-mA range.	Current is: ______ mA	—
3. Measure the V_A with a voltmeter.	V_A measures: ______ volts	—
4. Use Ohm's law ($I = V/R$) and calculate I from the measured value of V and the indicated R value.	V_A = ______ volts R = ______ ohms	I calculated = ______ mA
5. Increase V_A to twice its original value and note the new current reading.	V_A = ______ volts I now = ______ mA	Doubling V_A caused I to: ______. From this we conclude that with R constant (unchanged), current is directly proportional to ______.
6. Reduce V_A to 2 volts and note the new current reading.	I now = ______ mA	Reducing V_A to 2 volts caused the current to (*increase, decrease*) ______ proportionately. This shows again that current stays "in step" (is directly proportional) with voltage when R is unchanged. This means that if R is held constant and V is increased, I will (*increase, decrease*) ______; if V is decreased, I will (*increase, decrease*) ______.

PROJECT 6

OHM'S LAW
Relationship of I and R with V Constant

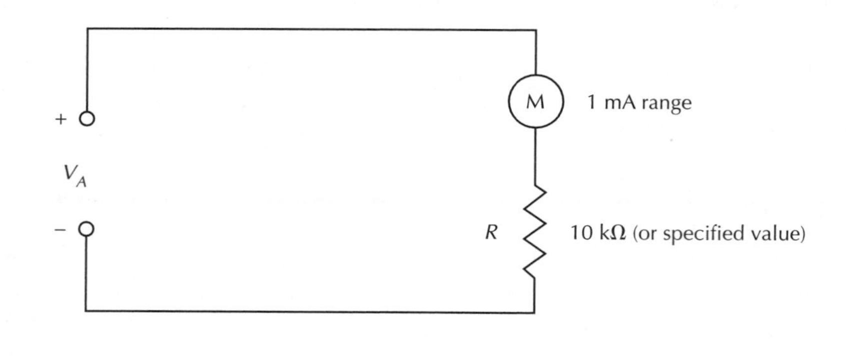

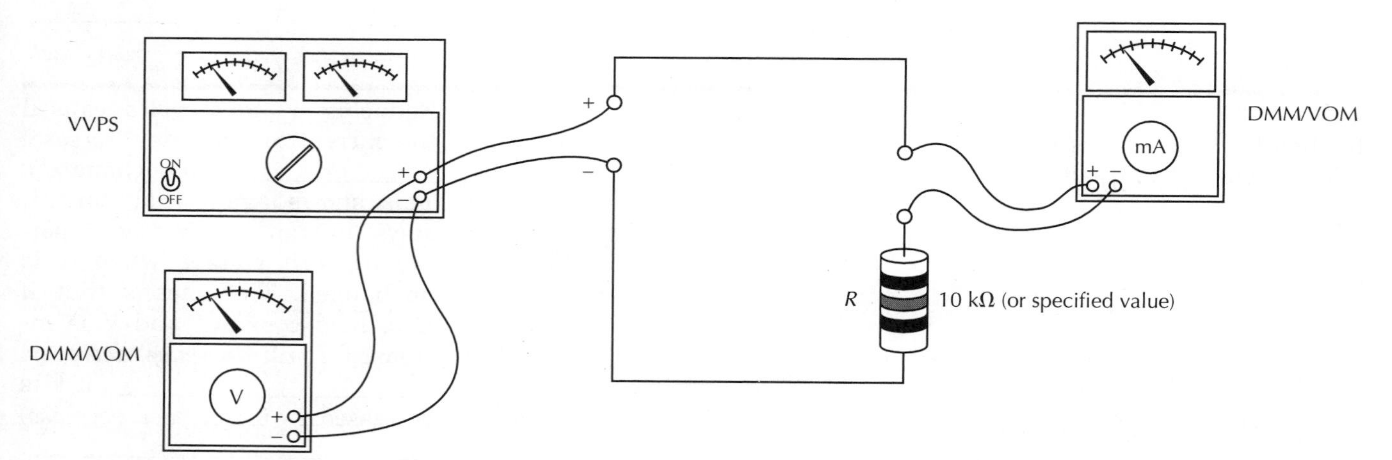

FIGURE 7
Standard schematic diagram (top) and pictorial diagram (bottom)

PROJECT PURPOSE

To demonstrate Ohm's law and the inverse relationship of current to resistance for a given voltage. To provide continued practice in circuit connection and measurement.

PARTS NEEDED

- ☐ DMM/VOM (2)
- ☐ VVPS (dc)
- ☐ CIS
- ☐ Resistors:
  - 10 kΩ
  - 27 kΩ
  - 47 kΩ
  - 100 kΩ

SAFETY HINTS

Be sure power is off when connecting meters or changing components.

ACTIVITY	OBSERVATION	CONCLUSION
1. Connect the initial circuit as shown in Figure 7.	—	—
2. Adjust V_A to obtain 1 mA of current.	Current is: ____________ mA	—
3. Measure V_A with the voltmeter and BE SURE *NOT* TO CHANGE V_A FOR THE REST OF THE STEPS IN THIS SECTION.	V_A measures: ____________ volts	—
4. Use Ohm's law ($R = V/I$) and calculate R from the measured values of V and I.	V = ____________ volts I = ____________ mA	R calculated = ____________ ohms
5. Remove the 10-kΩ resistor from the circuit and replace it with 100 kΩ and note the new current reading.	R now = ____________ Ω I now = ____________ mA	Keeping V_A constant at ________ volts and increasing R by 10 times to a value of ______ ohms caused the current to (*increase*, *decrease*) ____________ to one ________ of its original value. From this we conclude that I is inversely proportional to R. This means that if R is increased, I will (*increase*, *decrease*) ____________ proportionately; if R is decreased, I will (*increase*, *decrease*) __________ by the same factor as R was decreased.
6. Remove the 100-kΩ resistor from the circuit and replace it with 47 kΩ and record the new current reading. **NOTE:** Keep V_A the same as it was.	V_A = ____________ volts R = ____________ ohms I now = ____________ mA	The circuit resistance for this step is *approximately* (*1/4, 1/2*) ______ that of the previous step; the applied voltage is the same; and the resulting current is approximately (*2, 4*) __________ times that of step 5. This again tends to prove that I is (*directly*, *inversely*) ____________ proportional to R, with V held constant.

PROJECT 6 CONTINUED

OHM'S LAW Relationship of *I* and *R* with *V* Constant *(Continued)*

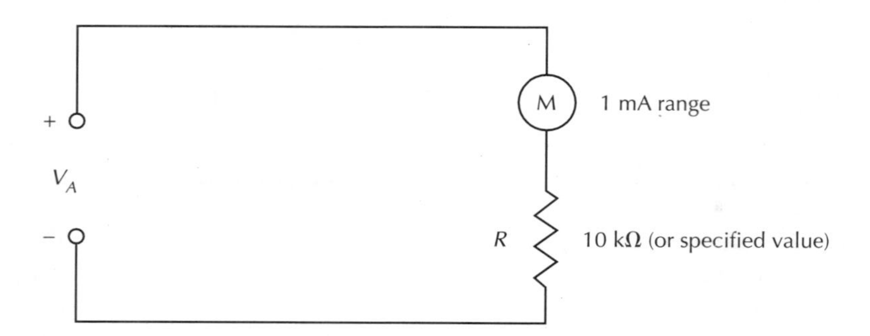

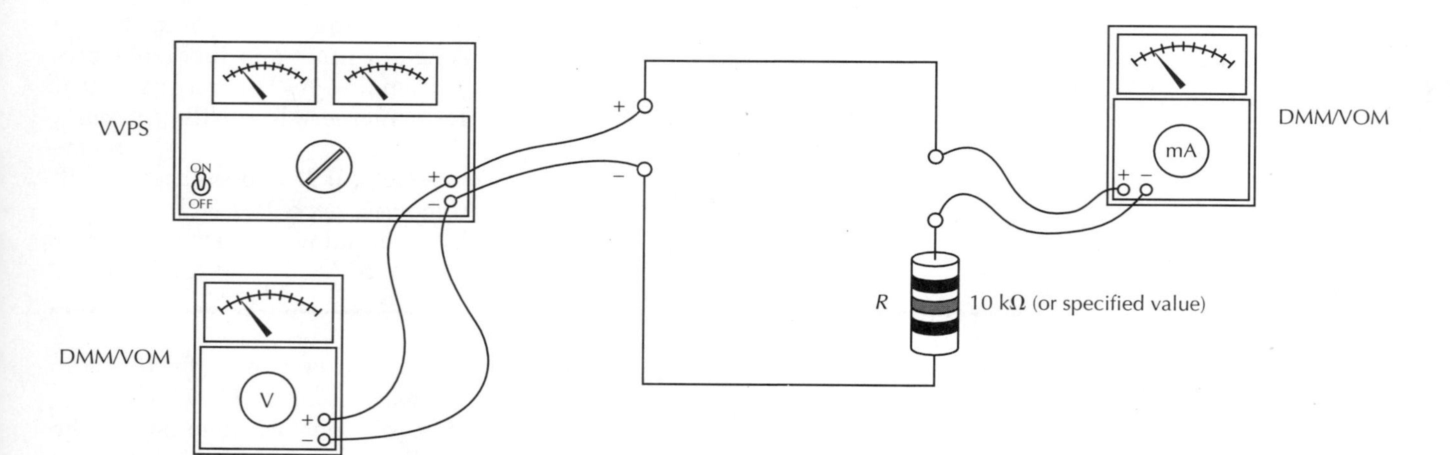

FIGURE 7
Standard schematic diagram (top) and pictorial diagram (bottom)

PROJECT PURPOSE

To demonstrate Ohm's law and the inverse relationship of current to resistance for a given voltage. To provide continued practice in circuit connection and measurement.

PARTS NEEDED

- ☐ DMM/VOM (2)
- ☐ VVPS (dc)
- ☐ CIS
- ☐ Resistors:
  - 10 kΩ
  - 27 kΩ
  - 47 kΩ
  - 100 kΩ

SAFETY HINTS

Be sure power is off when connecting meters or changing components.

EXTRA CREDIT STEP(S)

ACTIVITY	OBSERVATION	CONCLUSION
7. Remove the 47 kΩ resistor from the circuit and replace it with 27 kΩ. Record the new current reading.	V_A = ____________ volts R = ____________ kΩ I = ____________ mA	Did the circuit current change from the previous step? __________. Was the R larger or smaller than in the previous step? ______________. In your own words, explain what this data proves: ______________________ ______________________ ______________________

PROJECT 7

OHM'S LAW
Relationship of Power to V with R Constant

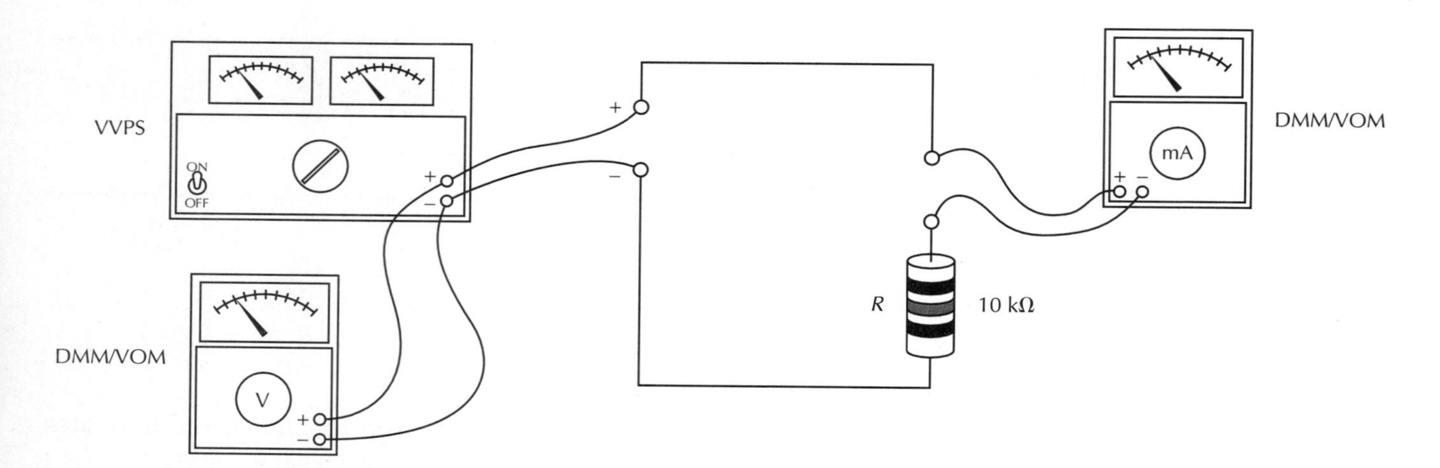

FIGURE 8
Standard schematic diagram (top) and pictorial diagram (bottom)

PROJECT PURPOSE

To demonstrate the use of the power formula form which shows that power is related to voltage squared, for a given resistance. To provide continued practice in circuit connection and measurement.

PARTS NEEDED

- ☐ DMM/VOM (2)
- ☐ VVPS (dc)
- ☐ CIS
- ☐ 10-kΩ resistor

SAFETY HINTS

Be sure power is off when connecting meters.

ACTIVITY	OBSERVATION	CONCLUSION
1. Connect the initial circuit as shown in Figure 8.	—	—
2. Adjust V_A to obtain 0.5 mA of current and measure V_A with the voltmeter.	I = ______ mA V_A = ______ volts	—
3. Use the $V \times I$ power formula and calculate the power dissipated by R.	V = ______ volts I = ______ mA	P calculated = ______ mW
4. Change V_A to obtain 1 mA of current. Measure V_A and calculate P using the $V \times I$ formula.	V now = ______ volts I now = ______ mA	P calculated now = ______ mW. Doubling V caused the current to increase by how many times? ______ Thus, the product of $V \times I$ increased ______ times (when we doubled the voltage and kept R constant). From this we conclude that power is proportional to the ______ of the voltage when R is not changed.
5. Calculate the power for the measured values of step 4 using the $P = V^2/R$ formula.	V = ______ volts R = ______ ohms	P calculated = ______ mW
6. Change V_A to 2.5 volts. Measure the appropriate voltage and current values and calculate P by both of the above formulas.	V now = ______ volts I now = ______ mA	P calculated by both methods is approximately ______ mW. Compared to the 10-volt V_A condition, we now have (*1/2, 1/4, 1/8*) ______ the V_A and (*1/4, 1/8, 1/16, 1/32*) ______ the power dissipation. This again illustrates that P is related to the (*square*, $\sqrt{\ }$) ______ of the voltage when R remains unchanged.

PROJECT

7

CONTINUED

OHM'S LAW

Relationship of Power to *V* with *R* Constant *(Continued)*

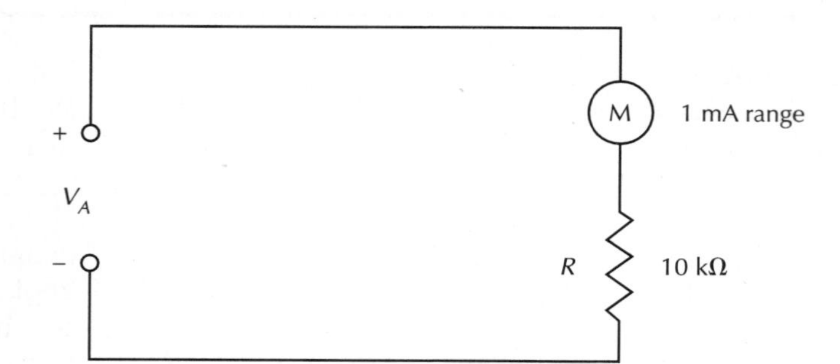

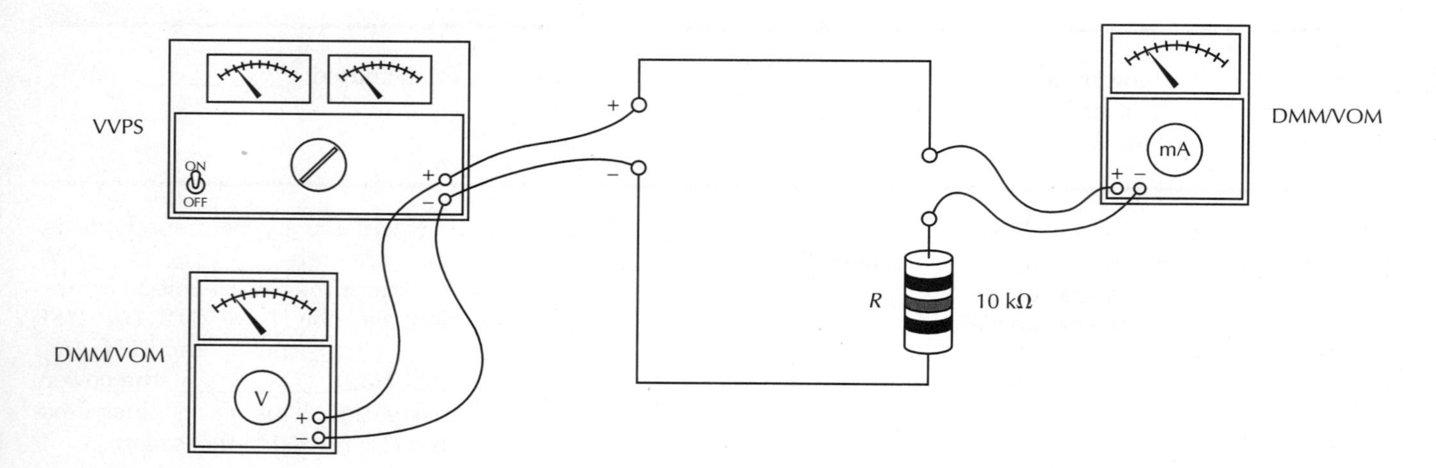

FIGURE 8
Standard schematic diagram (top) and pictorial diagram (bottom)

PROJECT PURPOSE

To demonstrate the use of the power formula form which shows that power is related to voltage squared, for a given resistance. To provide continued practice in circuit connection and measurement.

PARTS NEEDED

- ☐ DMM/VOM (2)
- ☐ VVPS (dc)
- ☐ CIS
- ☐ 10-kΩ resistor

SAFETY HINTS

Be sure power is off when connecting meters.

EXTRA CREDIT STEP(S)

ACTIVITY	OBSERVATION	CONCLUSION
7. Assume a V_A value of 4 volts. Calculate and predict what the circuit current and power would be for that value of applied voltage.	Predicted I = ________ mA Predicted P = ________ mW	—
8. Change the circuit V_A to 4 volts; measure I, and calculate P.	Measured I = ________ mA Calculated P = ________ mW	Was the current with 4 volts applied voltage higher or lower than when there was 2.5 volts applied? ____________. Was the calcualted power higher or lower? ____________. How many times higher or lower? ____________. Is the change coherent with the concept of power being proportional to V^2? ____________.

PROJECT

8

OHM'S LAW
Relationship of Power to *I* with *R* Constant

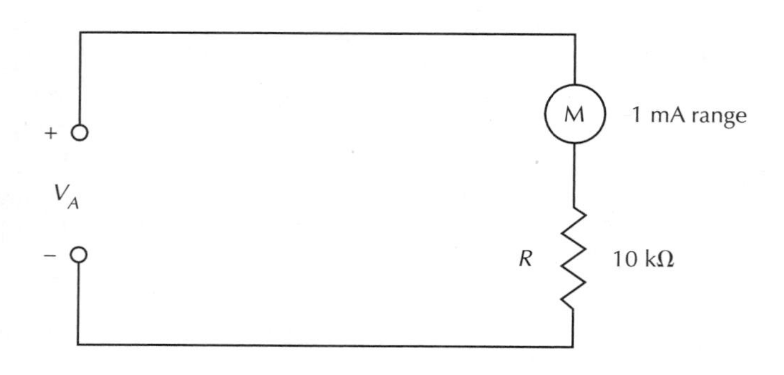

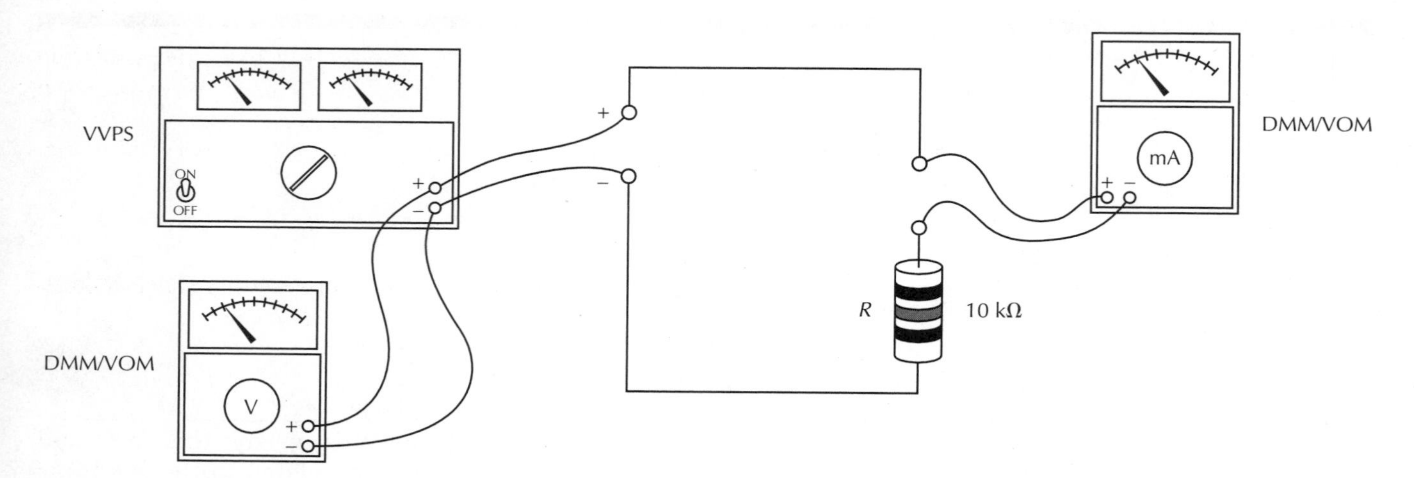

FIGURE 9
Standard schematic diagram (top) and pictorial diagram (bottom)

PROJECT PURPOSE

To verify the fact that power is related to the current squared, for a given resistance. To provide continued practice in circuit connection and measurement.

PARTS NEEDED

- ☐ DMM/VOM (2)
- ☐ VVPS (dc)
- ☐ CIS
- ☐ 10-kΩ resistor

SAFETY HINTS

Be sure power is off when connecting meters.

ACTIVITY	OBSERVATION	CONCLUSION
1. Connect the initial circuit as shown in Figure 9.	—	—
2. Adjust V_A to 7 volts and note the current.	I = ______ mA	—
3. Use the measured values of V and I and calculate P by the formula $P = V \times I$.	V = ______ volts I = ______ mA	P calculated = ______ mW
4. Decrease V_A until I is 1/2 its original value and calculate P using both the $V \times I$ and I^2R formulas.	V now = ______ volts I now = ______ mA	P calculated by both methods is approximately ______ mW. This is (*1/8, 1/4, 1/2*) ______ the power dissipated when I was double the value of the current for this step. From this we can conclude that P is proportional to I^2 when R in the circuit is unchanged. This means that if the circuit current has doubled, the power dissipation has increased (*1, 2, 3, 4*) ______ times; or, if circuit current were decreased to one-third its original value, the power must decrease to (*1/3, 1/6, 1/9*) ______ its original value.

EXTRA CREDIT STEP(S)

5. Change V_A until it is approximately one-third the original 7-volt value. (approximately 2.33 volts). Make appropriate measurements and calculations in order to fill in the blanks in the Observation column.	V_A now = ______ volts I now = ______ mA $V \times I$ now = ______ mW I^2R now = ______ mW V^2/R now = ______ mW	Is the power approximately one-ninth that when 7 volts was applied? ______. Does this verify the fact that power is related to the square of the circuit current with R constant? ______.

OHM'S LAW

Complete the following review questions, indicating the appropriate response by placing a check in the box next to the correct answer.

1. If V increases and R remains the same, then I will
   - ☐ increase
   - ☐ decrease
   - ☐ remain the same

2. If I increases and R remained the same, then V must have
   - ☐ increased
   - ☐ decreased
   - ☐ remained the same

3. If R increases and V remains the same, then I will
   - ☐ increase
   - ☐ decrease
   - ☐ remain the same

4. If V is doubled and R is halved, then I will
   - ☐ double
   - ☐ halve
   - ☐ quadruple
   - ☐ remain the same

5. If V is doubled, and R remains the same, then P will
   - ☐ double
   - ☐ halve
   - ☐ quadruple
   - ☐ remain the same

6. If I is halved, and R remains the same, then P will
   - ☐ double
   - ☐ halve
   - ☐ quadruple
   - ☐ decrease to one-quarter
   - ☐ remain the same

7. If V is doubled and R is halved, then P will
   - ☐ decrease 4 times
   - ☐ increase 8 times
   - ☐ increase 16 times
   - ☐ remain the same

OHM'S LAW

8. Increasing the voltage applied to a circuit will cause:

   a. Current to
      - ☐ increase
      - ☐ decrease
      - ☐ remain the same

   b. Resistance to
      - ☐ increase
      - ☐ decrease
      - ☐ remain the same

   c. Power dissipated to
      - ☐ increase
      - ☐ decrease
      - ☐ remain the same

9. Decreasing the resistance in a circuit will cause:

   a. Current to
      - ☐ increase
      - ☐ decrease
      - ☐ remain the same

   b. Voltage applied to
      - ☐ increase
      - ☐ decrease
      - ☐ remain the same

   c. Power dissipated to
      - ☐ increase
      - ☐ decrease
      - ☐ remain the same

10. In an electrical circuit

   a. current is directly proportional to
      - ☐ V
      - ☐ R

   b. and inversely proportional to
      - ☐ V
      - ☐ R

   c. while power is proportional to the square of the
      - ☐ V
      - ☐ R

SERIES CIRCUITS

Objectives

You will connect several dc resistive series circuits and make measurements and observations regarding their important electrical characteristics.

In completing these projects, you will connect circuits, make measurements, perform calculations, draw conclusions, and be able to answer questions about the following items related to series circuits.

- Total resistance
- Voltage distribution
- Circuit current
- Power dissipation(s)
- Effects of opens
- Effects of shorts
- Application of Ohm's law
- Application of Kirchhoff's voltage law

PROJECT/TOPIC CORRELATION INFORMATION

PROJECT		TEXT CHAPTER	SECTION	RELATED TEXT TOPIC(S)
9	Total Resistance in Series Circuits	4	4-2	Resistance in Series Circuits
10	Current in Series Circuits	4	4-1	Definition and Characteristics of a Series Circuit
11	Voltage Distribution in Series Circuits	4	4-3	Voltage in Series Circuits
12	Power Distribution in Series Circuits	4	4-5	Power in Series Circuits
13	Effects of an Open in Series Circuits	4	4-6	Effects of Opens in Series Circuits and Troubleshooting Hints
14	Effects of a Short in Series Circuits	4	4-7	Effects of Shorts in Series Circuits and Troubleshooting Hints

PROJECT 9

SERIES CIRCUITS
Total Resistance in Series Circuits

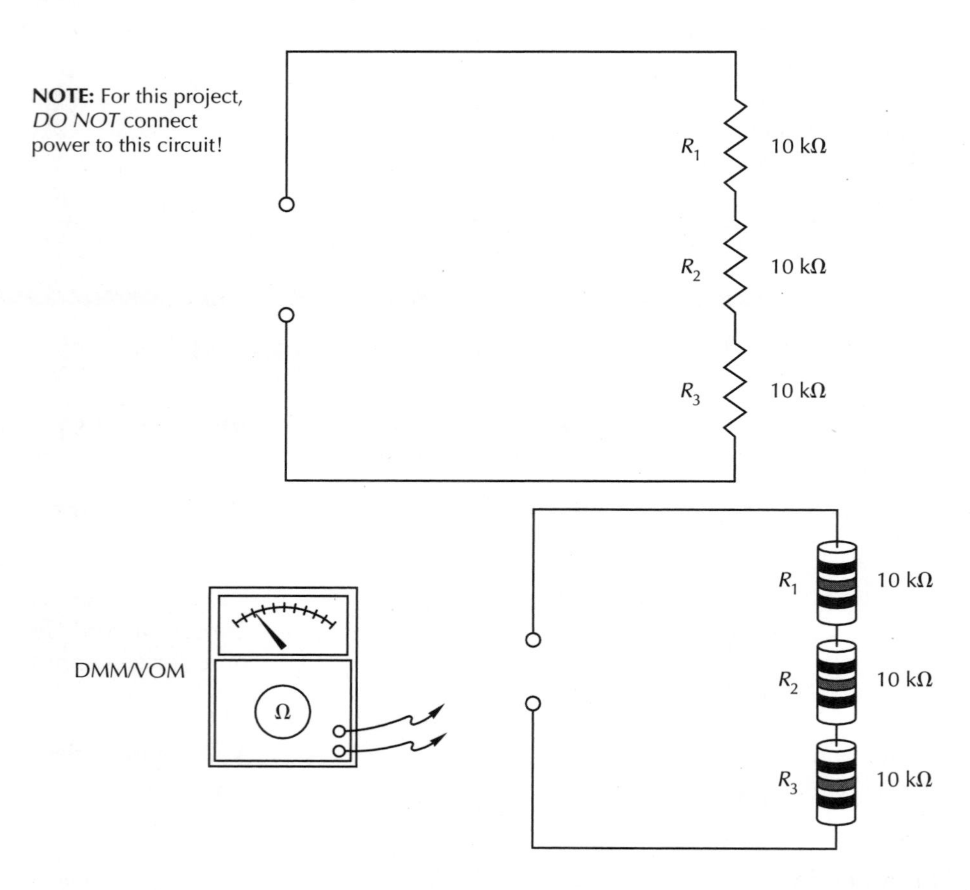

FIGURE 10
Standard schematic diagram (top) and pictorial diagram (bottom)

PROJECT PURPOSE

To confirm the series-circuit total resistance (R_T) formula by varying circuit conditions and making resistance measurements.

PARTS NEEDED

- ☐ DMM/VOM
- ☐ CIS
- ☐ Resistors:
  - 10 kΩ (3)
  - 27 kΩ
  - 47 kΩ
  - 100 kΩ (2)

SAFETY HINTS

DO NOT use a power supply for this project! (No power is to be applied when using an ohmmeter.)

ACTIVITY	OBSERVATION	CONCLUSION
1. Connect the initial circuit shown in Figure 10. CAUTION! DO NOT CONNECT POWER TO THE CIRCUIT FOR THIS PROJECT!	—	—
2. Use an ohmmeter and measure the total resistance of the circuit.	R total = ______ ohms	—
3. Use an ohmmeter and measure the resistance of each individual resistor and record your observations.	R_1 = ______ ohms R_2 = ______ ohms R_3 = ______ ohms	—
4. Add the resistances of the individual resistors and note the result.	$R_1 + R_2 + R_3$ = ______ ohms	The total resistance of a series circuit equals the (*product, sum*) ______ of all the individual resistances; therefore, R_T = ______.
5. Predict what the new R_T would be if R_1 were changed to a 27-kΩ resistor.	Pred. R_T = ______ ohms	—
6. Change R_1 to 27 kΩ and measure the new R_T.	New R_T = ______ ohms	Changing any element's resistance in a series circuit while the rest of the elements are unchanged will cause the circuit's total resistance to (*remain the same, change*) ______. If any element's R increases, then R_T will (*increase, decrease*) ______. If any element's R decreases, then R_T will (*increase, decrease*) ______.

EXTRA CREDIT STEP(S)

7. Change the circuit so that R_1 = 47 kΩ, R_2 = 100 kΩ, and R_3 = 100 kΩ. Predict R_T's value; then measure R_T to verify your prediction.	Predicted R_T = ______ kΩ Measured R_T = ______ kΩ	Did R_T equal the sum of the individual resistances in this case? ______. What might cause a difference in predicted and measrued values? ______ ______ ______ ______

SERIES CIRCUITS
Current in Series Circuits

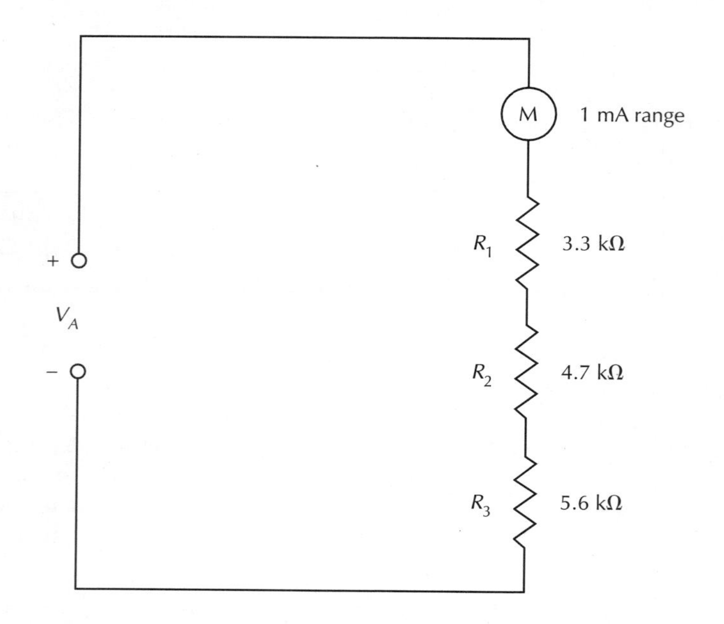

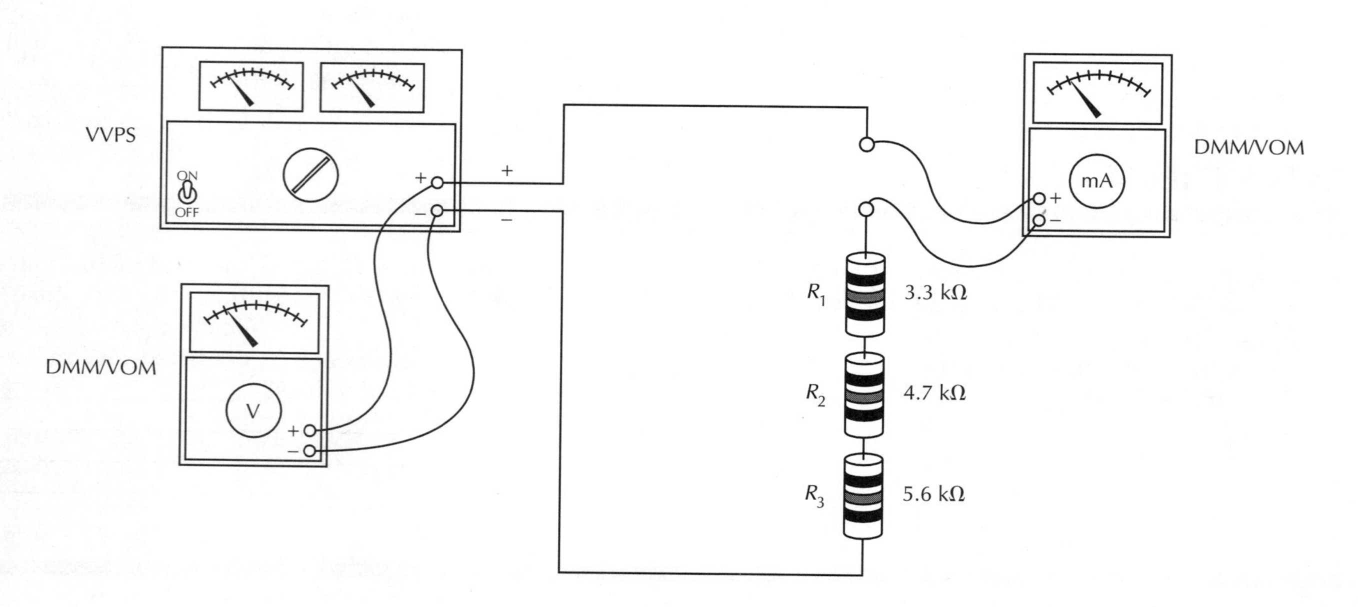

FIGURE 11
Standard schematic diagram (top) and pictorial diagram (bottom)

PROJECT PURPOSE

To verify that the current throughout a series circuit is the same current via "hands-on" experience in measuring current at various points throughout a series circuit.

PARTS NEEDED

- ☐ DMM/VOM (2)
- ☐ VVPS (dc)
- ☐ CIS
- ☐ Resistors:
  - 3.3 kΩ
  - 4.7 kΩ
  - 5.6 kΩ
  - 10 kΩ

SAFETY HINTS

Be sure power is off when connecting meters or changing components.

ACTIVITY	OBSERVATION	CONCLUSION
1. Connect the initial circuit shown in Figure 11.	—	—
2. Apply 9.5 volts to the circuit and note the current.	Number of paths for current in circuit = ____________. Current = __________ mA	—
3. "Swap" positions of R_1 and the current meter and note the current reading.	Current reading is now _____ mA.	—
4. "Swap" positions of the current meter and each of the remaining resistors, and note the current reading each time.	Current reading in every case was ____________ mA.	No matter where the current meter was placed in the circuit, the current reading was the same. This indicates that the current through all parts of a series circuit is the __________ current.
5. Change R_1 to a 10-kΩ resistor. Move the meter to various spots in the circuit and note the current reading.	Current reading in all cases was ____________ mA.	Increasing any R in a series circuit affects the current through all parts of the circuit. Would changing R by decreasing it cause I to change? ________ Increase or decrease? ____________.
6. Now change V_A to 19 volts. Move the meter to various spots in the circuit and note the current reading.	Current reading in all cases was ____________ mA.	Changing V_A for the series circuit caused I to change through all parts of the circuit. It changed by (*a different, the same*) __________ amount in all parts because the current through all parts of a series circuit is (*a different, the same*) __________ current. This is because there is only ______ path for current through series elements.

SERIES CIRCUITS
Voltage Distribution in Series Circuits

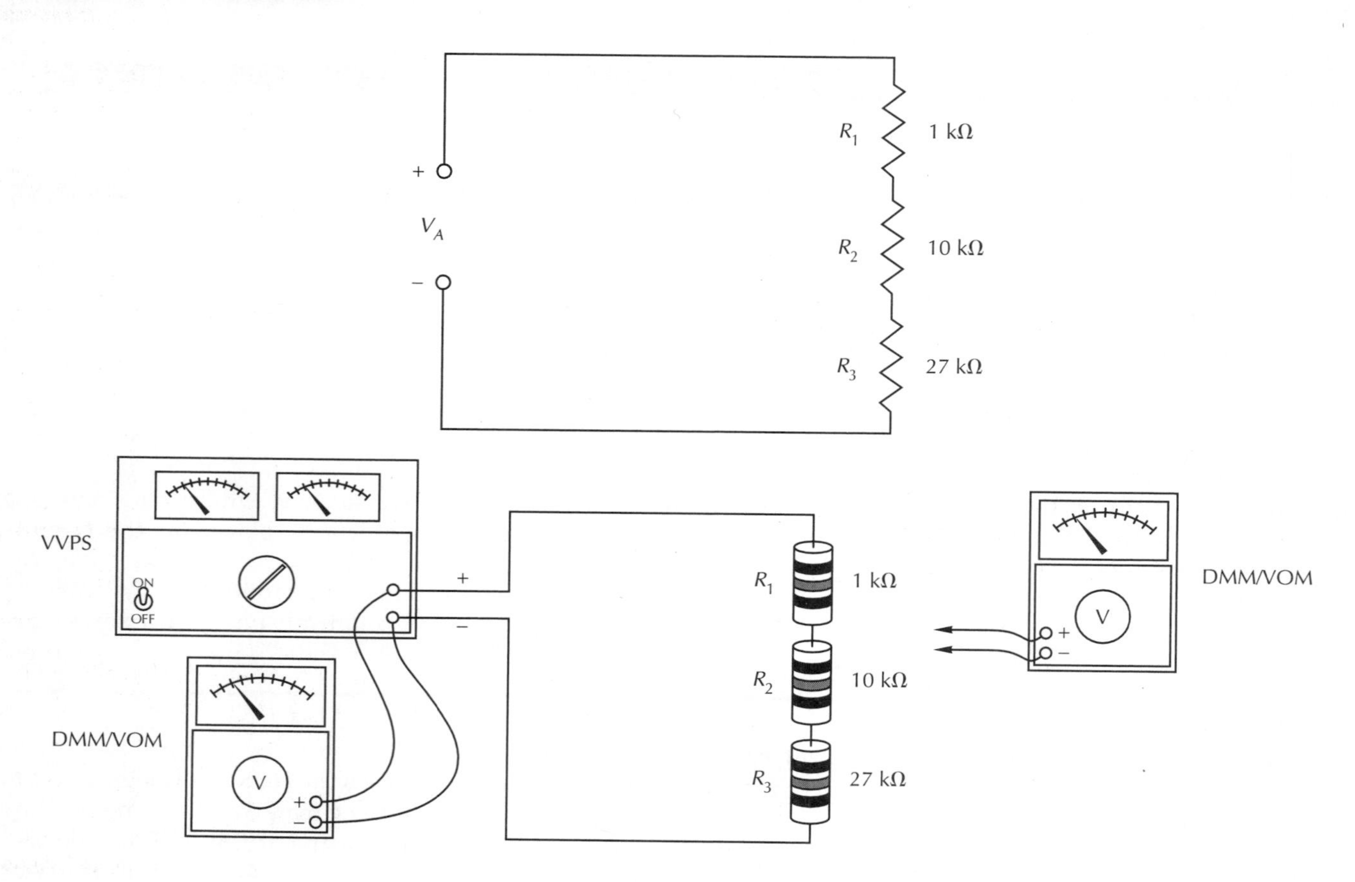

FIGURE 12
Standard schematic diagram (top) and pictorial diagram (bottom)

PROJECT PURPOSE

To demonstrate the proportional relationship of *R*s and *V*s in series circuits through circuit measurements and calculations.

PARTS NEEDED

- ☐ DMM/VOM (2)
- ☐ VVPS (dc)
- ☐ CIS
- ☐ Resistors:
  1 kΩ
  10 kΩ
  27 kΩ

SAFETY HINTS

Be sure power is off when connecting meters.

ACTIVITY	OBSERVATION	CONCLUSION
1. Connect the initial circuit shown in Figure 12.	—	—
2. Apply 9.5 volts to the circuit and measure each of the individual voltage drops. Also calculate the circuit current.	V_A = ________ volts V_1 = ________ volts V_2 = ________ volts V_3 = ________ volts I calc. = ________ mA	Since the I is the same through all the resistors, the voltage drop across any given resistor is directly related to its R compared to the total circuit (*number of Rs, resistance*) ________.
3. Add all the individual voltage drops and note the sum.	$V_1 + V_2 + V_3$ equals ________ volts or V ________.	In essence, Kirchhoff's voltage law states that the arithmetic sum of voltage drops around any circuit closed loop must equal V applied. Does it? ________.
4. Calculate what fraction of the applied voltage is dropped by each of the resistors. Express your answer as a fraction. (Example: 1/38, 10/38, etc.)	*Example:* V_1 = 1/38th V_A V_2 = ________ V_A V_3 = ________ V_A	Each resistor dropped the same fraction of V applied as its ________ value is of the total ________.
5. Compute the fractional relationship of V_1 to V_2 and V_3. Express answers as fractions.	*Example:* V_1 = 1/10th V_2 V_1 = ________ V_3	Because I is the same through all elements in a series circuit and since V = ________ × ________, the voltage drops across the resistors are related to each other by the same factor as their ________.
6. Predict what value V_2 would be if V_A were 19 volts. Change V_A to 19 volts and measure V_2.	V_2 pred. = ________ volts V_2 meas. = ________ volts	V_2 is now what fraction of V_A? ________. Is this the same fraction as when V_A = 9.5 volts? ________. Changing V (*does, does not*) ________ change the distribution percentages of V_A. If one of the Rs were changed, would the distribution percentages change? ________. We may conclude that in series circuits, the largest R will drop the (*least, most*) ________ voltage, and the smallest R the (*least, most*) ________ voltage.

PROJECT
11
CONTINUED

SERIES CIRCUITS
Voltage Distribution in Series Circuits *(Continued)*

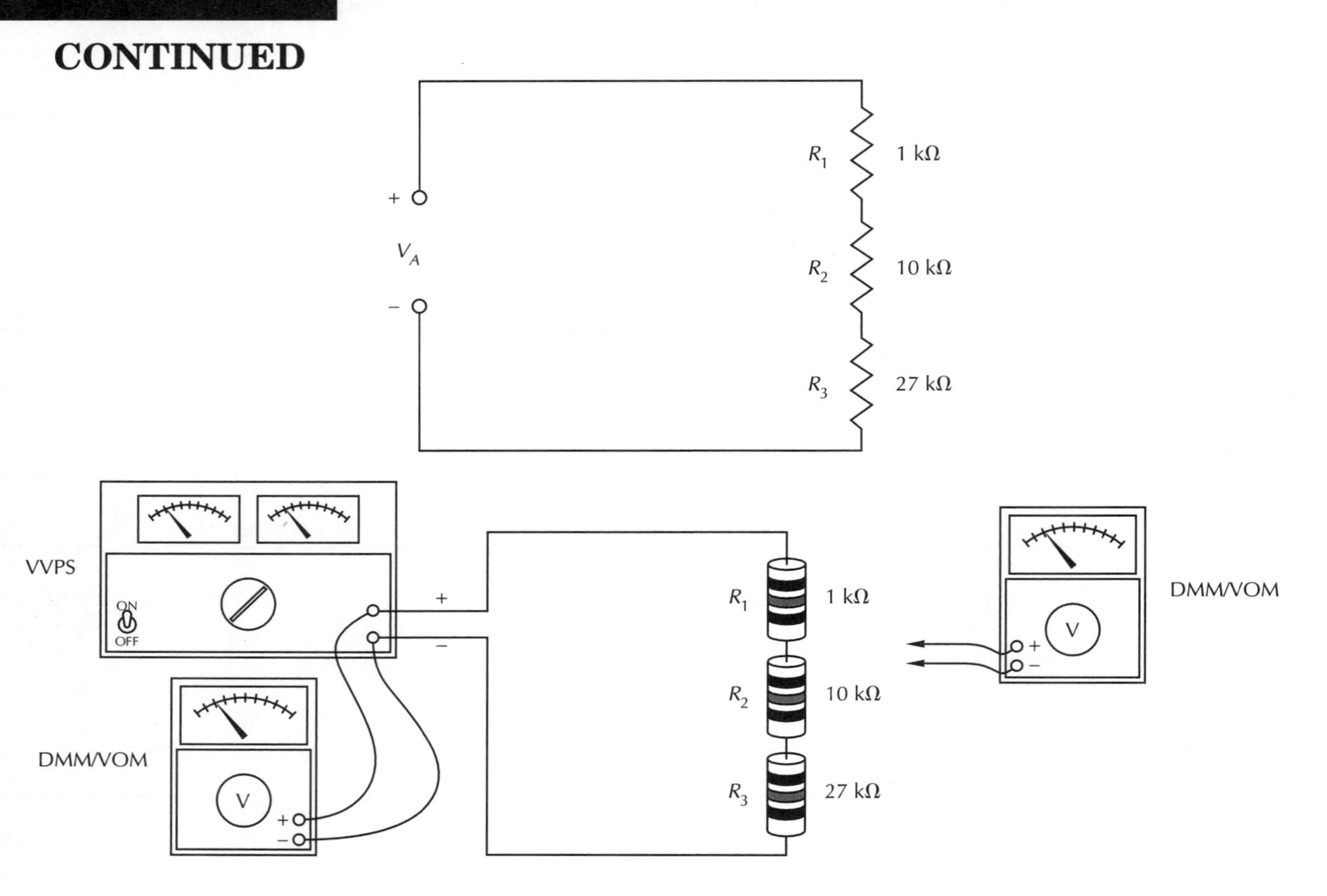

FIGURE 12
Standard schematic diagram (top) and pictorial diagram (bottom)

PROJECT PURPOSE
To demonstrate the proportional relationship of *R*s and *V*s in series circuits through circuit measurements and calculations.

PARTS NEEDED
- ☐ DMM/VOM (2)
- ☐ VVPS (dc)
- ☐ CIS
- ☐ Resistors:
  - 1 kΩ
  - 10 kΩ
  - 27 kΩ
  - 100 kΩ

SAFETY HINTS
Be sure power is off when connecting meters.

ACTIVITY	OBSERVATION	CONCLUSION
7. Use the voltage-divider rule and calculate V_1 and V_3 assuming $V_T = 19$ V. $(V_X = \frac{R_X}{R_T} \times V_T)$	V_1 calc. = ________ volts V_3 calc. = ________ volts	—
8. Measure V_1 and V_3 with 19 V applied to circuit.	V_1 meas. = ________ volts V_3 meas. = ________ volts	Do the measured values for V_1 and V_3 confirm the calculations using the voltage-divider rule? (*yes*, *no*) ____

EXTRA CREDIT STEP(S)

9. Turn off the power supply and replace R_3 with a 100-kΩ resistor.	—	—
10. Adjust voltage applied to the circuit to approximately 28 volts and measure the resistor voltage drops.	V_1 = ________ volts V_2 = ________ volts V_3 = ________ volts	Is V_3 equal to about 10 times V_2? ________? Is V_2 about 10 times greater than V_1? ________. Is the ratio of V_3's voltage to V_1's voltage about equal to their *R* ratio? ________? What might cause the ratios discussed to not be exactly 10:1 in each case? ________ ________ ________ ________

PROJECT 12

SERIES CIRCUITS
Power Distribution in Series Circuits

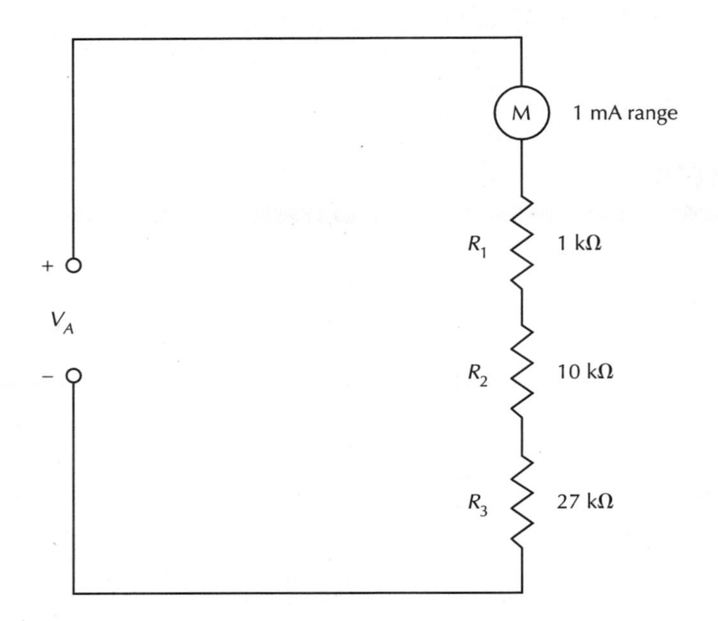

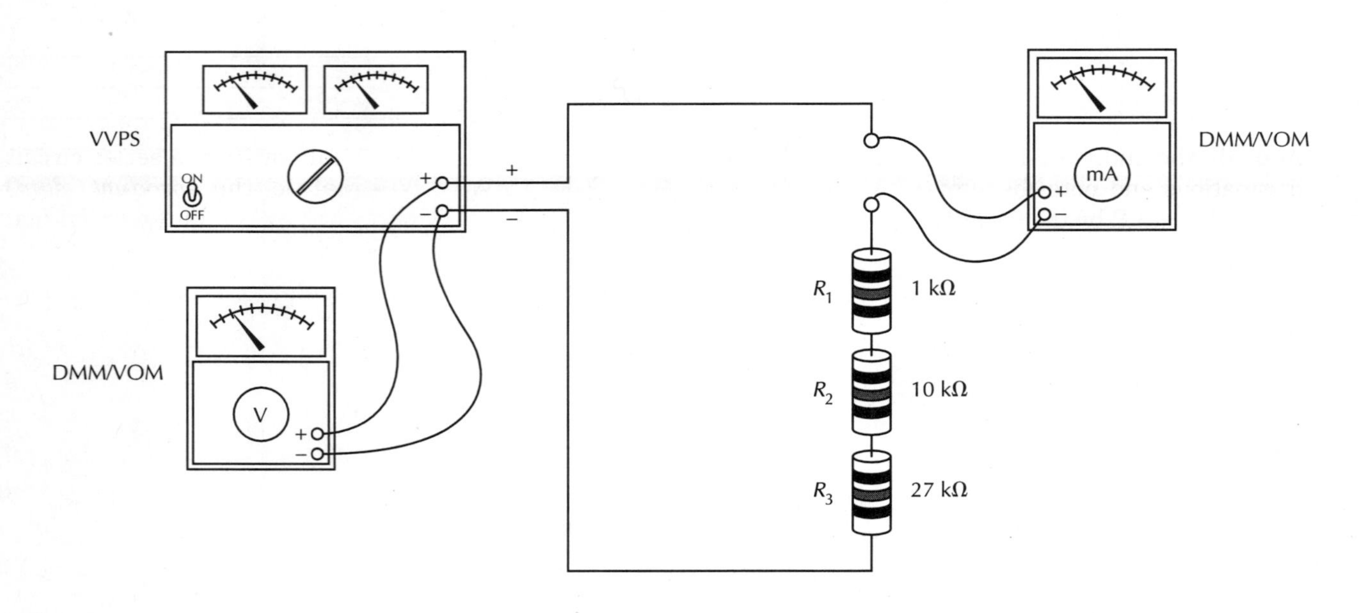

FIGURE 13
Standard schematic diagram (top) and pictorial diagram (bottom)

PROJECT PURPOSE

To illustrate that power distribution in a series circuit is directly related to resistance distribution and that total power equals the sum of all the individual power dissipations.

PARTS NEEDED

- ☐ DMM/VOM (2)
- ☐ VVPS (dc)
- ☐ CIS
- ☐ Resistors:
  - 1 kΩ
  - 10 kΩ
  - 27 kΩ

SAFETY HINTS

Be sure power is off when connecting meters.

ACTIVITY	OBSERVATION	CONCLUSION
1. Connect the initial circuit shown in Figure 13.	—	—
2. Apply 19 volts (V_A) to the circuit. Measure the current and the individual voltage drops and calculate the power dissipated by each of the resistors.	I = ______ mA V_1 = ______ volts V_2 = ______ volts V_3 = ______ volts	P_1 calculated = ______ mW P_2 calculated = ______ mW P_3 calculated = ______ mW We may conclude that since the I is the same through all resistors, the I^2R or (*voltage, power*) ______ dissipated by each resistor is directly related to its (*size, resistance*) ______ value. Furthermore, the power distribution throughout the circuit is the same as the (I, R) ______ distribution. This means that the largest value R will dissipate the (*most, least*) ______ amount of power; the smallest R, the (*most, least*) ______ power.
3. Add all the individual power dissipations and note the sum. Also, calculate P_T by the formula $P_T = V_T \times I_T$.	$P_1 + P_2 + P_3$ equals ______ mW $V_T \times I_T$ = ______ mW	The total power in a series circuit is equal to the (*product, sum*) ______ of all the individual power dissipations.
4. What ratio does P_1 have to P_3? Express as a ratio. (For example: 1:10, 2:5, etc.)	Ratio = ______	The ratios of power dissipated by two resistors in a series circuit is the same as their ______ ratio.

SERIES CIRCUITS
Power Distribution in Series Circuits *(Continued)*

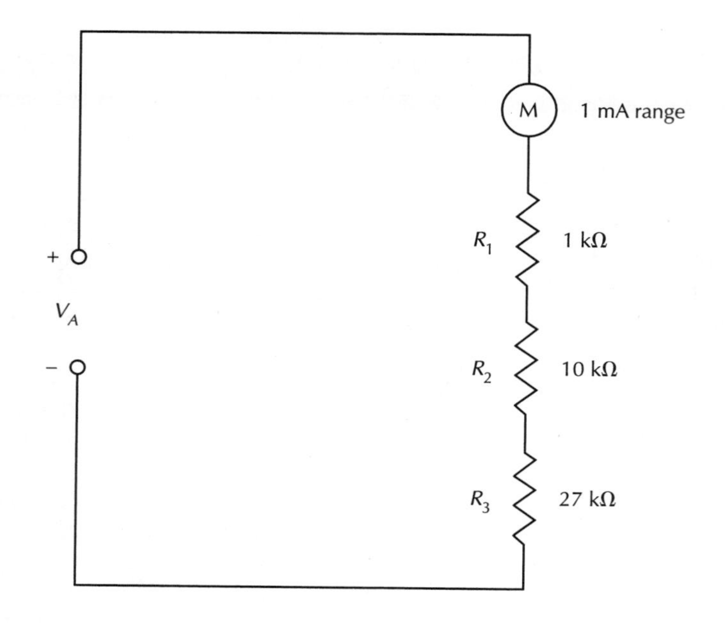

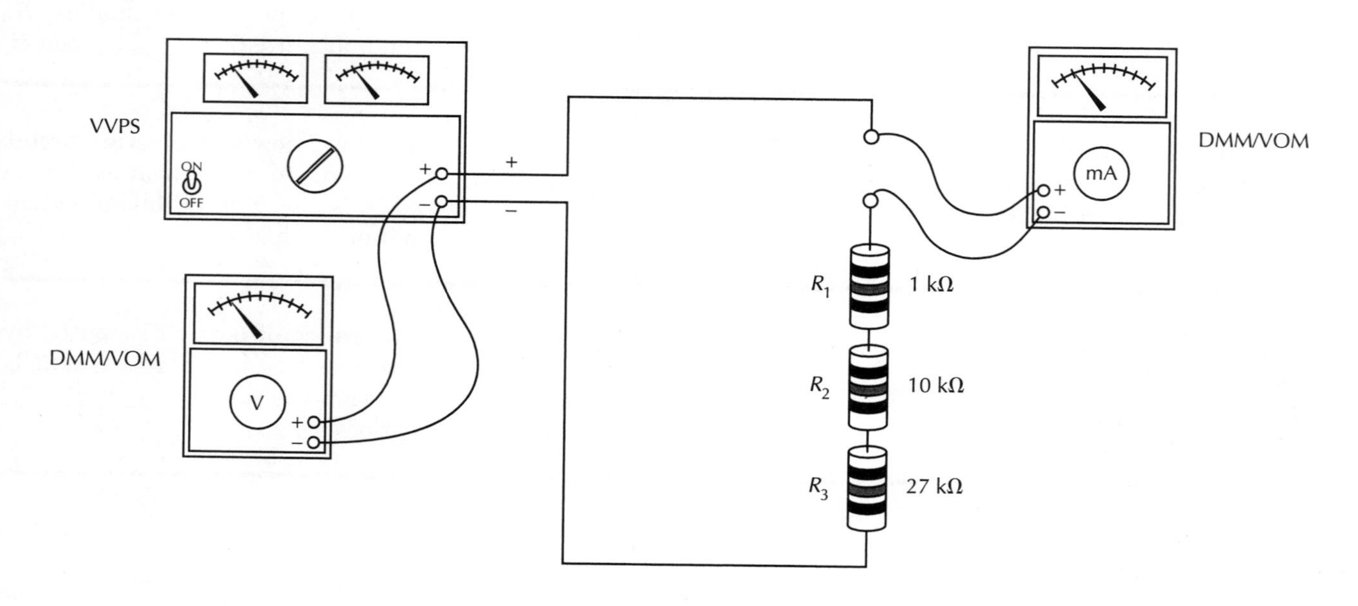

FIGURE 13
Standard schematic diagram (top) and pictorial diagram (bottom)

ACTIVITY	OBSERVATION	CONCLUSION
5. Indicate what would happen to P_T and to the individual power dissipations if V_A were cut in half.	P_T = ______________ as much as before. Individual Ps would also be ________________ original.	For a given R, decreasing V to one-half will also cause the circuit current to (*increase, decrease*) ______________ to ____________. Therefore, the product of $V \times I$ will be (*1/2, 1/4*) __________ the original value. If Rs remain unchanged but V is changed, the percentage of P_T dissipated by any given R will (*change, not change*) __________ but the actual value of power dissipated will (*change, not change*) ______________________.

EXTRA CREDIT STEP(S)

6. Change the V_A to a value of 9.5 volts. Measure and calculate values, as required to fill in the Observation column blanks.	V_A = ________________ volts I = ________________ mA P_T = ________________ mW V_1 = ________________ volts P_1 = ________________ mW V_2 = ________________ volts P_2 = ________________ mW V_3 = ________________ volts P_3 = ________________ mW	When V_A was reduced to half its original value, current decreased to ______________________ its original value and total power dissipated by the circuit decreased to ______________________ its original value. Does the power dissipated by each resistor also change by this same factor? ______________. Does this verify the data in step 5? ____________.

PROJECT 13

SERIES CIRCUITS
Effects of an Open in Series Circuits

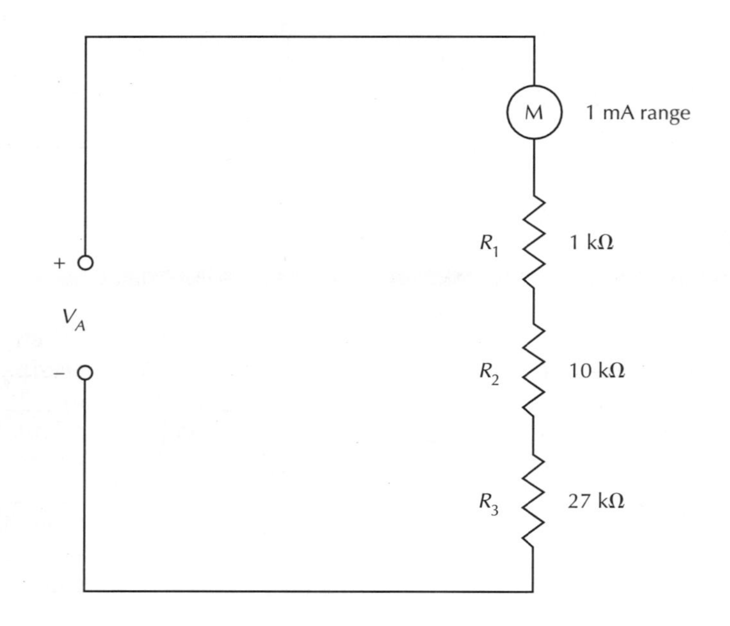

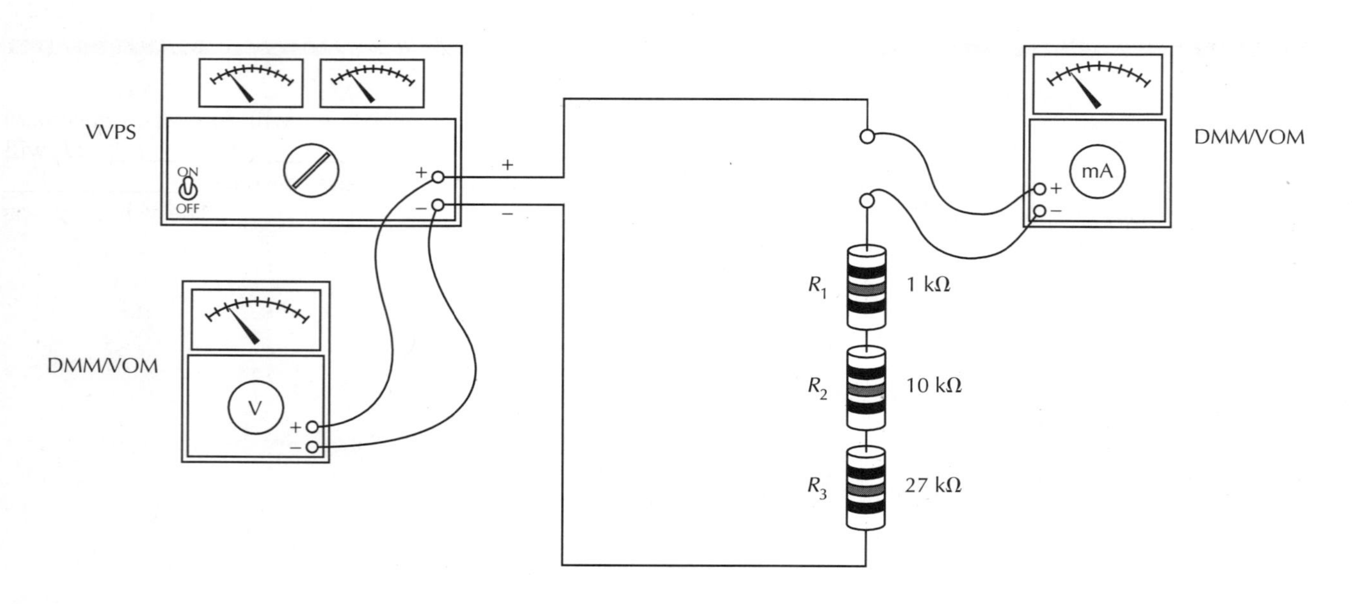

FIGURE 14
Standard schematic diagram (top) and pictorial diagram (bottom)

PROJECT PURPOSE

To provide "hands-on" experience regarding circuit parameter changes that occur when an open develops in a series circuit and to verify that voltage applied appears across the open portion of the circuit.

PARTS NEEDED

- ☐ DMM/VOM (2)
- ☐ VVPS (dc)
- ☐ CIS
- ☐ Resistors:
  - 1 kΩ
  - 10 kΩ
  - 27 kΩ

SAFETY HINTS

Be sure power is off when connecting meters.

ACTIVITY	OBSERVATION	CONCLUSION
1. Connect the initial circuit as shown in Figure 14.	—	—
2. Adjust V applied to 19 volts. Measure and record the current and individual voltage drops.	I = ________ mA V_1 = ________ volts V_2 = ________ volts V_3 = ________ volts	—
3. To simulate an R becoming open in this series circuit, remove R_2 and leave the circuit open between R_1 and R_3. Measure and record the circuit current and the individual voltage drops.	I = ________ mA V_1 = ________ volts V_2 = ________ volts (across open) V_3 = ________ volts	Since in a series circuit there is only one path for current flow, opening any element within the series circuit will cause: (*continuity*, *discontinuity*) ________. The R_T of the circuit then appears to be infinitely (*high*, *low*) ________. The voltage drop across R_1 was ________ volts because with zero current, the $I \times R$ drop must be ________. We conclude that if any part of a series circuit opens, R_T will (*increase*, *decrease*) ________ to ________; I_T will (*increase*, *decrease*) ________ to ________. The voltage drops across the unopened elements will (*increase*, *decrease*) ________ to ________, and the potential difference across the open portion of the circuit will (*increase*, *decrease*) ________ to ________.

PROJECT 13 CONTINUED

SERIES CIRCUITS
Effects of an Open in Series Circuits *(Continued)*

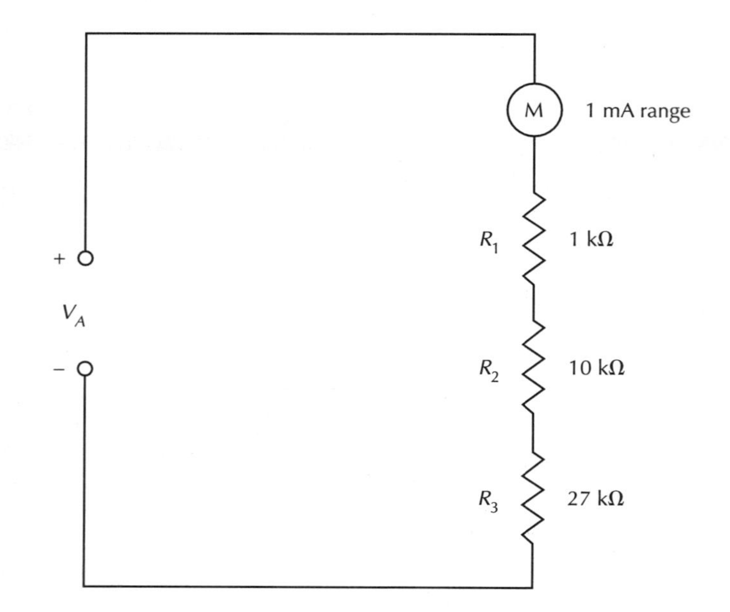

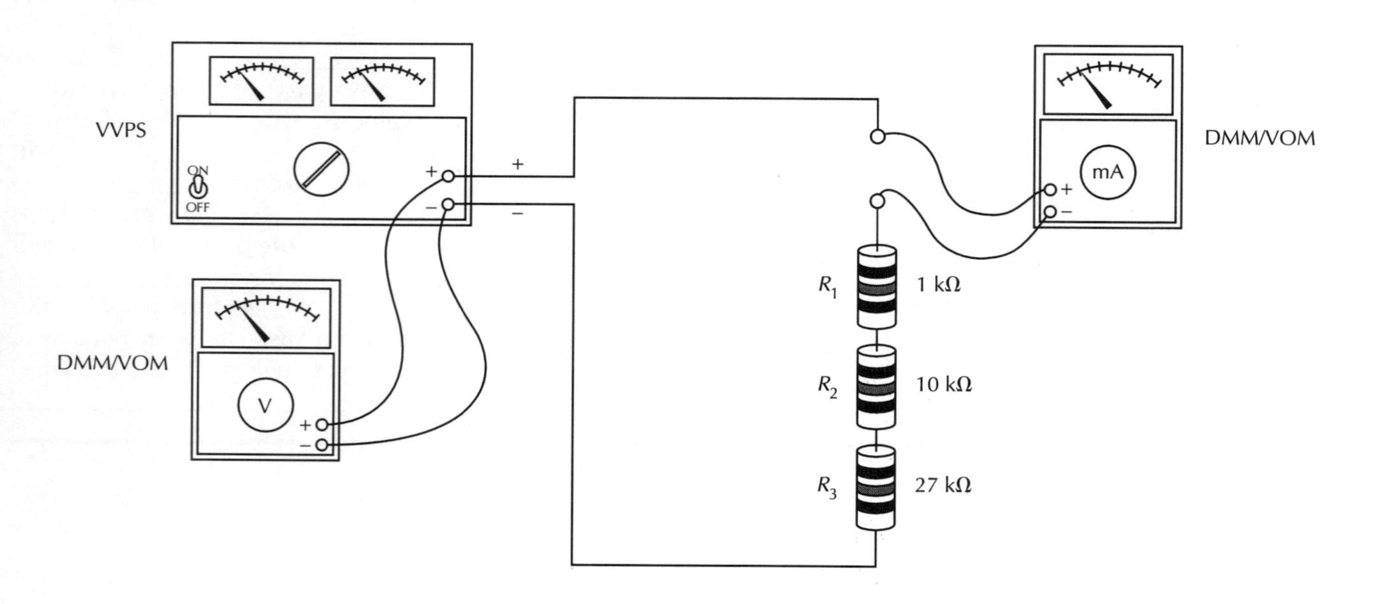

FIGURE 14
Standard schematic diagram (top) and pictorial diagram (bottom)

ACTIVITY	OBSERVATION	CONCLUSION
4. Assume R_2 were replaced in the circuit and R_3 removed. List the predicted results in the Observation column.	I = ________ mA V_1 = ________ volts V_2 = ________ volts V_3 = ________ volts	—
5. Make the change suggested in step 4 above, make appropriate measurements, and note the results.	I = ________ mA V_1 = ________ volts V_2 = ________ volts V_3 = ________ volts	Do the results of these measurements verify the conclusions of step 3? ________. The main difference noted in the parameters for this step compared to step 3 is that V applied now appears across R ________, rather than R ________.

SERIES CIRCUITS
Effects of a Short in Series Circuits

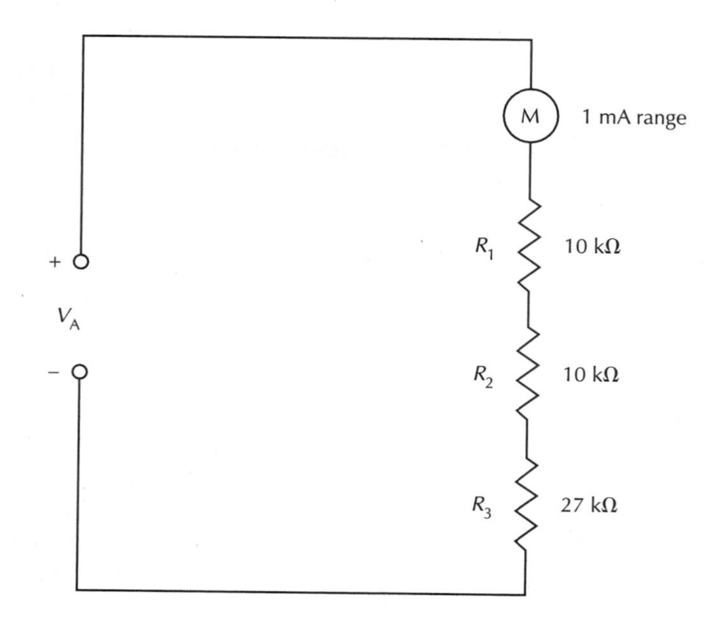

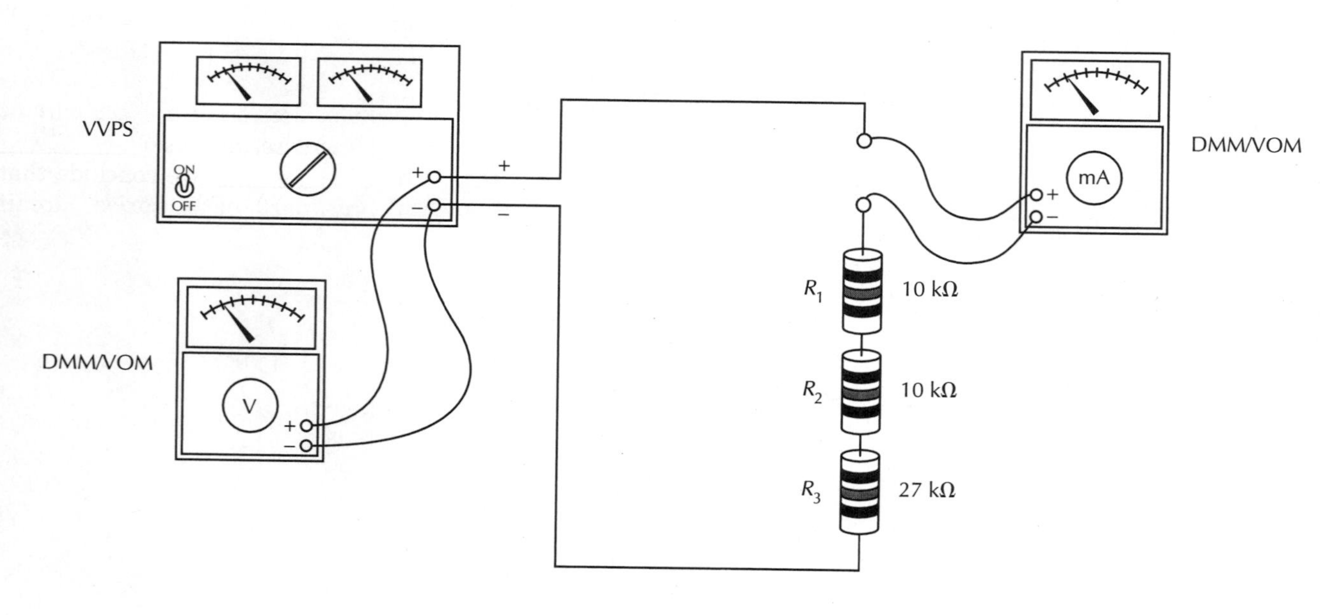

FIGURE 15
Standard schematic diagram (top) and pictorial diagram (bottom)

PROJECT PURPOSE

To provide "hands-on" experience regarding circuit parameter changes that occur when a short develops in part of a series circuit.

PARTS NEEDED

- ☐ DMM/VOM (2)
- ☐ VVPS (dc)
- ☐ CIS
- ☐ Resistors:
  10 kΩ (2)
  27 kΩ

SAFETY HINTS

DO NOT SHORT OUT ALL THREE RESISTORS AT ONCE!

Be sure power is off when connecting meters.

ACTIVITY

1. Connect the initial circuit as shown in Figure 15.
2. Adjust V applied to 20 volts. Measure and record the current and individual voltage drops.
3. To simulate an R "shorting out," remove R_2 and replace it with a jumper wire. Measure and note the circuit I and individual voltage drops with the shorted element replacing R_2.

OBSERVATION

—

I = ____________ mA
V_1 = ____________ volts
V_2 = ____________ volts
V_3 = ____________ volts

I = ____________ mA
V_1 = ____________ volts
V_2 = ____________ volts
(across short)
V_3 = ____________ volts

CONCLUSION

—

—

Shorting out R_2 has caused the circuit R_T to (*increase*, *decrease*) ____________ to ____________ ohms. This then caused I_T to (*increase*, *decrease*) ____________. The new (*higher*, *lower*) ____________ current caused the $I \times R$ drops across the unshorted elements (Rs) in the circuit to (*increase*, *decrease*) ____________. The R between the shorted terminals is effectively ____________ ohms. Therefore, the $I \times R$ drop across the shorted element or section of a series circuit will (*increase*, *decrease*) ____________ to ____________. We conclude that if any part of a series circuit "shorts," R_T will (*increase*, *decrease*) ____________; I_T will (*increase*, *decrease*) ____________; the voltage drops across the unshorted elements will (*increase*, *decrease*) ____________; and the V across the shorted element or section of the circuit will (*increase*, *decrease*) ____________ to ____________. The total power supplied to the circuit will (*increase*, *decrease*) ____________ with the shorted condition.

SERIES CIRCUITS

Complete the following review questions, indicating the appropriate response by placing a check in the box next to the correct answer.

1. The total resistance in a series circuit is equal to:

   ☐ sum of all the *R*s
   ☐ largest *R* minus smallest *R*
   ☐ neither of these

2. In a series circuit there is:

   ☐ only one path for current
   ☐ as many paths as there are components

3. The current in a series circuit is:

   ☐ the same through all parts
   ☐ different through each component

4. The highest voltage drop in a series circuit appears across:

   ☐ the smallest *R*
   ☐ the highest *R*
   ☐ the average value *R*

5. If R_1 is 10 times larger in value than R_2:

   ☐ V_1 = 10 times V_2
   ☐ V_1 = one-tenth V_2
   ☐ neither of these

6. In a series circuit, the smallest *R* will dissipate:

   ☐ the most power
   ☐ no power
   ☐ the least power
   ☐ none of these

7. In a series circuit, the applied voltage equals:

   ☐ the largest *V* drop minus the smallest *V* drop
   ☐ sum of all *V* drops

8. In a series circuit, the percentage of the circuit applied voltage dropped across any given resistor is directly related and dependent upon the ratio of the given *R* to:

   ☐ *R* total
   ☐ *I* total
   ☐ *V* total
   ☐ *P* total

9. In a series circuit, if part of the circuit becomes open, then:

   a. R_T will
      - ☐ increase
      - ☐ decrease
      - ☐ remain the same

   b. V_A will
      - ☐ increase
      - ☐ decrease
      - ☐ remain the same

   c. I_T will
      - ☐ increase
      - ☐ decrease
      - ☐ remain the same

   d. The voltage across the opened portion of the circuit will:
      - ☐ increase
      - ☐ decrease
      - ☐ remain the same

   e. The voltage drops across the unopened elements will:
      - ☐ increase
      - ☐ decrease
      - ☐ remain the same

10. In a series circuit, if part of the circuit becomes shorted, then:

    a. R_T will
       - ☐ increase
       - ☐ decrease
       - ☐ remain the same

    b. V_A will
       - ☐ increase
       - ☐ decrease
       - ☐ remain the same

    c. I_T will
       - ☐ increase
       - ☐ decrease
       - ☐ remain the same

    d. The voltage across the shorted portion of the circuit will:
       - ☐ increase
       - ☐ decrease
       - ☐ remain the same

    e. The voltage drops across the other components in the circuit will:
       - ☐ increase
       - ☐ decrease
       - ☐ remain the same

STUDENT LOG FOR OPTIONAL TROUBLESHOOTING EXERCISE

Students should write down the starting point symptom given to them by the instructor, and then proceed to log each step used in the troubleshooting process. For example, the first step after learning the starting point symptom information is to identify the area to be bracketed as the "area of uncertainty," and should include all circuitry components that "might" cause the trouble producing the symptoms. The next step is to make a decision about the first test, etc.

Starting point symptom:

____

Components and circuitry included in "initial brackets" ("area of uncertainty")

____

First test description (what type test and where?)

____

Components and circuitry *still* included in "bracketed" area after first test

____

Second test description (what type test and where?)

____

Components and circuitry *still* included in "bracketed" area after second test

____

Third test description (what type test and where?)

____

Components and circuitry *still* included in "bracketed" area after third test

____

Fourth test description (what type test and where?)

____

Components and circuitry *still* included in "bracketed" area after fourth test

____

Fifth test description (what type test and where?)

____

NOTE: *Keep repeating the testing, bracketing, and logging procedure until you are quite sure you have found the trouble. Call the instructor to check your work; then, replace the identified faulty component to see if the circuit operates properly!*

Suspected Trouble:

____

Trouble Verified by Instructor and by Circuit Operation: Yes ___ No ___

PARALLEL CIRCUITS

Objectives

You will connect several dc resistive parallel circuits and make measurements and observations regarding their important electrical characteristics.

In completing these projects, you will connect circuits, make measurements, perform calculations, draw conclusions, and be able to answer questions about the following items related to parallel circuits.

- Total resistance
- Voltage distribution
- Current distribution
- Power dissipation(s)
- Effects of opens
- Effects of shorts
- Application of Ohm's law
- Application of Kirchhoff's voltage law
- Application of Kirchhoff's current law

PROJECT/TOPIC CORRELATION INFORMATION

PROJECT		TEXT CHAPTER	SECTION	RELATED TEXT TOPIC(S)
15	Equivalent Resistance in Parallel Circuits	5	5-5, 5-6	Resistance in Parallel Circuits
16	Current in Parallel Circuits	5	5-3	Current in Parallel Circuits
17	Voltage in Parallel Circuits	5	5-1, 5-2	Voltage in Parallel Circuits
18	Power Distribution in Parallel Circuits	5	5-7	Power in Parallel Circuits
19	Effects of an Open in Parallel Circuits	5	5-8	Effects of Opens in Parallel Circuits and Troubleshooting Hints
20	Effects of a Short in Parallel Circuits	5	5-9	Effects of Shorts in Parallel Circuits and Troubleshooting Hints

PROJECT 15

PARALLEL CIRCUITS
Equivalent Resistance in Parallel Circuits

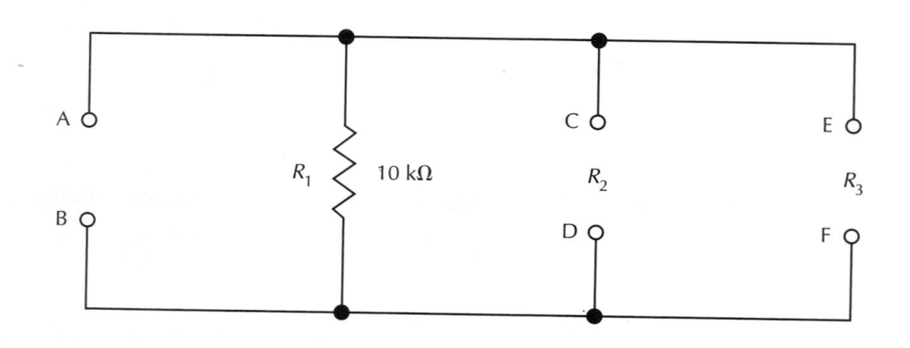

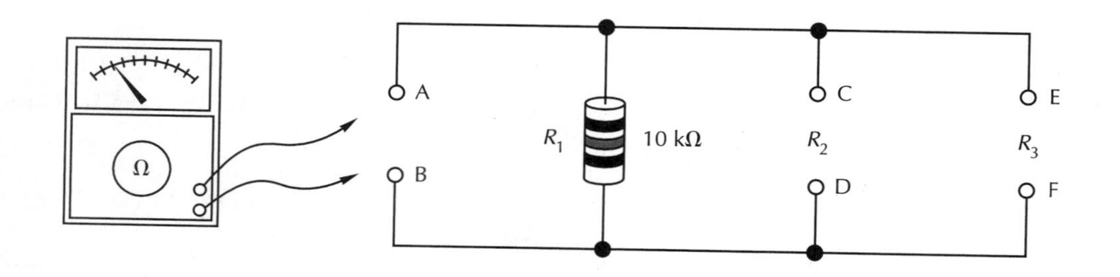

FIGURE 16
Standard schematic diagram (top) and pictorial diagram (bottom)

PROJECT PURPOSE

To verify that the equivalent resistance of parallel resistances is less than the least value resistance in parallel. To provide further practice in using parallel resistance formulas and confirmation of these through circuit measurements.

PARTS NEEDED

- ☐ DMM/VOM
- ☐ CIS
- ☐ Resistors:
  - 10 kΩ (3)
  - 47 kΩ
  - 100 kΩ

SAFETY HINTS

DO NOT use a power supply in the first six (6) steps of this project!

ACTIVITY	OBSERVATION	CONCLUSION
1. Connect the initial circuit as shown in Figure 16.	—	—
2. Measure R_e at points A and B.	R_e = ____________ ohms	—
3. Use the popular product-over-the-sum formula and calculate R_e if a 10-kΩ R were inserted between points C and D on the trainer. $R_e = R_1 \times R_2 \div (R_1 + R_2)$ After calculating R_e, measure it with a DMM/VOM at points A and B with a second 10-kΩ R inserted at points C and D.	R_e calculated = ____________ ohms R_e measured = ____________ ohms	When two resistors of equal value are connected in parallel, R_e is equal to one-half the R of (*one*, *both*) ____________ branch(es).
4. Assume a third 10-kΩ R is to be added at points E and F and calculate R_e. Use the reciprocal method, or, $R_e = R_{e^1} \times R_3 \div (R_{e^1} + R_3)$ After calculating, connect a third 10-kΩ R between points E and F and measure R_e to verify your calculations.	R_e calculated = ____________ ohms R_e measured = ____________ ohms	When three resistors of equal value are connected in parallel, R_e is equal to (*1/2, 1/3, 1/4*) ________ the R of one branch. In parallel circuits, the total resistance of the circuit is always less than the (*highest*, *lowest*) ________ branch R. If two or more *unequal* Rs are in parallel, can we divide one branch R by the number of branches to find R_e? ____________.
5. Change R_2 to a 47-kΩ R and R_3 to a 100-kΩ R. Use the assumed voltage method of solving for R_e as follows: a. Assume 47 volts applied. b. Solve for each branch I. c. Calculate I_T. d. Find R_e by Ohm's law. $\left(R_e = \frac{V_T}{I_T}\right)$ After calculating R_e, use the DMM or VOM and measure R_e to verify your calculations.	I_1 = ____________ mA I_2 = ____________ mA I_3 = ____________ mA I_T = ____________ mA R_e = ____________ ohms R_e measured = ____________ ohms	The assumed voltage method of finding R_e is sometimes easier to use than the product-over-the-sum method, especially when several branches of unequal values are involved. Many times, if an appropriate value of V applied is assumed, one can solve for the branch currents in his/her head, then add the branch currents easily to solve for (R_T, I_T) ____________. Then all that remains is a simple (*multiplication*, *division*) ____________ problem to find R_e. **NOTE:** Using a calculator, the reciprocal method is also very easy!

PROJECT 15 CONTINUED

PARALLEL CIRCUITS
Equivalent Resistance in Parallel Circuits *(Continued)*

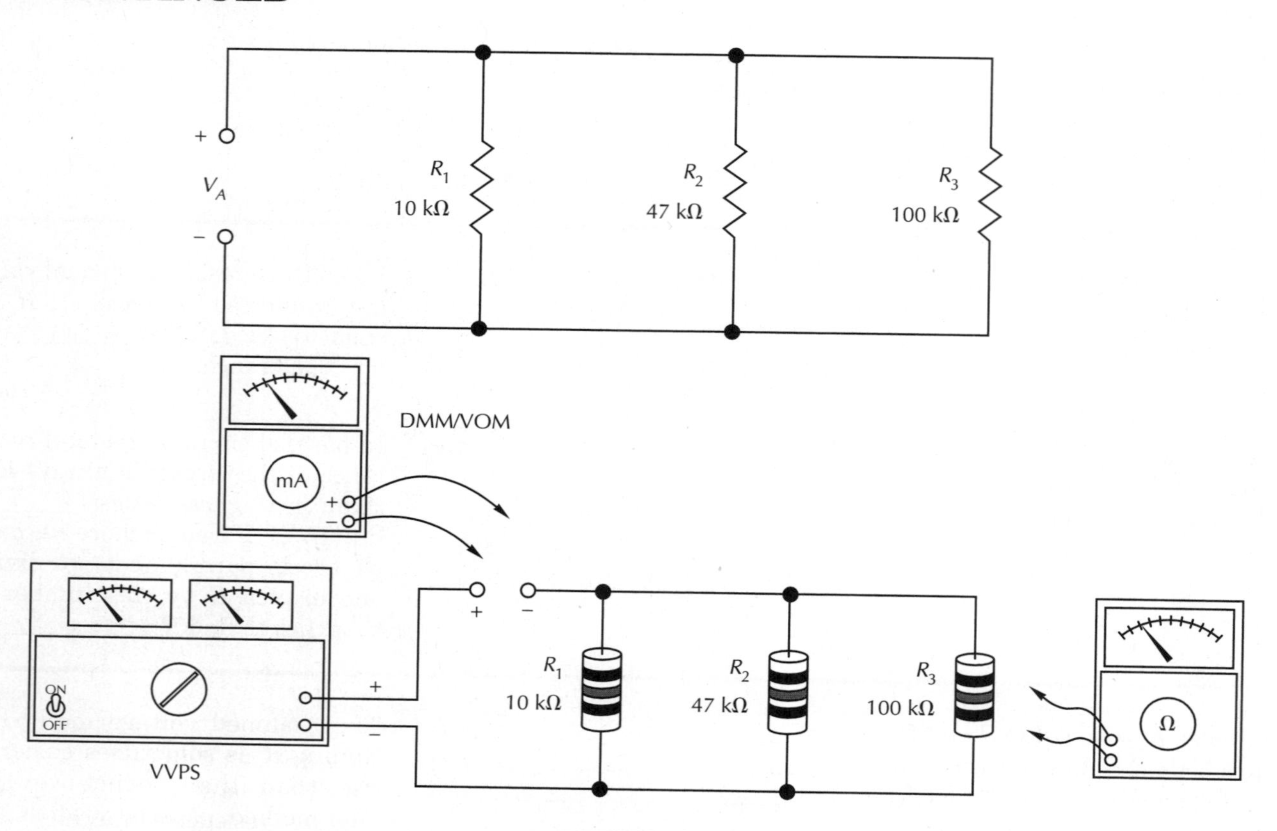

FIGURE 17
Standard schematic diagram (top) and pictorial diagram (bottom)

PROJECT PURPOSE

To provide further practice in using parallel resistance formulas and confirmation of these through circuit measurements.

PARTS NEEDED

- ☐ DMM/VOM (2)
- ☐ VVPS (dc)
- ☐ CIS
- ☐ Resistors:
  - 10 kΩ
  - 47 kΩ
  - 100 kΩ
  - one R to be determined by student calculation

SAFETY HINTS

Be sure power is off when making circuit changes and/or when connecting meters.

ACTIVITY	OBSERVATION	CONCLUSION
6. With R_1 = 10 kΩ, R_2 = 47 kΩ, and R_3 = 100 kΩ as shown in Figure 17, what will the parameters (electrical circuit values) be if we apply 7.6 volts? (Use 7.6 as the assumed voltage and calculate as before.)	I_1 = ______ mA I_2 = ______ mA I_3 = ______ mA I_T = ______ mA R_e = ______ ohms	Changing V applied does not change (R_T, I_T) ______ $_T$. The voltage that is "assumed" does make a difference in difficulty of computing the final result (R_e). From the preceding steps, it would seem wise to choose a value of assumed voltage that is easily divided by the resistance values for solving branch currents, since no matter what voltage is assumed, it does not change the actual circuit (I, R) ______.
7. With the circuit described in step 6 connected, insert the milliammeter to read I_T. Apply 7.6 volts to the circuit and measure I_T. After measuring, REMOVE the power supply and current meter from the circuit. Replace the current meter with a jumper wire.	I_T = ______ mA	The assumed voltage method of calculating R_e (*has been, has not been*) ______ verified.
8. A practical problem that sometimes confronts a technician is what value R must be put in parallel with the existing Rs in order to arrive at a desired equivalent resistance. A simple formula that helps solve this is: $R_u = \frac{R_k \times R_e}{R_k - R_e}$ where: R_u is R unknown, R_k is R known, and R_e is the desired resultant equivalent R. Assume you have a 27-kΩ R and want an R_e of 17 kΩ. Use the above formula, solve for R_u, then connect the circuit on the matrix using an available resistor as close to the calculated R_u as possible and measure R_e to verify results.	R_u = ______ ohms	The formula for solving for the unknown needed parallel resistance to arrive at a given equivalent resistance seems to work, because the results of the practical circuit were close to the theoretical result. It has been proven again that R_e turns out to be (*less, more*) ______ than the least resistance branch.

PROJECT 16

PARALLEL CIRCUITS
Current in Parallel Circuits

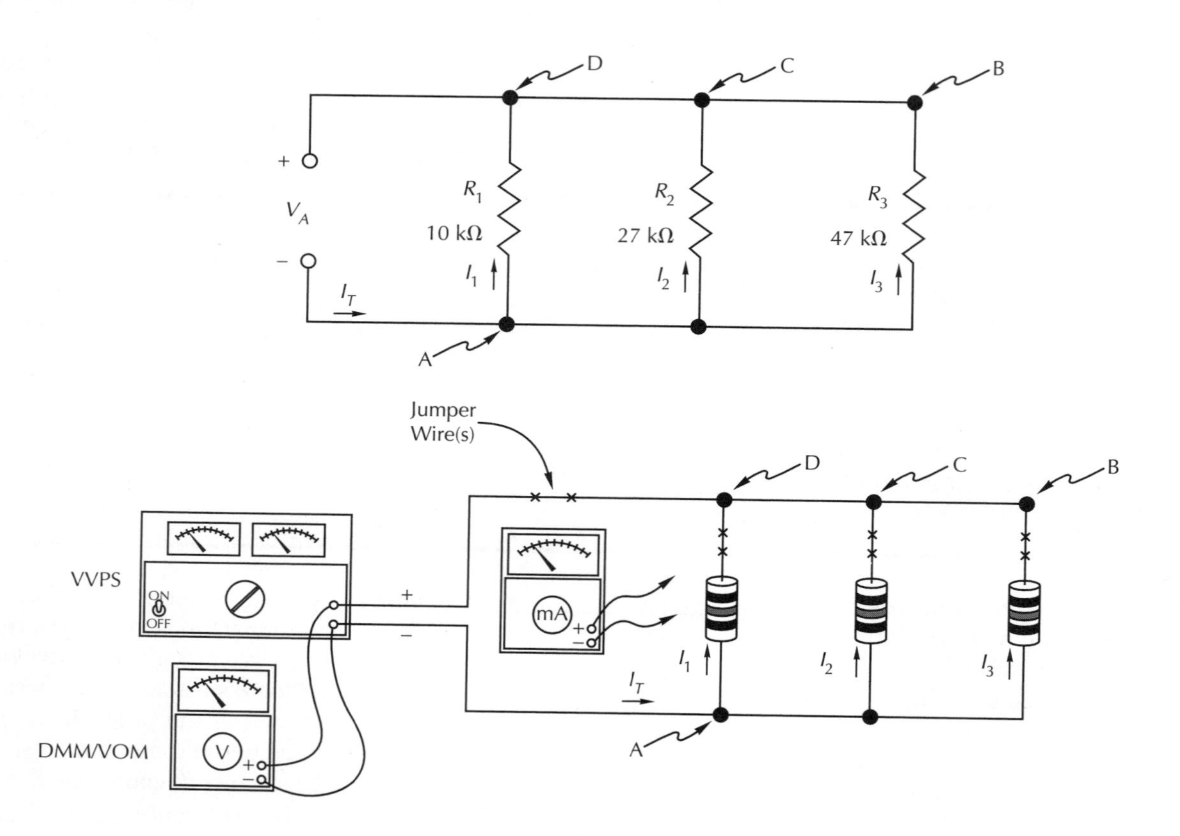

FIGURE 18
Standard schematic diagram (top) and pictorial diagram (bottom)

PROJECT PURPOSE

To confirm that current through parallel branches is inverse to each branch's resistance value and that total current equals the sum of all branch currents. To provide practical verification of Kirchhoff's current law.

PARTS NEEDED

- ☐ DMM/VOM (2)
- ☐ VVPS (dc)
- ☐ CIS
- ☐ Resistors:
  - 10 kΩ
  - 27 kΩ
  - 47 kΩ
  - 100 kΩ

SAFETY HINTS

Be sure power is off when making circuit changes and/or when connecting meters.

NOTE: Due to using standard-value resistors, some approximating is called for in this project to simply demonstrate concepts.

ACTIVITY	OBSERVATION	CONCLUSION
1. Connect the initial circuit as shown in Figure 18.	—	—
2. Replace the appropriate jumper with the current meter to read I_1. Apply 5 volts to the circuit and note I_1. Then move the meters and jumpers as required to measure I_2, I_3, and I_T.	I_1 = ______ mA I_2 = ______ mA I_3 = ______ mA I_T = ______ mA	R_2 is roughly ______ times larger in resistance than R_1. The current through R_2 is roughly ______ the current through R_1. R_3 is roughly ______ times larger in R value than R_1 and its current is roughly ______ of the current through R_1. From this we conclude that the current through parallel branches is (*directly*, *inversely*) ______ proportional to the branch Rs. Also, we observe from the measured currents that total circuit current equals the (*product*, *sum*) ______ of the ______ currents.
3. Change R_2 from 27 kΩ to 100 kΩ. Measure and record all the circuit currents one at a time. Make V applied 5 V in each case.	I_1 = ______ mA I_2 = ______ mA I_3 = ______ mA I_T = ______ mA	When R_2 was changed from 27 kΩ to 100 kΩ, keeping the same V_A, did the current through R_1 or R_3 change? ______. Did I_T change? ______. If so, was the change in I_T the same as the change in I_2? ______. This again proves that I_T is simply the sum of all the ______ currents. Since R_2 is 10 times larger than R_1, its current should be ______ of I_1. Is it? ______.
4. Refer again to the circuit diagram, and make the appropriate statements in the Conclusion column that will verify Kirchhoff's current law that states the current away from any point in a circuit must equal the current to that point.	—	The currents going away from point A in the circuit are labeled as ______ and ______ and ______. These equal the current coming to point A, and this current is designated by the symbol ______. The current *to and away from* point B is labeled as ______ on the diagram. The currents to and away from point C include ______ and ______; and the currents to and away from point D include ______ and ______ and ______.

PROJECT 17

PARALLEL CIRCUITS
Voltage in Parallel Circuits

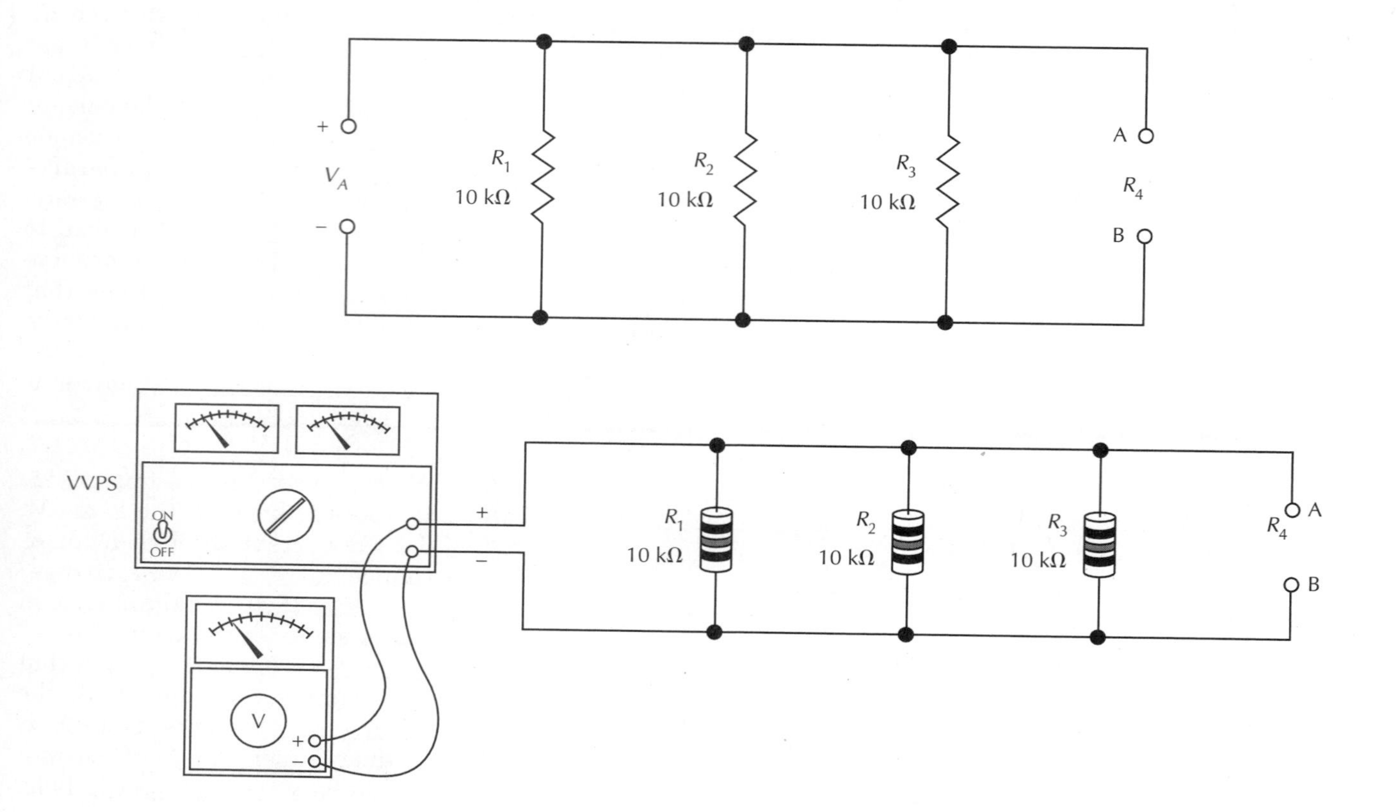

FIGURE 19
Standard schematic diagram (top) and pictorial diagram (bottom)

PROJECT PURPOSE

To demonstrate that the voltage across parallel branches is equal, and that in a simple parallel circuit, each branch voltage equals the circuit applied voltage.

PARTS NEEDED

- ☐ DMM/VOM (2)
- ☐ VVPS (dc)
- ☐ CIS
- ☐ Resistors:
  10 kΩ (4)
  100 kΩ

SAFETY HINTS

Be sure power is off when making circuit changes and/or when connecting meters.

ACTIVITY	OBSERVATION	CONCLUSION
1. Connect the initial circuit as shown in Figure 19.	—	—
2. Apply 10 volts V_A to the circuit and measure all the circuit voltages. Record in the Observation column.	V_A = ______ volts V_1 = ______ volts V_2 = ______ volts V_3 = ______ volts	We can conclude from the measurements that the voltage across all the resistors in parallel is the ______. Also, in a simple, purely parallel circuit configuration, the voltage across each branch equals V ______.
3. Change the value of R_2 from 10 kΩ to 100 kΩ and measure all the circuit voltages. Record in the Observation column.	V_A = ______ volts V_1 = ______ volts V_2 = ______ volts V_3 = ______ volts	Did all the voltages measure the same as before? ______. This indicates that changing the value of branch *R*s in a parallel circuit (*does, does not*) ______ alter the branch voltages. As a matter of fact, the branch voltages are not separate voltages, but really all the ______ *V*. Does changing the value of a branch *R* change the circuit total current? ______. Does it change the current through the unchanged *R* branch(es)? ______. Does it change the current through the changed *R* branch? ______.
4. Add a fourth branch to the parallel circuit by inserting a 10-kΩ resistor between points A and B. Measure the circuit voltages with 10 volts applied.	V_A = ______ volts V_1 = ______ volts V_2 = ______ volts V_3 = ______ volts V_4 = ______ volts	From the results of this step, we conclude that adding additional branches to a parallel circuit (*does, does not*) ______ alter the voltage across the parallel branches. (This assumes that the source of voltage is capable of meeting the current requirements of the circuit without its output voltage being altered.)

PROJECT 18

PARALLEL CIRCUITS
Power Distribution in Parallel Circuits

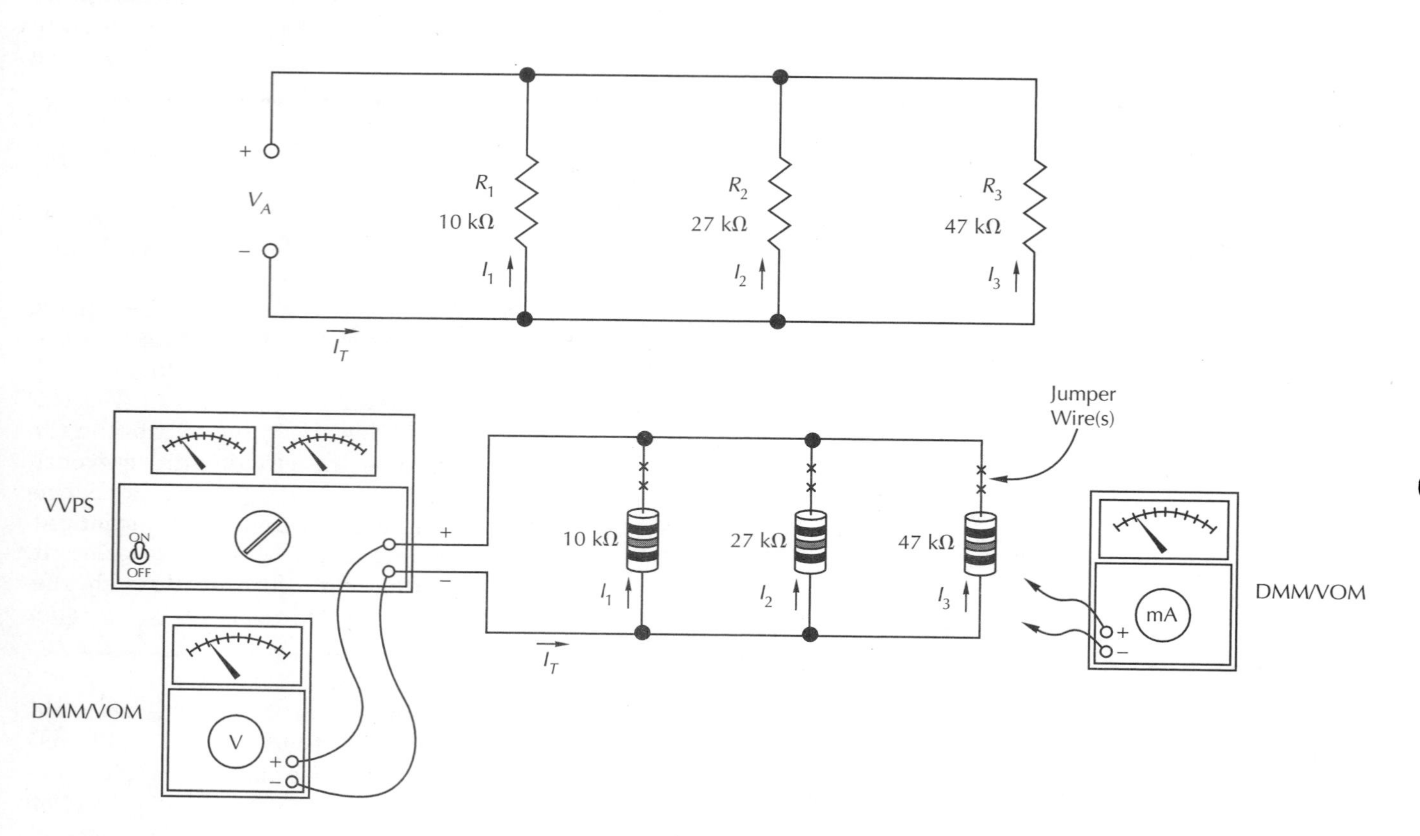

FIGURE 20
Standard schematic diagram (top) and pictorial diagram (bottom)

PROJECT PURPOSE

To illustrate that in parallel circuits power dissipation by any given branch is inverse to that branch's resistance value. To also demonstrate that total power in parallel circuits equals the sum of the branch power dissipations.

PARTS NEEDED

- ☐ DMM/VOM (2)
- ☐ VVPS (dc)
- ☐ CIS
- ☐ Resistors:
  - 10 kΩ (2)
  - 27 kΩ
  - 47 kΩ

SAFETY HINTS

Be sure power is off when making circuit changes and/or when connecting meters.

ACTIVITY	OBSERVATION	CONCLUSION
1. Connect the initial circuit as shown in Figure 20.	—	—
2. Apply 10 volts to the circuit, replace the appropriate "branch jumpers" one at a time with the milliammeter in order to measure each of the branch currents (I_1, I_2, and I_3), and record the results in the Observation column.	I_1 = ______ mA I_2 = ______ mA I_3 = ______ mA	Using the law that total current in a parallel circuit is equal to the (*product, sum*) ______ of the ______ currents, the total current for the circuit must be ______ mA. Since the total power dissipated in any resistive circuit can be calculated as V_T times ______, then the total power dissipated by this circuit and being supplied by the source is ______ mW.
3. Use the appropriate power formula(s) and calculate the individual power dissipations of R_1, R_2, and R_3. Record your answers.	P_1 = ______ mW P_2 = ______ mW P_3 = ______ mW	Does the sum of $P_1 + P_2 + P_3$ equal P_T? ______. Which resistor dissipates the most power? ______. This resistor is the (*largest, smallest*) ______ value *R* in the circuit. Which resistor dissipated the least power? ______. This resistor is the (*largest, smallest*) ______ value *R* in the circuit. From this we conclude that in a parallel circuit the smaller the branch *R*, the (*lesser, greater*) ______ power it will dissipate, because all *V*s are the same, and the *I*s are inversely proportional to the branch *R*s. Therefore, the $V \times I$ product will be greater if the resistor is of (*low, high*) ______ value.
4. Change R_2 from a 27-kΩ to a 10-kΩ resistor. Measure the branch currents and calculate the individual power dissipations and the P_T with 10 volts *V* applied.	I_1 = ______ mA I_2 = ______ mA I_3 = ______ mA P_1 = ______ mW P_2 = ______ mW P_3 = ______ mW P_T = ______ mW	Changing the resistance of one branch in a parallel circuit will cause the power dissipated by that branch to change and the ______ power to change. But the power dissipated by the other branches will remain the same as long as their resistance values are not changed.

PROJECT 19

PARALLEL CIRCUITS
Effects of an Open in Parallel Circuits

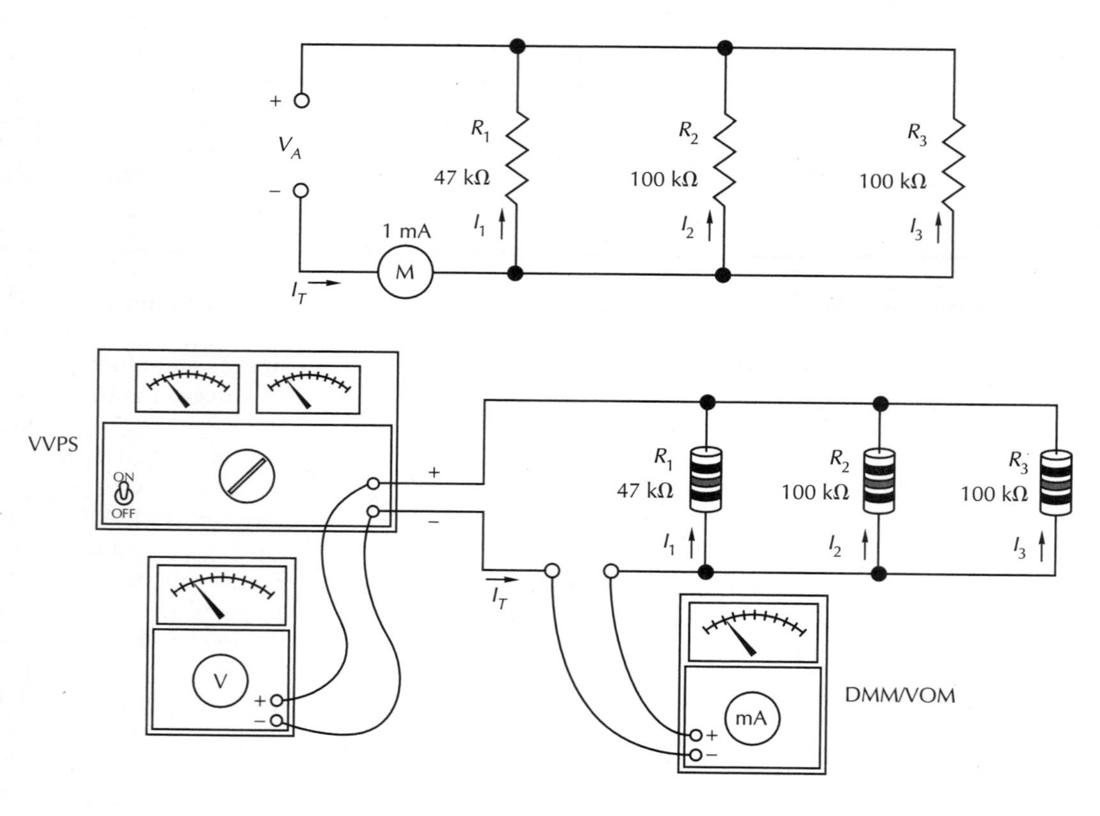

FIGURE 21
Standard schematic diagram (top) and pictorial diagram (bottom)

PROJECT PURPOSE
To provide experience through circuit measurements and observations regarding what happens to circuit parameters when an open occurs in a parallel circuit. To verify that total circuit current will decrease by the amount of current that was passing through the defective branch, prior to its opening.

PARTS NEEDED
- ☐ DMM/VOM (2)
- ☐ VVPS (dc)
- ☐ CIS
- ☐ Resistors:
  47 kΩ
  100 kΩ (2)

SAFETY HINTS
Be sure power is off when making circuit changes and/or when connecting meters.

ACTIVITY	OBSERVATION	CONCLUSION
1. Connect the initial circuit as shown in Figure 21.	—	—
2. Apply 20 volts to the circuit and note the total current. Use Ohm's law and calculate R_T and the branch currents. Measure I_T to verify.	I_T = ______ mA R_T = ______ ohms I_1 = ______ mA I_2 = ______ mA I_3 = ______ mA	R_T is less than the (*highest, lowest*) ______ resistance branch. I_T is equal to the ______ of the branch currents. The ratios of the branch currents to each other is inverse to the ratios of their ______ to each other.
3. Remove R_1 from the circuit to simulate an open in that branch. Apply 20 volts to the circuit and measure I_T, I_1, I_2, and I_3. Calculate R_T.	I_T = ______ mA R_T = ______ ohms I_1 = ______ mA I_2 = ______ mA I_3 = ______ mA	It can be concluded that if any branch of a parallel circuit opens, R_T will (*increase, decrease*) ______; therefore, I_T will (*increase, decrease*) ______ since V remained the same. Did the current through R_2 and R_3 change when R_1 opened? ______ The reason R_T increased was that when R_1 opened there was one less ______ path. The current through the opened branch (*increased, decreased*) ______ to ______ mA; whereas, the current through the unopened branches (*increased, decreased, remained the same*) ______. The voltage across all branches (*did, did not*) ______ change when the R_1 branch was opened.
4. Use the proper formula(s) and calculate the branch power dissipations and P_T for the normal circuit condition of step 2.	P_T = ______ mW P_1 = ______ mW P_2 = ______ mW P_3 = ______ mW	—
5. Use the proper formula(s) and calculate the branch power dissipations and P_T for the circuit conditions when R_1 was open in step 3.	P_T = ______ mW P_1 = ______ mW P_2 = ______ mW P_3 = ______ mW	Opening one branch of a parallel circuit (*does, does not*) ______ affect the power dissipated by the other branches. However, total P will (*increase, decrease*) ______ by the amount of power that was being dissipated by the opened branch previous to its opening.

PROJECT 20

PARALLEL CIRCUITS
Effects of a Short in Parallel Circuits

NOTE: Do not connect power to this curcuit for this project!

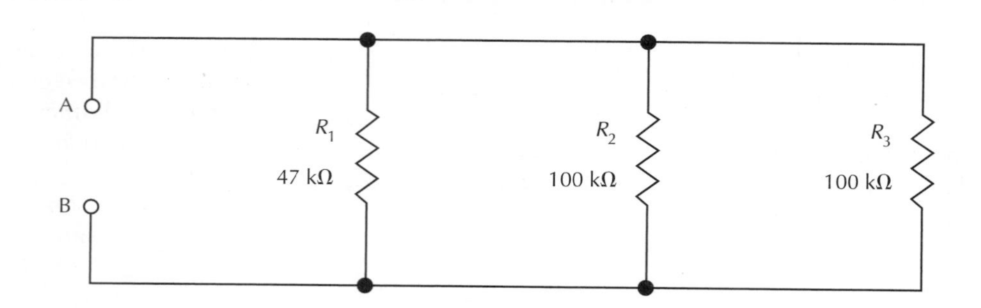

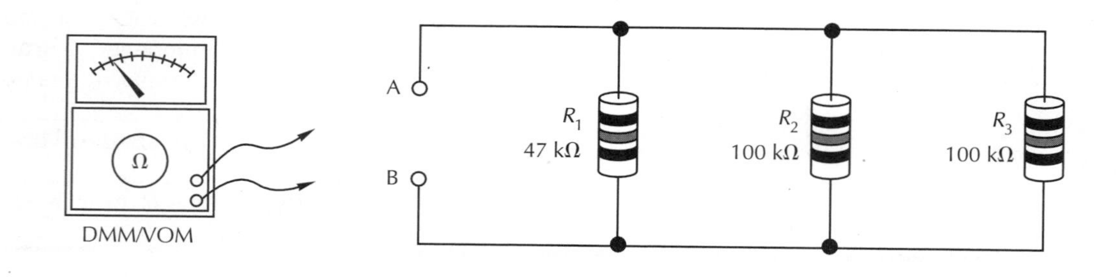

FIGURE 22
Standard schematic diagram (top) and pictorial diagram (bottom)

PROJECT PURPOSE

To provide experience through "POWER-OFF" ohmmeter measurements regarding what happens to parallel circuit parameters, should one or more branches become shorted.

PARTS NEEDED

- ☐ DMM/VOM
- ☐ CIS
- ☐ Resistors:
  47 kΩ
  100 kΩ (2)

SAFETY HINTS

DO NOT use a power supply for this project!

ACTIVITY	OBSERVATION	CONCLUSION
1. Connect the initial circuit as shown in Figure 22. CAUTION: DO NOT USE THE POWER SUPPLY OR MILLIAMMETER FOR THIS DEMONSTRATION OR PROJECT.	—	—
2. Use the DMM/VOM and measure R_e at points A and B.	R_e = ______ ohms	—
3. To simulate R_1 becoming shorted, remove R_1 and replace it with a jumper wire; then measure R_e again.	R_e = ______ ohms	A short across one branch of a parallel circuit causes R_e to (*increase, decrease*) ______ effectively to ______ ohms. If we were using a power supply that could supply infinite current, the current through R_2 would be ______ mA; through R_3 would be ______ mA; and through the shorted branch would be ______ mA. Since $V = I \times R$, what is the *IR* drop across the shorted branch? ______ volts. In parallel, the voltage across all branches in parallel is the ______ voltage; therefore, V_2 and V_3 for the conditions described above would be ______ volts.
4. Replace R_1 into the circuit and simulate a short in branch R_2 by replacing R_2 with a jumper wire. Measure R_e again.	R_e = ______ ohms	A short of *any* branch of a parallel circuit will cause R_e to (*increase, decrease*) ______ and I_T to (*increase, decrease*) ______ (before the power supply fuse blows). Also, current through the "unshorted" branches would (*increase, decrease*) ______, and current through the "shorted" branch would (*increase, decrease*) ______. The voltage across all branches would (*increase, decrease*) ______. It should be noted that in the strictest sense, if any branch of a parallel circuit is "shorted," all branches are shorted. However, only one branch "contains" the actual shorted element that is "shunting" the whole circuit with the undesired low resistance path, whose resistance approaches (∞, *0*) ___ ohms.

PARALLEL CIRCUITS

Complete the following review questions, indicating the appropriate response by placing a check in the box next to the correct answer.

1. A greater change of R_e occurs if the resistance that is added in shunt with the existing circuit has a resistance value that is:

   - ☐ high
   - ☐ low

2. Adding another resistor in parallel with an existing circuit will cause the circuit's total current to:

   - ☐ increase
   - ☐ decrease
   - ☐ remain the same

3. Adding another resistor in parallel with an existing parallel circuit will cause the currents through the original branches to:

   - ☐ increase
   - ☐ decrease
   - ☐ remain the same

4. In a simple two-branch parallel circuit consisting of a 2-ohm and a 3-ohm resistor in parallel, three-fifths of the total circuit current will flow through:

   - ☐ the 3-ohm *R*
   - ☐ the 2-ohm *R*
   - ☐ neither *R*

5. The total resistance of the circuit in question 4 is:

   - ☐ 8 ohms
   - ☐ 2 ohms
   - ☐ 1.2 ohms
   - ☐ 2.1 ohms
   - ☐ 3 ohms
   - ☐ none of these

6. In a circuit consisting of 100-kΩ, 47-kΩ, and 1-kΩ branches, the total circuit resistance must be:

   - ☐ more than 1 kΩ
   - ☐ less than 1 kΩ
   - ☐ more than 100 kΩ
   - ☐ more than 47 kΩ

PARALLEL CIRCUITS

7. The important fact to remember about parallel circuits is that the voltage across all parallel branches is:

   ☐ divided equally
   ☐ different
   ☐ the same
   ☐ none of these

8. When one branch of a parallel circuit opens:

   a. V_A will
      ☐ increase
      ☐ decrease
      ☐ remain the same

   b. I_T will
      ☐ increase
      ☐ decrease
      ☐ remain the same

   c. The unopened branch currents will
      ☐ increase
      ☐ decrease
      ☐ remain the same

   d. The opened branch currents will
      ☐ increase
      ☐ decrease
      ☐ remain the same

   e. R_T will
      ☐ increase
      ☐ decrease
      ☐ remain the same

9. If any branch of a parallel circuit "shorts," it will cause:

   ☐ excessive current
   ☐ excessive voltage
   ☐ excessive resistance
   ☐ none of these

10. In a parallel circuit, the resistor that dissipates the most power is:

   ☐ the lowest value R
   ☐ the highest value R
   ☐ neither of these

STUDENT LOG FOR OPTIONAL TROUBLESHOOTING EXERCISE

Students should write down the starting point symptom given to them by the instructor, and then proceed to log each step used in the troubleshooting process. For example, the first step after learning the starting point symptom information is to identify the area to be bracketed as the "area of uncertainty," and should include all circuitry components that "might" cause the trouble producing the symptoms. The next step is to make a decision about the first test, etc.

Starting point symptom:

____

Components and circuitry included in "initial brackets" ("area of uncertainty")

____

First test description (what type test and where?)

____

Components and circuitry *still* included in "bracketed" area after first test

____

Second test description (what type test and where?)

____

Components and circuitry *still* included in "bracketed" area after second test

____

Third test description (what type test and where?)

____

Components and circuitry *still* included in "bracketed" area after third test

____

Fourth test description (what type test and where?)

____

Components and circuitry *still* included in "bracketed" area after fourth test

____

Fifth test description (what type test and where?)

____

NOTE: *Keep repeating the testing, bracketing, and logging procedure until you are quite sure you have found the trouble. Call the instructor to check your work; then, replace the identified faulty component to see if the circuit operates properly!*

Suspected Trouble:

____

Trouble Verified by Instructor and by Circuit Operation: Yes ___ No ___

SERIES-PARALLEL CIRCUITS

Objectives

You will connect several dc resistive series-parallel circuits and make measurements and observations regarding their important electrical characteristics.

In completing these projects, you will connect circuits, make measurements, perform calculations, draw conclusions, and be able to answer questions about the following items related to series-parallel circuits.

- Total resistance
- Voltage distribution
- Current distribution
- Power dissipation(s)
- Effects of opens
- Effects of shorts
- Application of Ohm's law
- Application of Kirchhoff's voltage law
- Application of Kirchhoff's current law

PROJECT/TOPIC CORRELATION INFORMATION

PROJECT		TEXT CHAPTER	SECTION	RELATED TEXT TOPIC(S)
21	Total Resistance in Series-Parallel Circuits	6	6-3	Total Resistance in Series-Parallel Circuits
22	Current in Series-Parallel Circuits	6	6-4	Current in Series-Parallel Circuits
23	Voltage Distribution in Series-Parallel Circuits	6	6-5	Voltage in Series-Parallel Circuits
24	Power Distribution in Series-Parallel Circuits	6	6-6	Power in Series-Parallel Circuits
25	Effects of an Open in Series-Parallel Circuits	6	6-7	Effects of Opens in Series-Parallel Circuits and Troubleshooting Hints
26	Effects of a Short in Series-Parallel Circuits	6	6-8	Effects of Shorts in Series-Parallel Circuits and Troubleshooting Hints

PROJECT
21

SERIES-PARALLEL CIRCUITS
Total Resistance in Series-Parallel Circuits

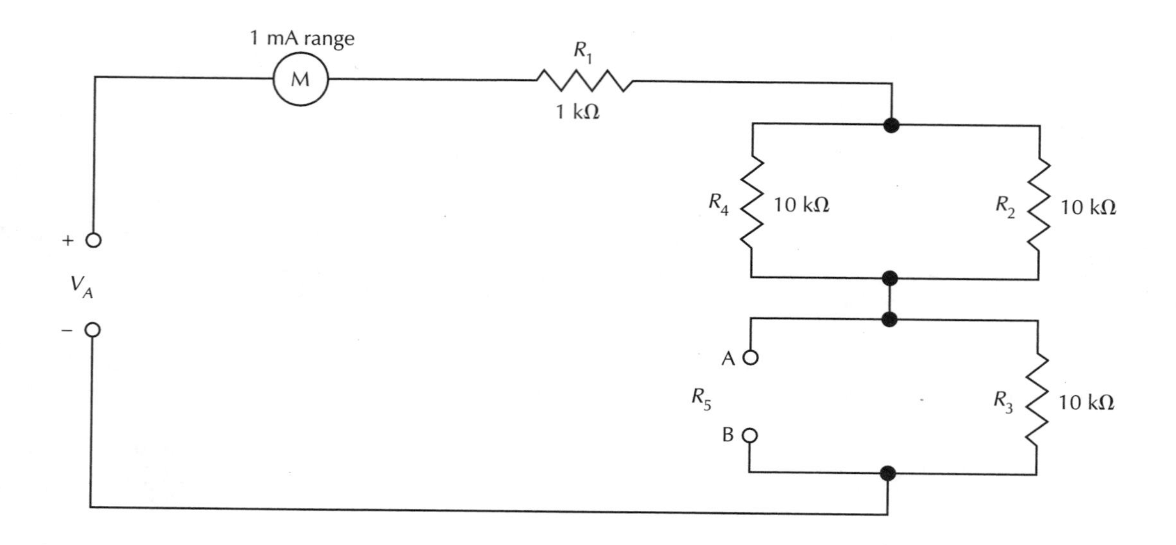

FIGURE 23

NOTE: From this point on, you will notice that there are no pictorial diagrams. You will be expected to wire your circuits and make measurements following the schematic diagram only.

PROJECT PURPOSE

To use and verify series-parallel circuit resistance analysis techniques by circuit measurements and/or related calculations.

PARTS NEEDED

- ☐ DMM/VOM (2)
- ☐ VVPS (dc)
- ☐ CIS
- ☐ Resistors:
  - 1 kΩ
  - 10 kΩ (4)

SAFETY HINTS

NOTE: From this point on, we will not repeat the standard safety hints you have been seeing thus far but will reserve this box for special safety hints, as appropriate.

NOTE: The term R_T generally refers to total circuit resistance, whereas the term R_e (R equivalent) may refer to the total resultant resistance of specific parallel resistors in just a portion of the circuit. If the total circuit is a purely parallel circuit, the term R_e is the same as R_T, for that case.

ACTIVITY	OBSERVATION	CONCLUSION
1. Connect the initial circuit as shown in Figure 23.	—	—
2. Calculate the circuit R_T from the indicated resistance values as follows: Add $R_1 + R_e$ of R_2 and R_4 (in parallel) $+ R_3$.	R_T = __________ ohms	Is the total resistance of this circuit less than the least resistor? ________. Is the total resistance of this circuit equal to the sum of all the individual resistors? ________.
3. Apply just enough voltage to the circuit to obtain 1 mA of current and measure V_A, V_1, V_2, V_3, and V_4.	V_A = __________ volts V_1 = __________ volts V_2 = __________ volts V_3 = __________ volts V_4 = __________ volts	Since there is ______ mA of total circuit current and V_A is _______ volts, then by Ohm's law, R_T must be _______ ohms. Which resistors in this circuit are in parallel with one another? ______ and ______. Is the voltage across these two *R*s the same? _______. Which resistor(s) are in series with the main line (carry I_T)? _________.
4. DISCONNECT the *power supply and milliammeter* from the circuit and measure R_T (at appropriate points in the circuit) with an ohmmeter.	R_T = __________ ohms	It can be concluded that in series-parallel circuits the total resistance is equal to the sum of all the components in (*series, parallel*) __________ with the main line (components through which I_T passes) plus the equivalent resistance of parallel elements, whose R_e is effectively in (*series, parallel*) __________ with the main line components.
5. Using the logic just discussed, calculate what the circuit R_T would be if a 10-kΩ resistor were added to the circuit between points A and B. After calculating R_T, measure it with a DMM/VOM to verify your thinking (meter and power supply still not connected).	R_{calc} = __________ ohms R_{meas} = __________ ohms	R_T was calculated by adding the resistance of R _______ + the R_e of R ______ and R ______ + the R_e of R ______ and R ______. Would R_T change if the jumper were removed that runs between the top of R_3 and the top of R_5? __________

PROJECT 21 CONTINUED

SERIES-PARALLEL CIRCUITS
Total Resistance in Series-Parallel Circuits *(Continued)*

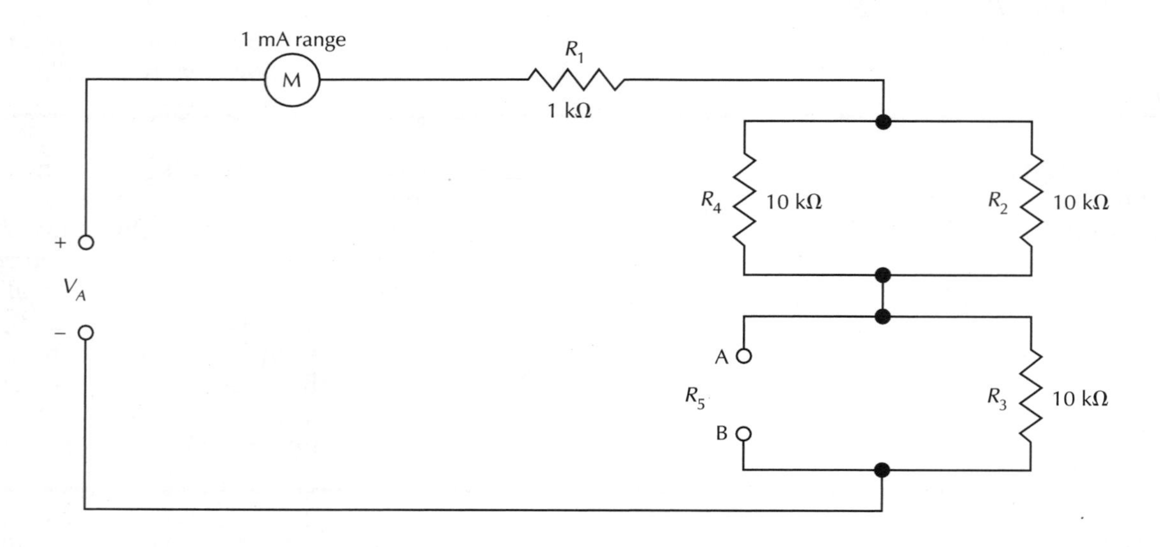

FIGURE 23

NOTE: From this point on, you will notice that there are no pictorial diagrams. You will be expected to wire your circuits and make measurements following the schematic diagram only.

PROJECT PURPOSE

To use and verify series-parallel circuit resistance analysis techniques by circuit measurements and/or related calculations.

PARTS NEEDED

- ☐ DMM/VOM (2)
- ☐ VVPS (dc)
- ☐ CIS
- ☐ Resistors:
  - 1 kΩ (2)
  - 10 kΩ (4)

SAFETY HINTS

NOTE: From this point on, we will not repeat the standard safety hints you have been seeing thus far but will reserve this box for special safety hints, as appropriate.

EXTRA CREDIT STEP(S)

6. Assume that the 10-kΩ resistor is left in the circuit between points A and B. Also assume that the circuit is to be rewired so that all four 10-kΩ resistors will be in parallel with each other, and that combination is to still be in series with R_1 (the 1-kΩ resistor).

   Predict the total resistance for the conditions just described.

   $R_{predicted}$ = ______________ kΩ

   Is the predicted resistance higher or lower than that measured in step 5? ______________.
   Why?______________.
   ______________.

7. Actually rewire the circuit so that the four 10-kΩ resistors are in parallel with each other, and in series with the 1-kΩ resistor; then, *without any power applied to the circuit*, measure and record the circuit total resistance.

   R_{meas} = ______________ kΩ

   Did the measured value reasonably agree with your predicted value in the preceding step? ________. If a 1-kΩ resistor had been placed across points A and B rather than the 10 kΩ, would total resistance have been higher or lower? ______________.
   Logic, and the measurements of this project seem to indicate that a low R value placed in parallel with an existing S-P circuit will affect the circuit's total resistance (*more, less*) ______________ than a larger R value being placed in parallel with the existing circuit.

8. Actually replace R_5 (across points A and B) with a 1-kΩ resistor; then, measure total resistance to see if it verifies your conclusions.

   R_{meas} = ______________ kΩ

   Are the conclusions verified? __________.

PROJECT
22

SERIES-PARALLEL CIRCUITS
Current in Series-Parallel Circuits

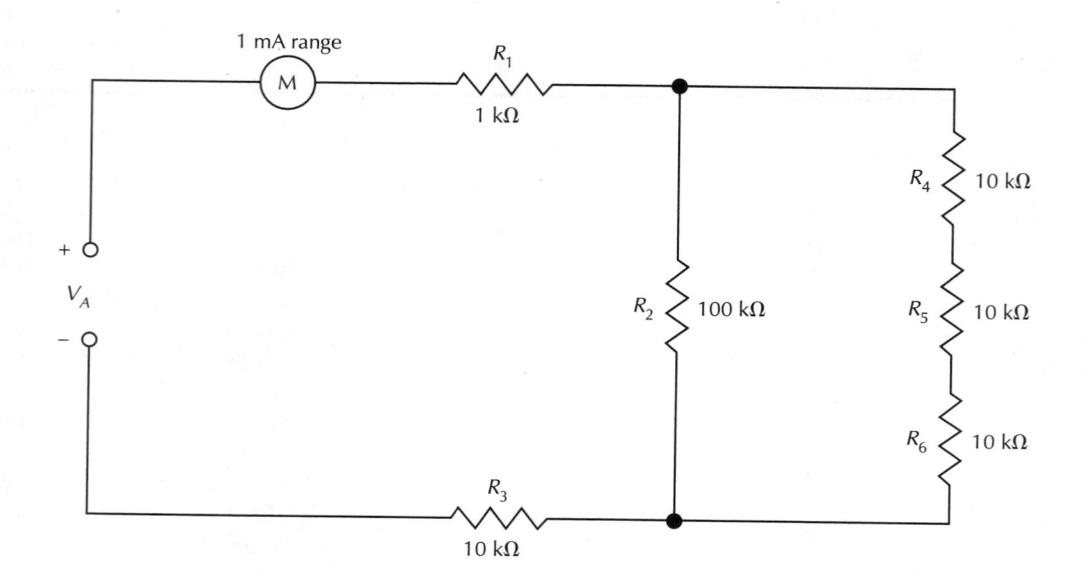

FIGURE 24

PROJECT PURPOSE

To confirm the characteristics of current division and summation through a series-parallel circuit, using measurements, observations, and both series and parallel circuit analysis techniques. To observe that a component's location, as well as its value, has significance regarding its electrical parameters when it is in a series-parallel circuit.

PARTS NEEDED

- ☐ DMM/VOM (2)
- ☐ VVPS (dc)
- ☐ CIS
- ☐ Resistors:
  - 1 kΩ (2)
  - 10 kΩ (4)
  - 27 kΩ
  - 100 kΩ

ACTIVITY	OBSERVATION	CONCLUSION
1. Connect the initial circuit as shown in Figure 24.	—	—
2. Apply just enough voltage to the circuit to obtain 0.5 mA of I_T. Measure all the individual voltage drops and calculate the current through each component as required to fill in the blanks in the Observation column.	V_1 = ______ volts V_2 = ______ volts V_3 = ______ volts V_4 = ______ volts V_5 = ______ volts V_6 = ______ volts V_A = ______ volts I_1 = ______ mA I_2 = ______ mA I_3 = ______ mA I_4 = ______ mA I_5 = ______ mA I_6 = ______ mA	The components that are in series with the main line and through which total current flows are: R ______ and R ______. The components that are in series with each other, but not the main line are: ______, ______, and ______. What value of resistance is in parallel with R_2? ______ ohms. Does the current divide through the parallel branches inverse to the resistance relationships? ______. It should be noted that the current and voltage distribution in a series-parallel circuit for any given component is dependent upon its location in the circuit. However, the series "sections" of the circuit can be analyzed by the same rules as used in ______ circuits, and the parallel sections of the circuit can be analyzed by the rules for ______ circuits. To finalize the circuit analysis, you then combine the results of the sectional analysis.
3. Analyze the circuit and predict what will happen to all the currents in the circuit if R_5 is changed to a 27-kΩ resistor. Will they: (*increase*, *decrease*, or *remain the same*)? Note your predictions in the Observation column.	I_1 will ______ I_2 will ______ I_3 will ______ I_4 will ______ I_5 will ______ I_6 will ______ I_T will ______	Increasing the value of any resistor in a series-parallel circuit (or any other circuit) will cause I_T to (*increase*, *decrease*) ______. It is possible for current to increase through a component in parallel with the changed value R because a larger percentage of I_T (even though I_T has decreased some) will pass through the unchanged branch.

PROJECT

22

CONTINUED

SERIES-PARALLEL CIRCUITS
Current in Series-Parallel Circuits *(Continued)*

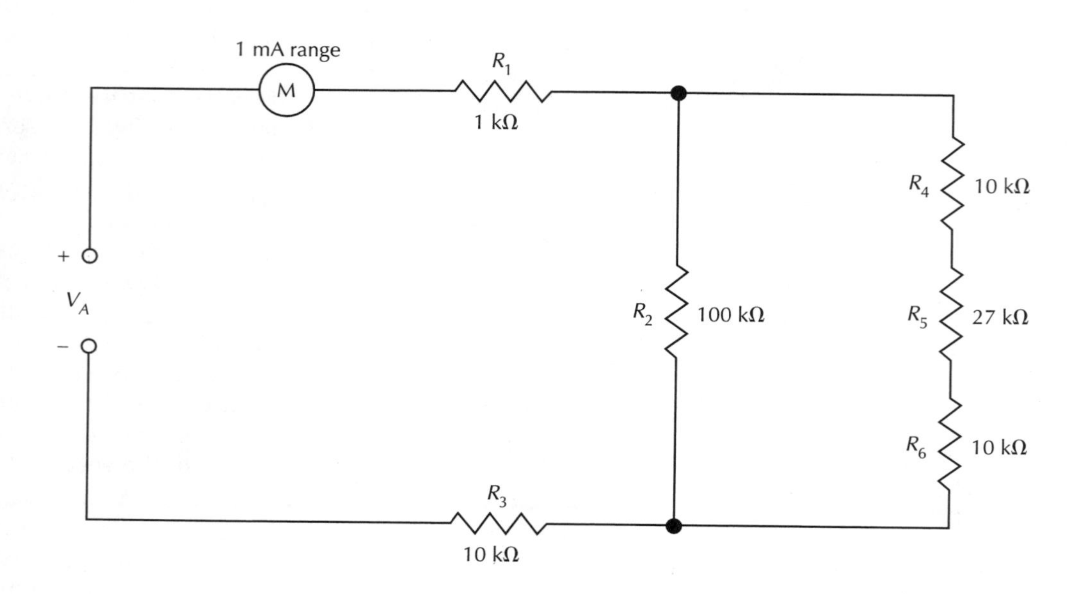

FIGURE 25

PROJECT PURPOSE

To confirm the characteristics of current division and summation through a series-parallel circuit, using measurements, observations, and both series and parallel circuit analysis techniques. To observe that a component's location, as well as its value, has significance regarding its electrical parameters when it is in a series-parallel circuit.

PARTS NEEDED

- ☐ DMM/VOM (2)
- ☐ VVPS (dc)
- ☐ CIS
- ☐ Resistors:
  - 1 kΩ (2)
  - 10 kΩ (4)
  - 27 kΩ
  - 100 kΩ

ACTIVITY	OBSERVATION	CONCLUSION
4. To verify your predictions of the previous step 3 in Figure 24, change R_5 from a 10-kΩ to a 27-kΩ resistor as shown in the diagram in Figure 25. Measure all the circuit voltages with 17 volts applied to the circuit. Calculate the current through each component using Ohm's law. Fill in the blanks as required in the Observation column. Compare these results with those of the previous step 2 in Figure 24 to see if the current(s) increased or decreased.	V_1 = ________ volts V_2 = ________ volts V_3 = ________ volts V_4 = ________ volts V_5 = ________ volts V_6 = ________ volts V_A = ________ volts I_1 = ________ mA I_2 = ________ mA I_3 = ________ mA I_4 = ________ mA I_5 = ________ mA I_6 = ________ mA I_T = ________ mA	Do the measurements agree with the predictions? ________. Another way of analyzing why I_2 increased even though I_T decreased is to consider the circuit to be a "simple" series circuit of R_1 in series with the R_e of R_2 in parallel with R_4, R_5, R_6, and R_3. Thus, the "equivalent" series circuit is $R_1 + R_e + R_3$. Now, because R_5 was increased from 10 kΩ to 27 kΩ, R_e will increase. Now R_e is a bigger percentage of R_T, and thus more of V_A will be dropped across R_e and hence R_2. Since R_2 has not itself changed value, yet V_2 is higher, the current through R_2 must also be higher.
5. Remove the current meter. Change R_5 from 27 kΩ to 1 kΩ (which is lower than the original circuit's 10 kΩ). As time permits, predict the circuit parameter "trends" (as opposed to specific values). After predicting, make any measurements required to verify or correct the predictions.	The currents that will increase are: ________ ________ The currents that will decrease are: ________ ________	Decreasing the value of any resistor in a series-parallel circuit (or any other circuit) will cause I_T to (*increase, decrease*) ________. The current through R ________ decreased because with a higher I_T the $I \times R$ drops increased across R ________ and R ________, which means less of V_A was left to be dropped by R ________. Since this resistor has not changed value and it dropped less V, then the current through it must have (*increased, decreased*) ________.

PROJECT 22 CONTINUED

SERIES-PARALLEL CIRCUITS
Current in Series-Parallel Circuits *(Continued)*

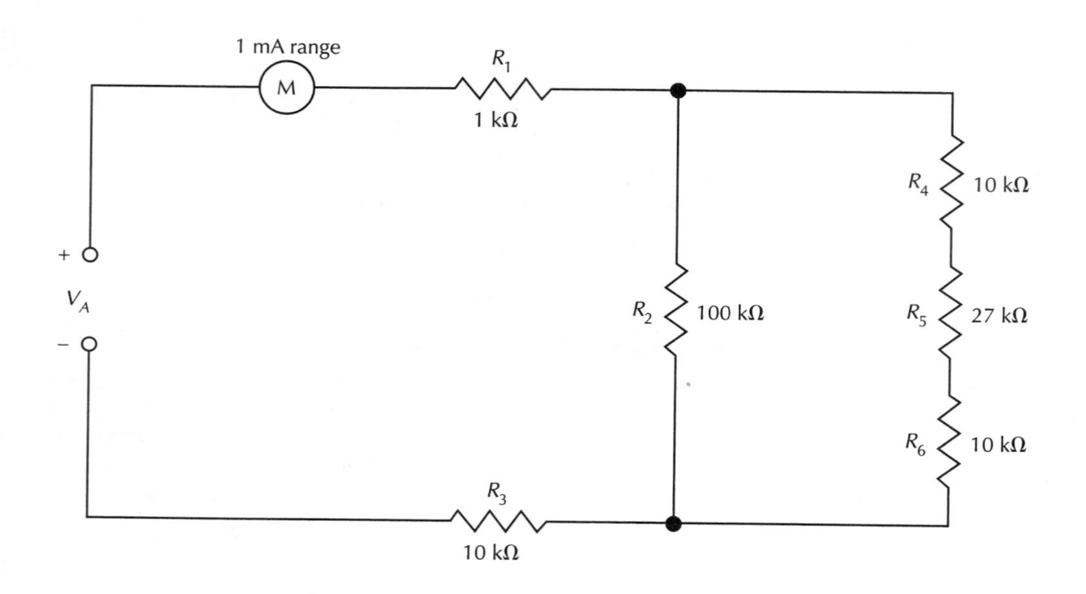

FIGURE 25

PROJECT PURPOSE

To confirm the characteristics of current division and summation through a series-parallel circuit, using measurements, observations, and both series and parallel circuit analysis techniques. To observe that a component's location, as well as its value, has significance regarding its electrical parameters when it is in a series-parallel circuit.

PARTS NEEDED

- ☐ DMM/VOM (2)
- ☐ VVPS (dc)
- ☐ CIS
- ☐ Resistors:
  - 1 kΩ (2)
  - 10 kΩ (4)
  - 27 kΩ
  - 100 kΩ

EXTRA CREDIT STEP(S)

6. Use the circuit of Figure 25; however, change R_1 from 1 kΩ to 12 kΩ. Use 17 volts applied voltage. Measure and calculate, as appropriate, to fill in blanks and draw conclusions.

I_T = ______________________ mA
I_2 = ______________________ mA
I_4 = ______________________ mA

Does R_1 carry total current? __________. Did the circuit total current change when R_1 was changed? _______. Did the value of current through R_2 change? _____. Did the percentage of total current through the R_2 branch change when R_1 was changed? _______. (Be careful when thinking about this question!). Did the percentage of total current carried by the branch with R_4 in it change when R_1 was changed? ______. This indicates that changing a series component that carries total current in a series-parallel circuit will cause a change in total circuit current, but (*will, will not*) ______________ affect the division of current through parallel branches in the circuit.

PROJECT

23

SERIES-PARALLEL CIRCUITS
Voltage Distribution in Series-Parallel Circuits

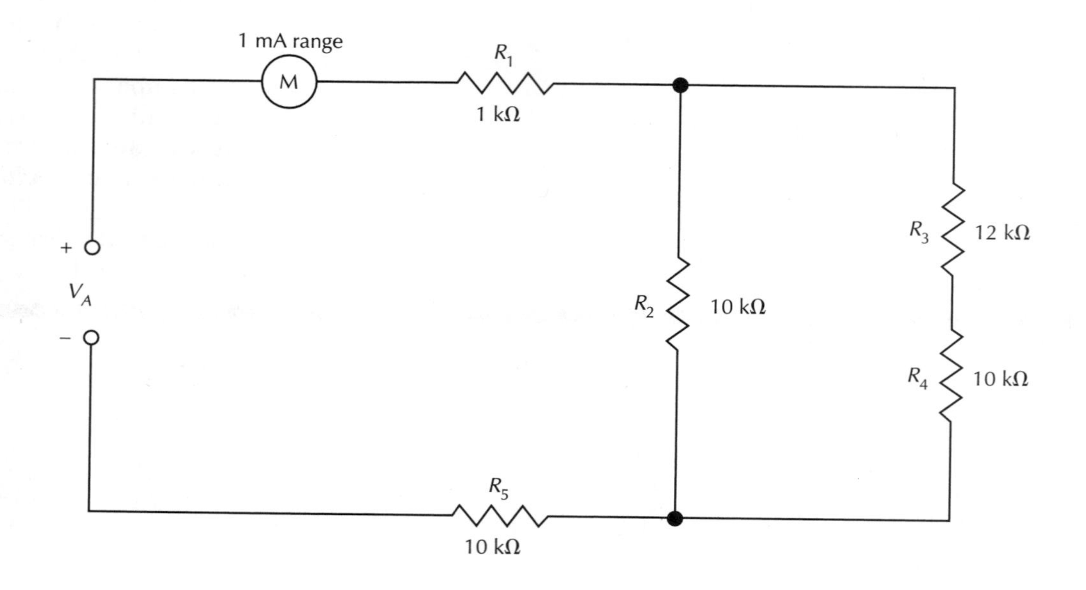

FIGURE 26

PROJECT PURPOSE

To observe voltage distribution characteristics in a series-parallel circuit, using measurements and observations. To prove that in a series-parallel circuit, a component's location, as well as its value affects its voltage drop.

PARTS NEEDED

- ☐ DMM/VOM (2)
- ☐ VVPS (dc)
- ☐ CIS
- ☐ Resistors:
  - 1 kΩ
  - 10 kΩ (3)
  - 12 kΩ

ACTIVITY	OBSERVATION	CONCLUSION
1. Connect the initial circuit as shown in Figure 26.	—	—
2. Assume an I_T of 0.5 mA and calculate all the circuit voltages. Note results of calculations. (**NOTE:** Solve for the R_e of branch R_2 in parallel with branch $(R_3 + R_4)$ so you can then use the current-divider rule(s) to find currents through these particular branch resistors. This will then allow you to find their individual resistor voltage drops, using the appropriate $I \times R$ values.)	V_1 = ______ volts V_2 = ______ volts V_3 = ______ volts V_4 = ______ volts V_5 = ______ volts	According to Kirchhoff's voltage law, the applied voltage to obtain these results would be ______ volts.
3. Apply just enough V_A to obtain 0.5 mA of I_T and measure all the circuit voltages.	V_1 = ______ volts V_2 = ______ volts V_3 = ______ volts V_4 = ______ volts V_5 = ______ volts V_A = ______ volts	Did the measurements verify the calculations of step 2? ______. (**NOTE:** The effects of meter loading will affect some of the readings slightly, if using a VOM.) Did the largest value R drop the most voltage? ______. Because of its location in the circuit, only a small portion of the total ______ passes through this component. The important consideration when analyzing voltage distribution in a series-parallel circuit is the electrical location of each component being considered. I_T only passes through those components that are in series with the (*largest branch, main line*) ______ ______ or source. Parallel sections of the circuit must be analyzed using parallel circuit rules. However, the "net R_e" of a parallel section can be considered as being in series with (*largest branch, main line*) ______ ______ components.

PROJECT
23
CONTINUED

SERIES-PARALLEL CIRCUITS
Voltage Distribution in Series-Parallel Circuits *(Continued)*

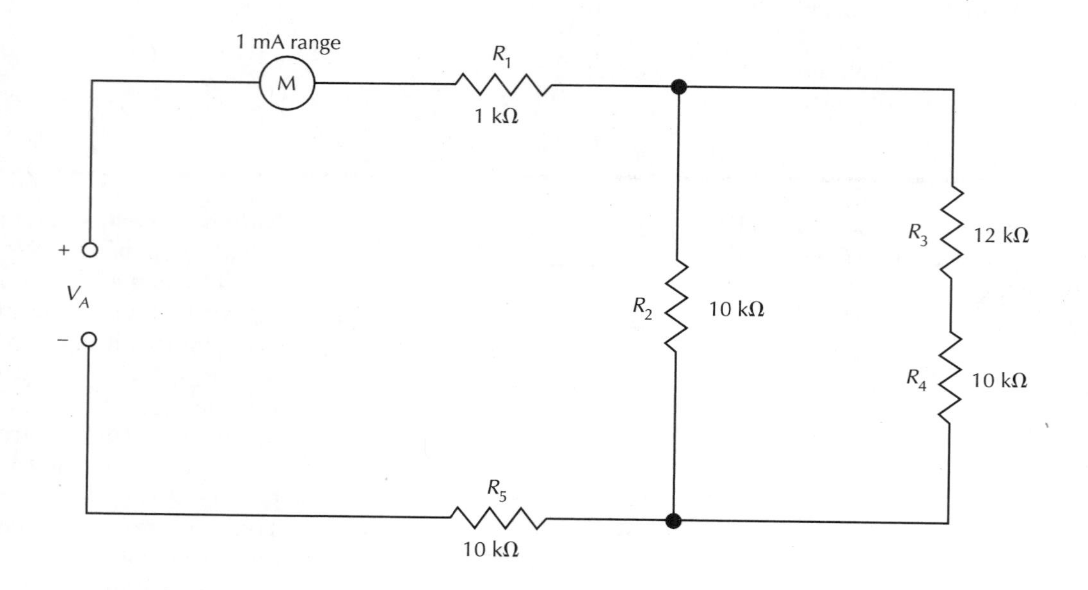

FIGURE 26

PROJECT PURPOSE

To observe voltage distribution characteristics in a series-parallel circuit, using measurements and observations. To prove that in a series-parallel circuit, a component's location, as well as its value affects its voltage drop.

PARTS NEEDED

- ☐ DMM/VOM (2)
- ☐ VVPS (dc)
- ☐ CIS
- ☐ Resistors:
  - 1 kΩ
  - 10 kΩ (3)
  - 12 kΩ

ACTIVITY	OBSERVATION	CONCLUSION
4. Predict what will happen to V_2 if R_3 is replaced with a jumper wire short. After making the prediction, make the suggested circuit change and note the results.	V_2 will ____________ V_2 did ____________	Decreasing the R in the parallel branch causes total current to (*increase, decrease*) __________, which causes the components that are in series with the main line I_R drops to (*increase, decrease*) __________. Thus, less of V_A is dropped across R_2.

PROJECT

24

SERIES-PARALLEL CIRCUITS
Power Distribution in Series-Parallel Circuits

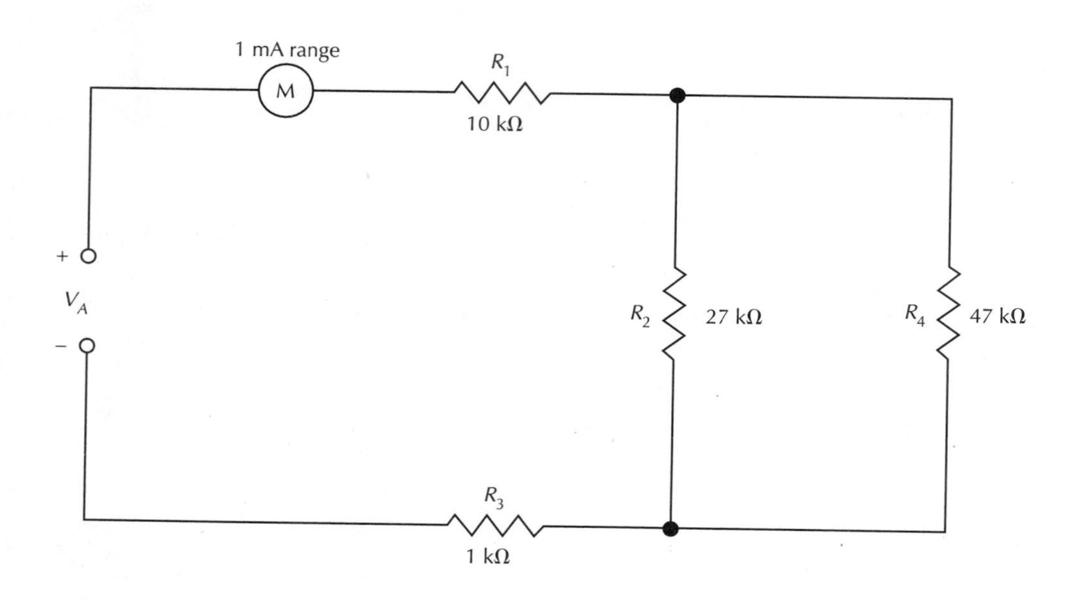

FIGURE 27

PROJECT PURPOSE

To connect a series-parallel circuit, make measurements and perform calculations that illustrate the power distribution characteristics of this type circuit. To illustrate that it is the "electrical location" of a given component, not the physical location of a component, which influences its electrical parameters.

PARTS NEEDED

- ☐ DMM/VOM (2)
- ☐ VVPS (dc)
- ☐ CIS
- ☐ Resistors:
  - 1 kΩ
  - 10 kΩ
  - 27 kΩ
  - 47 kΩ

ACTIVITY	OBSERVATION	CONCLUSION
1. Connect the initial circuit as shown in Figure 27. 2. Apply just enough V_A to obtain 0.5 mA of current through the meter.	— V_A = __________ volts (approx.)	— R_T must equal V_T/I_T or ______ kΩ.
3. Measure all voltage drops and calculate the power dissipated by each component.	V_1 = __________ volts V_2 = __________ volts V_3 = __________ volts V_4 = __________ volts P_1 = __________ mW P_2 = __________ mW P_3 = __________ mW P_4 = __________ mW P_T = __________ mW	The power dissipated by each component in this circuit is directly proportional to its ________ drop and the ________ through it. This statement is true for any circuit. For the parallel sections of the circuit, the larger the R, the (*more, less*) ________ the power dissipated by that component. For the components in series with the main line, the larger the R, the (*more, less*) ______ the power dissipated by that component. Which component in the circuit dissipated the most power? __________. Notice this is not the largest R in the circuit, as it would have to be in a simple series circuit if it dissipated the most power. Also, it is not the smallest R, as it would have to be in a simple parallel circuit if it dissipated the most power. The key to the power distribution for series-parallel circuits is again the electrical (*size, location*) ________ of the component being considered in the circuit.
4. "Swap" the positions of R_2 and R_4. Make measurements and determine if the powers dissipated by each of the Rs have changed. (Measure voltages and compare to step 3 results.)	Results are: (*same, different*) ____________________	It is not the "physical" location that is important but rather the ________ location of components that determines the distribution of power (and other parameters) in a series-parallel circuit.

PROJECT 25

SERIES-PARALLEL CIRCUITS
Effects of an Open in Series-Parallel Circuits

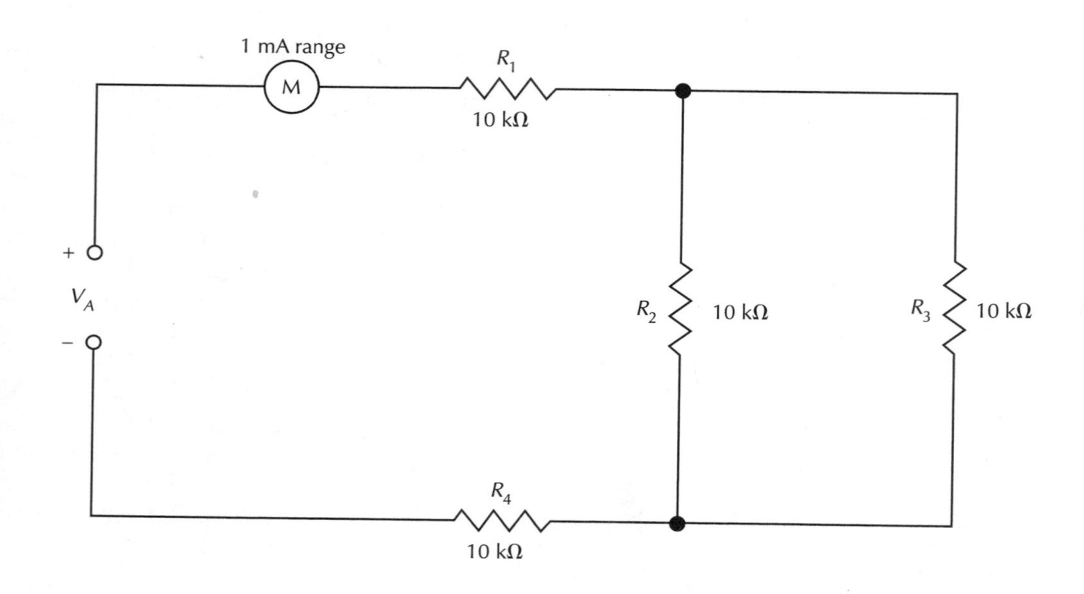

FIGURE 28

PROJECT PURPOSE

To demonstrate, through measurements and calculations, the various effects on electrical parameters of an open occurring in a series portion and in a parallel portion of a series-parallel circuit. To note the effect on the overall circuit total current when an open occurs.

PARTS NEEDED

- ☐ DMM/VOM (2)
- ☐ VVPS (dc)
- ☐ CIS
- ☐ Resistors:
  10 kΩ (4)

ACTIVITY	OBSERVATION	CONCLUSION
1. Connect the initial circuit as shown in Figure 28.	—	—
2. Increase V applied until 1 mA of current is indicated by the milliammeter. Measure and note the circuit voltages.	V_A = ____________ volts V_1 = ____________ volts V_2 = ____________ volts V_3 = ____________ volts V_4 = ____________ volts	—
3. Calculate the current through each component using Ohm's law.	I_1 = ____________ mA I_2 = ____________ mA I_3 = ____________ mA I_4 = ____________ mA	Total current flows through R ____ and R ______. The current divides equally between R ____ and R ____ as they are ________ resistances, and are in ____________, thus have the same voltage drop.
4. Remove R_1 from the circuit to simulate R_1 "opening." Measure the circuit voltages.	V_A = ____________ volts V_1 = ____________ volts (across the open) V_2 = ____________ volts V_3 = ____________ volts V_4 = ____________ volts	Opening a component in series with the source caused I_T to (*increase, decrease*) ________ to ____ mA. The voltage drops across all the good components decreased to ________ volts. The potential difference across the open became equal to V______. What electrical law does this verify? ________ voltage law.
5. Replace R_1 into the circuit. Now remove R_3 to simulate it "opening." Measure the circuit voltages.	V_A = ____________ volts V_1 = ____________ volts V_2 = ____________ volts V_3 = ____________ (across the open) V_4 = ____________ volts	Opening a parallel branch in a series-parallel circuit causes I_T to (*increase, decrease*) ________ but not to ______ mA. The voltage drops across the components in series with the main line (*increase, decrease*) ________, whereas, the voltage across the components directly in parallel with the open (*increase, decrease*) ________. No matter whether an open occurs in the series or parallel sections of the series-parallel circuit, the circuit total resistance will (*increase, decrease*) ____________.

PROJECT
25
CONTINUED

SERIES-PARALLEL CIRCUITS
Effects of an Open in Series-Parallel Circuits *(Continued)*

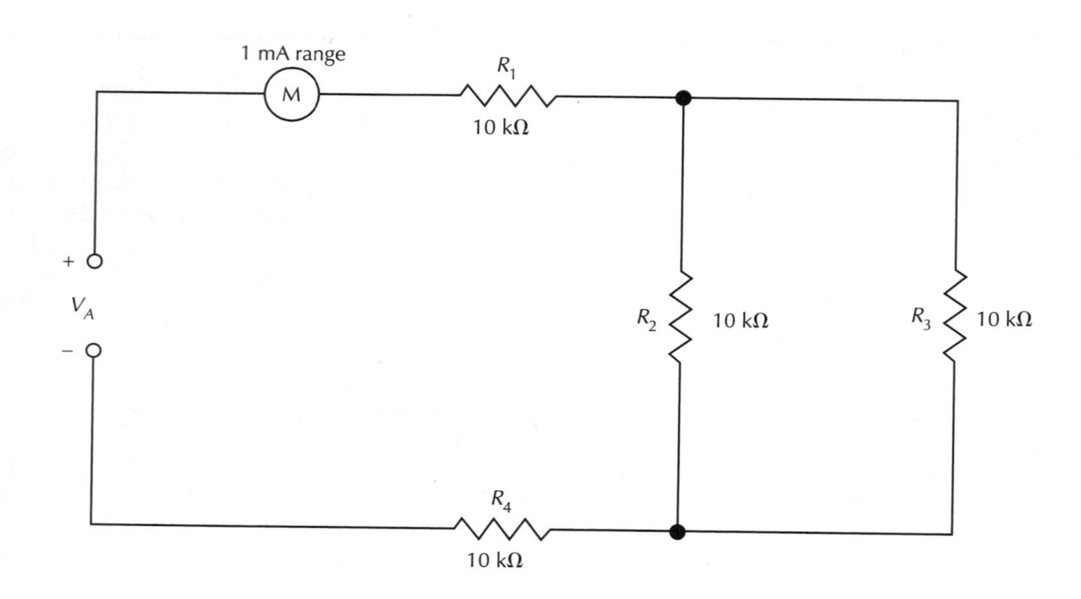

FIGURE 28

PROJECT PURPOSE

To demonstrate, through measurements and calculations, the various effects on electrical parameters of an open occurring in a series portion and in a parallel portion of a series-parallel circuit. To note the effect on the overall circuit total current when an open occurs.

PARTS NEEDED

- ☐ DMM/VOM (2)
- ☐ VVPS (dc)
- ☐ CIS
- ☐ Resistors: 10 kΩ (4)

EXTRA CREDIT STEP(S)

6. Use the circuit of Figure 28. With the power off, remove R_4 from the circuit to simulate R_4 opening. Turn the power back on and measure and record V_A and the voltage across the open terminals where R_4 was previously connected.

V_A measures ________________ volts
V_4 measures ________________ volts

Did the voltage across the "opened resistor" exactly equal V_A? ________. If not, was it higher or lower than V_A?______________. Explain what might cause this to be true.
____
____
____
____________________________________.

PROJECT 26

SERIES-PARALLEL CIRCUITS
Effects of a Short in Series-Parallel Circuits

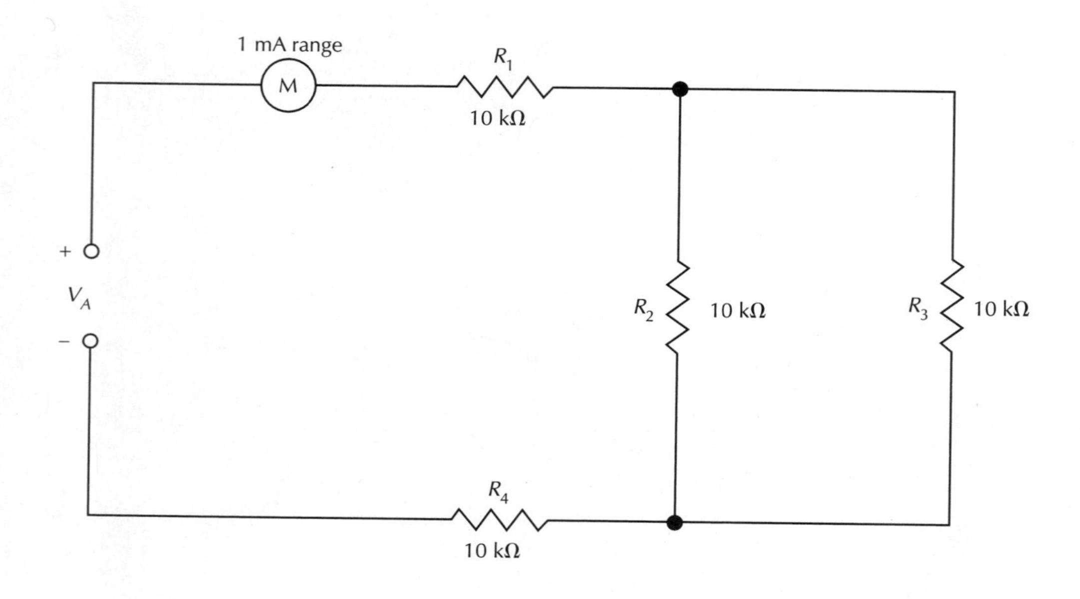

FIGURE 29

PROJECT PURPOSE

To demonstrate, through measurements and calculations, the various effects on electrical parameters of a short occurring in a series portion and in a parallel portion of a series-parallel circuit. To note the effect on the overall circuit total current when a short occurs.

PARTS NEEDED

- ☐ DMM/VOM (2)
- ☐ VVPS (dc)
- ☐ CIS
- ☐ Resistors: 10 kΩ (4)

ACTIVITY	OBSERVATION	CONCLUSION
1. Connect the initial circuit as shown in Figure 29.	—	—
2. Increase V applied until 0.5 mA of current is indicated by the milliammeter. Measure and note the circuit voltages.	V_A = ________ volts V_1 = ________ volts V_2 = ________ volts V_3 = ________ volts V_4 = ________ volts	
3. Calculate the current through each component using Ohm's law.	I_1 = ________ mA I_2 = ________ mA I_3 = ________ mA I_4 = ________ mA	Total current flows through R_1 and R ____. One (*third, half, fourth*) ________ of total current flows through R_2 and one half through R (*1, 2, 3, 4*) ________.
4. Simulate R_1 shorting by replacing it with a jumper wire. Measure the circuit voltages and record.	V_A = ________ volts V_1 = ________ volts (across the short) V_2 = ________ volts V_3 = ________ volts V_4 = ________ volts	Shorting a component in series with the source caused I_T to (*increase, decrease*) ________; the voltage drops across all the "unshorted" or normal components to (*increase, decrease*) ________; and the voltage across the shorted element to (*increase, decrease*) ________ to ________ volts. Do Kirchhoff's voltage and current laws still hold true when there is a short in the circuit? ________.
5. Replace R_1 into the circuit. Now replace R_3 with a short. Measure and record the circuit voltages.	V_A = ________ volts V_1 = ________ volts V_2 = ________ volts V_3 = ________ volts V_4 = ________ volts	Shorting a parallel branch caused I_T to (*increase, decrease*) ________ to ________ mA. The voltage across the shorted branch and any branch in parallel with it (*increases, decreases*) ________ to ________ volts. The voltages across all other components (*increased, decreased*) ________ because total current (*increased, decreased*) ________. A short anywhere in any type of circuit will cause the circuit total resistance to (*increase, decrease*) ________. Therefore, the circuit total current will (*increase, decrease*) ________.

SERIES-PARALLEL CIRCUITS

Complete the following review questions, indicating the appropriate response by placing a check in the box next to the correct answer.

1. The total resistance (R_T) of a series-parallel circuit is always greater than the sum of the Rs through which total current flows.

   ☐ True
   ☐ False

2. Total current passes through every component in a series-parallel circuit.

   ☐ True
   ☐ False

3. Considering only the components in series with the "main line," the resistor that will dissipate the most power is:

   ☐ the largest R
   ☐ the smallest R
   ☐ neither of these

4. Considering only the components in the parallel sections of the series-parallel circuit, the component that will dissipate the most power is:

   ☐ the largest R
   ☐ the smallest R
   ☐ neither of these

5. In a series-parallel circuit, does the largest resistor always have the largest $I \times R$ drop?

   ☐ Yes
   ☐ No

6. In a series-parallel circuit, if a component in series with the source (in the main line) opens, the voltage across the other components will:

   ☐ increase
   ☐ decrease
   ☐ remain the same

7. If a resistor in a parallel section of a series-parallel circuit opens, the voltage across the unopened components will:

   ☐ increase in some cases and decrease in others
   ☐ decrease in all cases
   ☐ increase in all cases

8. If a component in series with the "main line" shorts, the voltage across all the other components will:

   - ☐ increase
   - ☐ decrease
   - ☐ remain the same

9. If a resistor in a parallel section of a series-parallel circuit shorts, the voltage across all the other components will:

   - ☐ increase in some cases and decrease in others
   - ☐ decrease in all cases
   - ☐ increase in all cases

10. To analyze a series-parallel circuit it is necessary to use both series circuit rules and parallel circuit rules.

   - ☐ True
   - ☐ False

STUDENT LOG FOR OPTIONAL TROUBLESHOOTING EXERCISE

Students should write down the starting point symptom given to them by the instructor, and then proceed to log each step used in the troubleshooting process. For example, the first step after learning the starting point symptom information is to identify the area to be bracketed as the "area of uncertainty," and should include all circuitry components that "might" cause the trouble producing the symptoms. The next step is to make a decision about the first test, etc.

Starting point symptom:

____

Components and circuitry included in "initial brackets" ("area of uncertainty")

____

First test description (what type test and where?)

____

Components and circuitry *still* included in "bracketed" area after first test

____

Second test description (what type test and where?)

____

Components and circuitry *still* included in "bracketed" area after second test

____

Third test description (what type test and where?)

____

Components and circuitry *still* included in "bracketed" area after third test

____

Fourth test description (what type test and where?)

____

Components and circuitry *still* included in "bracketed" area after fourth test

____

Fifth test description (what type test and where?)

____

NOTE: *Keep repeating the testing, bracketing, and logging procedure until you are quite sure you have found the trouble. Call the instructor to check your work; then, replace the identified faulty component to see if the circuit operates properly!*

Suspected Trouble:

____

Trouble Verified by Instructor and by Circuit Operation: Yes ___ No ___

BASIC NETWORK THEOREMS

Objectives

You will connect circuits illustrating Thevenin's, Norton's, and Maximum Power Transfer Theorems.

In completing these projects, you will connect circuits, make measurements, perform calculations, draw conclusions, and be able to answer questions about the following items related to these theorems.

- R_{TH}
- V_{TH}
- R_N
- I_N
- Current through a load
- Voltage across a load
- Efficiency of circuit at maximum power transfer

PROJECT/TOPIC CORRELATION INFORMATION

PROJECT	TEXT CHAPTER	SECTION	RELATED TEXT TOPIC(S)
27 Thevenin's Theorem	7	7-4	Thevenin's Theorem
28 Norton's Theorem	7	7-5 7-6	Norton's Theorem
29 Maximum Power Transfer Theorem	7	7-2	Maximum Power Transfer Theorem

PROJECT 27

BASIC NETWORK THEOREMS
Thevenin's Theorem

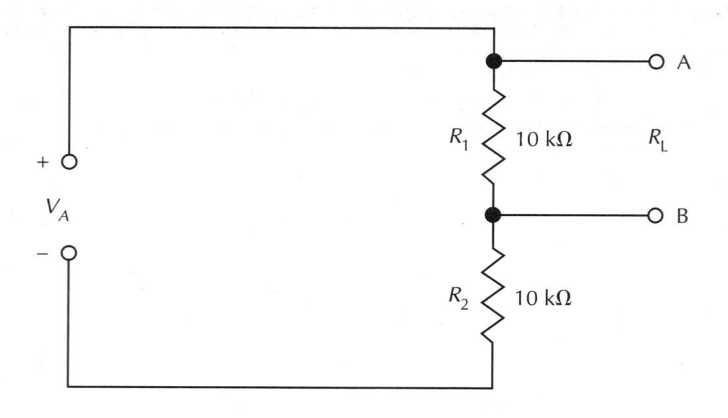

FIGURE 30

PROJECT PURPOSE

To practice analyzing a resistive network, using Thevenin's theorem. To calculate R_{TH} and V_{TH} and predict circuit parameters based on these values. To confirm these predictions by actual measurement.

PARTS NEEDED

- ☐ DMM/VOM (2)
- ☐ VVPS (dc)
- ☐ CIS
- ☐ Resistors:
  - 1 kΩ (3)
  - 10 kΩ (2)

For our purposes Thevenin's theorem might be stated as follows: The current through a load resistor (R_L) connected between any two points on an existing resistive network (circuit) can be calculated by dividing the voltage at those two points, prior to connecting R_L, by the sum of R that would be measured at those two points (with the voltage source(s) replaced by shorts) + R_L. Therefore: $I_L = V_{TH}/R_{TH} + R_L$.

In essence, Thevenin's theorem tells us that the entire network of resistors and voltage source(s) can be replaced by an equivalent circuit of a voltage source, called V_{TH}, and a single series resistor, called R_{TH}.

V_{TH} = the open-circuit voltage at the two points (V without R_L connected). R_{TH} = resistance seen when looking back into the circuit at the two points, with any sources replaced by shorts (and without R_L connected).

In order to quickly calculate I_L (current through R_L) when R_L is to be connected to any existing network (complicated or otherwise), we then consider R_L as being connected across the Thevenin equivalent circuit so the circuit consists of: V_{TH} supplying voltage to R_{TH} and R_L in series. Hence, $I_L = V_{TH}/R_{TH} + R_L$.

ACTIVITY	OBSERVATION	CONCLUSION
1 Connect the initial circuit as shown in Figure 30.	—	—
2. Apply the voltage distribution characteristics of series circuits and determine what the voltage would be between points A and B with a source V of 20 volts.	$V_{A\text{–}B}$ = ______ volts	In a series circuit consisting of two equal value resistors, each resistor will drop (*1/3, 1/2, 1/4*)______ of V applied. If the load resistor, R_L, is to be connected to points A and B, then V_{TH} = ______ volts.
3. Assume the source were zero ohms and calculate R_{TH}. (**NOTE:** R_1 and R_2 would then be effectively in parallel.)	R_{TH} = ______ ohms	The Thevenin equivalent circuit would therefore be drawn as follows: 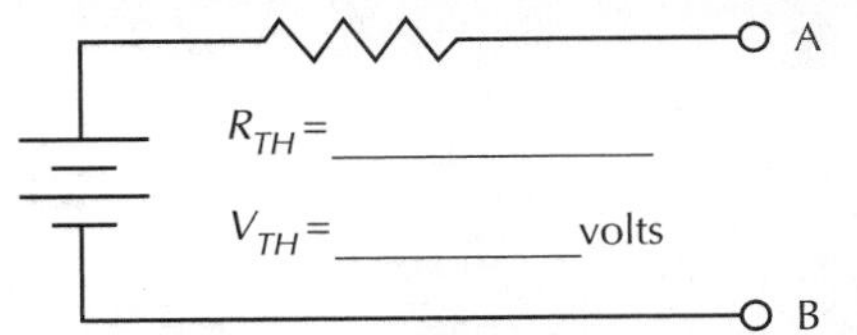
4. If R_L is to be a 10-kΩ resistor, calculate I_L and V_L.	I_L = ______ mA V_L = ______ volts	$V_L = I_L \times$ ______.
5. Actually apply 20 volts of V applied. Measure and record circuit parameters as indicated in the Observation column.	V_A = ______ volts V_L = ______ volts V_2 = ______ volts	With R_L connected, total circuit current is ______ mA. If R_L were disconnected, I_T would = ______ mA and V_2 would = ______ volts.

PROJECT 27 CONTINUED

BASIC NETWORK THEOREMS
Thevenin's Theorem *(Continued)*

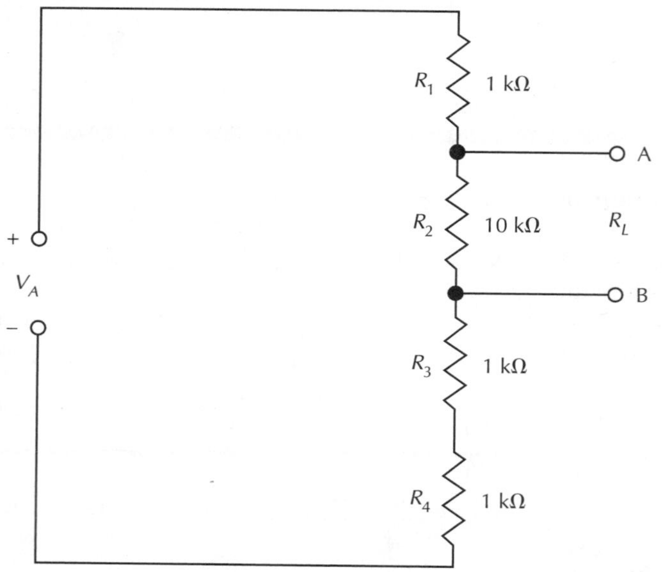

FIGURE 31

PROJECT PURPOSE

To practice analyzing a resistive network, using Thevenin's theorem. To calculate R_{TH} and V_{TH} and predict circuit parameters based on these values. To confirm these predictions by actual measurement.

PARTS NEEDED

- ☐ DMM/VOM (2)
- ☐ VVPS (dc)
- ☐ CIS
- ☐ Resistors:
  - 1 kΩ (3)
  - 4.7 kΩ
  - 10 kΩ (2)

ACTIVITY	OBSERVATION	CONCLUSION
6. Using the Thevenin equivalent circuit, as appropriate, calculate and project what the value of V_L and I_L would be if R_L was changed to a value of 4.7 kΩ.	V_L calc. = ______ volts I_L calc. = ______ mA	—
7. Change R_L to a value of 4.7 kΩ and make measurements needed to compare measurements with your projections.	V_L meas. = ______ volts	Did the measured and projected values for V_L reasonably compare? ______. Were the calculations made easier by virtue of the Thevenin equivalent circuit approach? ______.
8. Connect the initial circuit as shown in Figure 31.	—	—
9. Calculate V_{TH} and R_{TH} (assume 13 volts V applied).	V_{TH} = ______ volts R_{TH} = ______ ohms	The Thevenin equivalent circuit would therefore be drawn as follows: 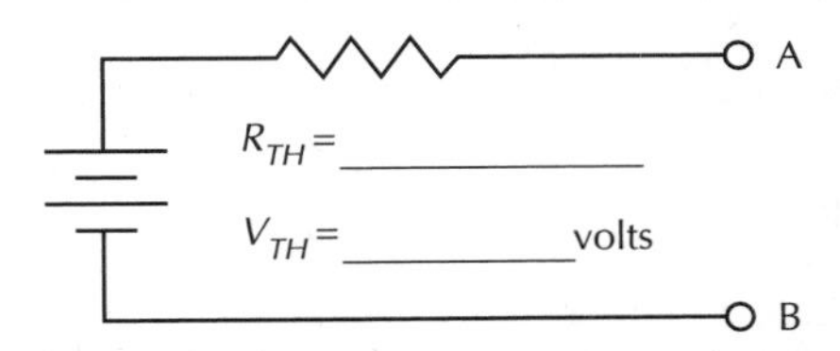
10. If R_L is to be a 10-kΩ resistor, calculate I_L and V_L.	I_L = ______ mA V_L = ______ volts	$I_L = V_{TH}/R_L + R_{TH}$ = ______ mA $V_L = I_L \times R_L$ = ______ × 10 kΩ = ______ volts
11. Apply 13 volts to the circuit. Measure the circuit parameters as indicated in the Observation column. .	V_A = ______ volts V_L = ______ volts V_4 = ______ volts	The main advantage of using Thevenin's theorem for solving I_L is not that it particularly simplifies the initial calculations for a single value of R_L, but rather, if R_L were going to be changed several times, V_{TH} and ______ could be used for each case. This means that each new change of R_L does not necessitate solving the whole network problem each time. All that is necessary is to substitute the new value of R_L in the formula: $I_L = V_{TH}/$ ______.

PROJECT 28

BASIC NETWORK THEOREMS
Norton's Theorem

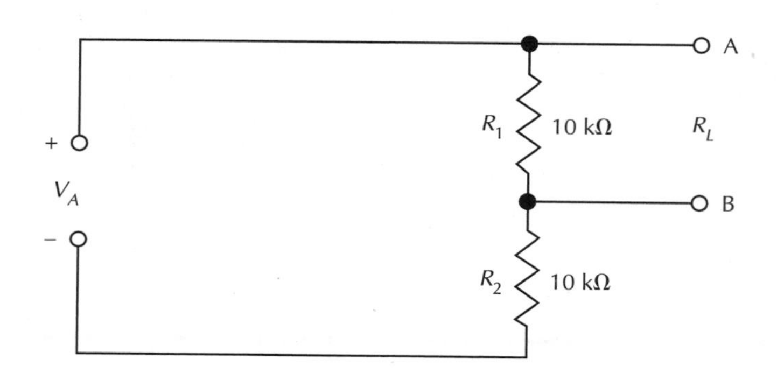

FIGURE 32

PROJECT PURPOSE

To practice analyzing a resistive network, using Norton's theorem. To calculate R_N and I_N and predict circuit parameters based on these values. To confirm these predictions by actual measurements.

PARTS NEEDED

- ☐ DMM/VOM (2)
- ☐ VVPS (dc)
- ☐ CIS
- ☐ Resistors:
  - 1 kΩ
  - 10 kΩ (4)

For our purposes Norton's theorem might be stated as follows: Any two-terminal network of resistors and source(s) may be represented by a single current source shunted by a single resistance. The size of the current source is determined by calculating the amount of current that will flow through the two terminals if they are shorted. The size of the resistance (shunt R) is equal to the resistance that would be seen looking back into the network from the two points (when they are not shorted and without an R_L).

In essence, the above statement means:

I_N (equivalent Norton current source) = current that would flow through a short at points A and B if A and B were shorted.

R_N = resistance seen when looking back into the circuit at the two points (to which R_L will be connected), with any sources replaced by shorts. (In effect, R_N is the same as R_{TH} would be.)

In order to quickly calculate I_L (current through R_L) when R_L is to be connected to points A and B, we then consider R_L as being connected across the Norton equivalent circuit so the circuit consists of: I_N supplying current to R_N and R_L in parallel and the current dividing according to the rules of parallel circuits. Hence: $I_L = (R_N/R_N + R_L) \times I_N$.

ACTIVITY	OBSERVATION	CONCLUSION
1. Connect the initial circuit as shown in Figure 32.	—	—
2. Assume a short across points A and B and determine what the current through this short would be if the source voltage is to be 15 volts.	I_N = ______ mA	In effect, R ______ is shorted out and the circuit total resistance is ______ ohms.
3. Assume the source were zero ohms and calculate R_N. **(NOTE:** R_1 and R_2 would then be effectively in parallel.)	R_N = ______ ohms	The Norton equivalent circuit would therefore be drawn as follows: 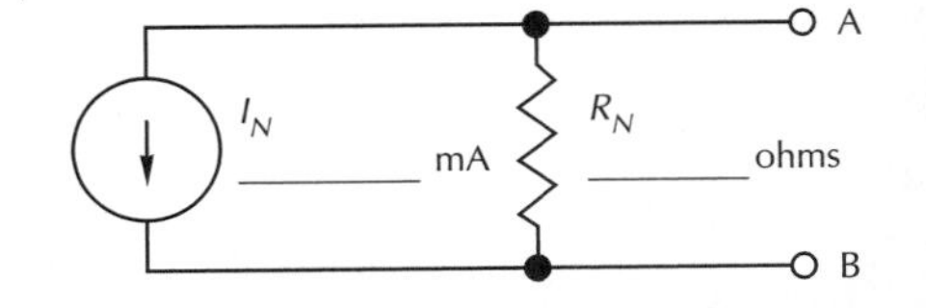
4. If R_L is to be a 10-kΩ resistor, calculate I_L and V_L. (Use the formula for I_L shown in the discussion at the top of the page.)	I_L = ______ mA V_L = ______ volts	$V_L = I_L \times$ ______.
5. Actually apply 15 volts to the circuit and measure the circuit parameters to verify predicted results.	V_A = ______ volts V_L = ______ volts V_2 = ______ volts	Does Norton's theorem actually work in analyzing and predicting I_L? ______.

PROJECT **28** CONTINUED

BASIC NETWORK THEOREMS
Norton's Theorem *(Continued)*

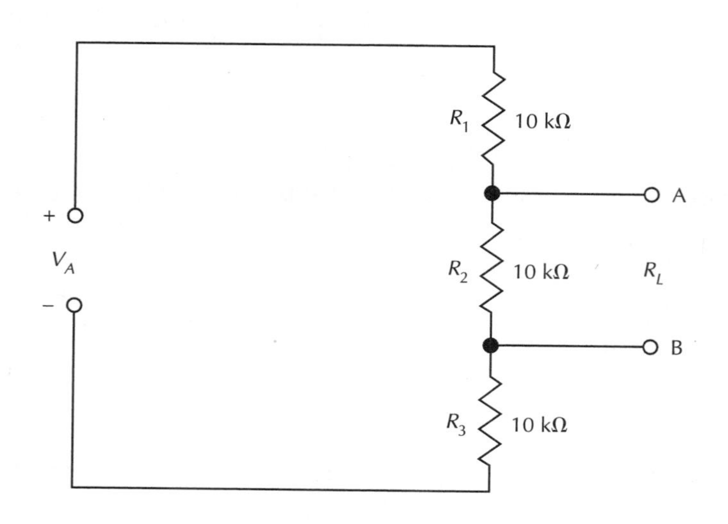

FIGURE 33

PROJECT PURPOSE

To provide further practice in using Norton's theorem by analyzing a second resistive circuit. Again, to use Norton's theorem and predict circuit parameters, and use measurements to verify the accuracy of the predictions.

PARTS NEEDED

- ☐ DMM/VOM (2)
- ☐ VVPS (dc)
- ☐ CIS
- ☐ Resistors:
  - 1 kΩ
  - 5.6 kΩ
  - 10 kΩ (4)

ACTIVITY	OBSERVATION	CONCLUSION
6. Using the Norton equivalent circuit, as appropriate, calculate and project what the value of V_L and I_L would be if R_L was changed to a value of 5.6 kΩ.	V_L calc. = ________ volts I_L calc. = ________ mA	—
7. Change R_L to a value of 5.6 kΩ and make measurements needed to compare measurements with your projections.	V_L meas. = ________ volts	Did the measured and projected values for V_L reasonably compare? ________. Were the calculations made easier by virtue of the Norton equivalent circuit approach? ________.
8. Connect the initial circuit as shown in Figure 33.	—	—
9. Calculate I_N and R_N (assume 20 volts V_A).	I_N = ________ mA R_N = ________ ohms	$I_N = V_A/$ ________ ohms = ________ mA R_N = 10 kΩ × ________ /10 kΩ + ________ = ________
10. If R_L is to be 1 kΩ, calculate I_L and V_L for a V_A of 20 volts.	I_L = ________ mA V_L = ________ volts	$I_L = \left(\frac{R_N}{R_N + \_\_\_\_}\right) \times I_N =$ ________ mA
11. Apply 20 volts to the circuit. Measure parameters and verify the predictions made by calculations using Norton's theorem.	V_A = ________ volts V_L = ________ volts V_3 = ________ volts	The predictions made using Norton's theorem (*were, were not*) ________ verified by the measurements.
12. If R_L were to be changed to a 10-kΩ resistor, use Norton's theorem and predict the parameters called for in the Observation column (assume 20 volts V_A).	I_N = ________ mA R_N = ________ ohms I_L = ________ mA V_L = ________ volts V_1 = ________ volts V_2 = ________ volts V_3 = ________ volts	Changing the value of R_L that will be used (*does, does not*) ________ change the Norton equivalent circuit. The Norton equivalent circuit for the circuit shown on the "setup" diagram can therefore be drawn as follows:

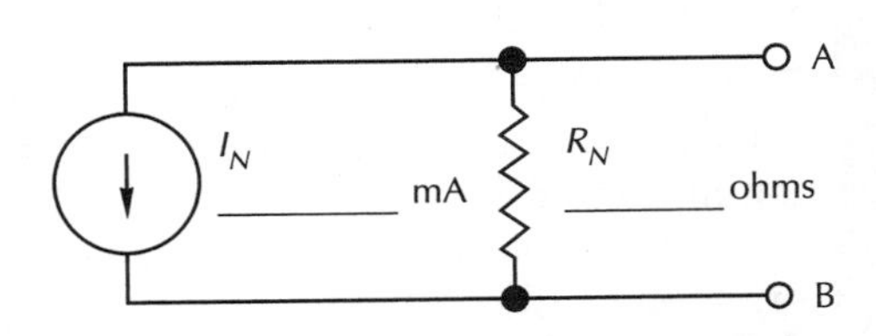

PROJECT

29

BASIC NETWORK THEOREMS
Maximum Power Transfer Theorem

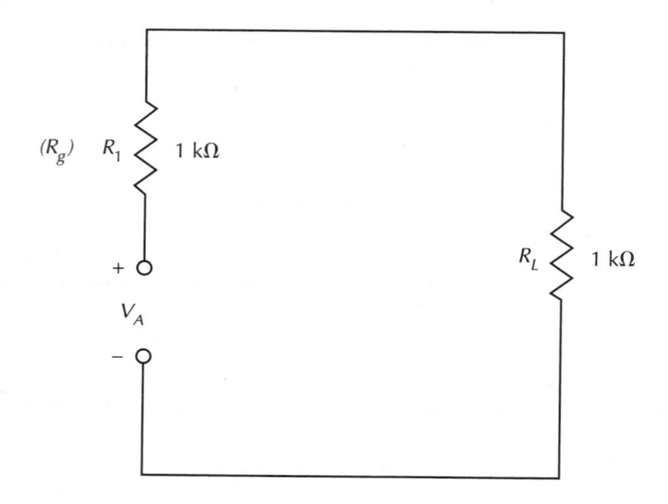

FIGURE 34

PROJECT PURPOSE

To verify the maximum power transfer theorem through circuit measurements and calculations. To observe the relationships of R_g and R_L values to both the amount of power delivered to the load, and the efficiency of power delivery to the load.

PARTS NEEDED

- ☐ DMM/VOM (2)
- ☐ VVPS (dc)
- ☐ CIS
- ☐ Resistors:
  - 1 kΩ (2)
  - 10 kΩ
  - 100 Ω

For our purposes, the maximum power transfer theorem might be stated as: Maximum power will be transferred from the generator (or source) to the load (R_L) when the resistance of the load equals the resistance of the generator (or source).

ACTIVITY	OBSERVATION	CONCLUSION
1. Connect the initial circuit as shown in Figure 34.	—	—
2. Apply 10 volts to the circuit. Measure V_L and calculate P_L.	V_L = ________ volts P_L = ________ mW	This circuit is solved as a simple (*series, parallel*) ________ circuit.
3. Assume that R_1 represents the resistance of the generator (R_g). Calculate the power transfer efficiency as follows: Eff. = $(P_{out}/P_{in}) \times 100$ where $P_{out} = P_L$ and $P_{in} = P_T$	P_{out} = ________ mW P_{in} = ________ mW Eff. = ________ %	When $R_L = R_g$ (*1/3, 1/4, 1/2*) ________ the total power provided by the source is transferred to the load (R_L).
4. Change R_L to 10 kΩ. Measure V_L and calculate P_L and efficiency (use 10 volts V_A).	V_L = ________ volts P_L = ________ mW P_{out} = ________ mW P_{in} = ________ mW Eff. = ________ %	Making R_L larger than R_g caused efficiency to (*increase, decrease*) ________. However, P_L was (*less, more*) ________ than when $R_L = R_g$. The reason for this is that with the higher R_L, the circuit total resistance was larger and thus the total power supplied to the circuit by the source was (*less, more*) ________ than when $R_L = R_g$, even though P_L was a (*larger, smaller*) ________ percentage of P_T, the power transferred to the 10-kΩ load was (*less, more*) ________ than to the 1-kΩ load.
5. Change R_L to a 100-ohm resistor. Measure V_L and calculate P_L and efficiency (use 10 volts V_A).	V_L = ________ volts P_L = ________ mW P_{out} = ________ mW P_{in} = ________ mW Eff. = ________ %	When R_L is smaller than R_g, the efficiency is (*more, less*) ________ than 50% as when $R_L = R_g$. Total power produced by the source with R_L smaller than R_g is greater than for maximum power transfer conditions. However, a smaller percentage of the power is dissipated by R ________.

BASIC NETWORK THEOREMS

Complete the following review questions, indicating the appropriate response by placing a check in the box next to the correct answer.

1. To find R_{TH}, we look back into the network from the two points to which R_L will be connected and assume the source(s) as:

   - ☐ open
   - ☐ shorted
   - ☐ neither of these

2. When solving for R_{TH}, we assume that R_L is:

   - ☐ connected to the circuit
   - ☐ not connected to the circuit

3. When solving for R_{TH}, we assume the part of the circuit connected to the two points to which R_L will be connected to be:

   - ☐ open
   - ☐ shorted
   - ☐ neither of these

4. A Thevenin equivalent circuit consists of:

   - ☐ a voltage source and series R
   - ☐ a current source and shunt R
   - ☐ neither of these

5. To find R_N, we look back into the network from the two points to which R_L will be connected and assume the source(s) as:

   - ☐ open
   - ☐ shorted
   - ☐ neither of these

6. When solving for I_N, we assume the part of the circuit connected to the two points to which R_L will be connected to be:

   - ☐ open
   - ☐ shorted
   - ☐ neither of these

7. A Norton equivalent circuit consists of:

   - ☐ a voltage source and series R
   - ☐ a current source and shunt R
   - ☐ neither of these

8. Maximum power transfer from source to load occurs when:

   - ☐ R_L is greater than R_g
   - ☐ R_L is less than R_g
   - ☐ $R_L = R_g$
   - ☐ all are the same

9. Highest efficiency occurs when:

   - ☐ R_L is greater than R_g
   - ☐ R_L is less than R_g
   - ☐ $R_L = R_g$
   - ☐ makes no difference

10. At maximum power transfer, the efficiency is:

    - ☐ 100%
    - ☐ 25%
    - ☐ 10%
    - ☐ 50%
    - ☐ none of these

NETWORK ANALYSIS TECHNIQUES

Objectives

In this series of three projects, you will connect several resistive circuits in order to make measurements and observations regarding network analysis techniques.

In completing these projects, you will connect circuits, make measurements, perform calculations, draw conclusions, and be able to answer questions about the following items relating to network analysis techniques:

- Appropriate labeling of schematic diagrams for mesh and nodal analysis
- Use of equations in solving for circuit parameters using mesh and nodal analysis
- Use of equations when performing wye and delta network conversions

PROJECT/TOPIC CORRELATION INFORMATION

PROJECT		TEXT CHAPTER	SECTION	RELATED TEXT TOPIC(S)
30	Loop/Mesh Analysis	8	8-1	Assumed Loop or Mesh Current Analysis Approach
31	Nodal Analysis	8	8-2	Nodal Analysis Approach
32	Wye-Delta and Delta-Wye Conversions	8	8-3	Conversions of Delta and Wye Networks

PROJECT 30

NETWORK ANALYSIS TECHNIQUES
Loop/Mesh Analysis

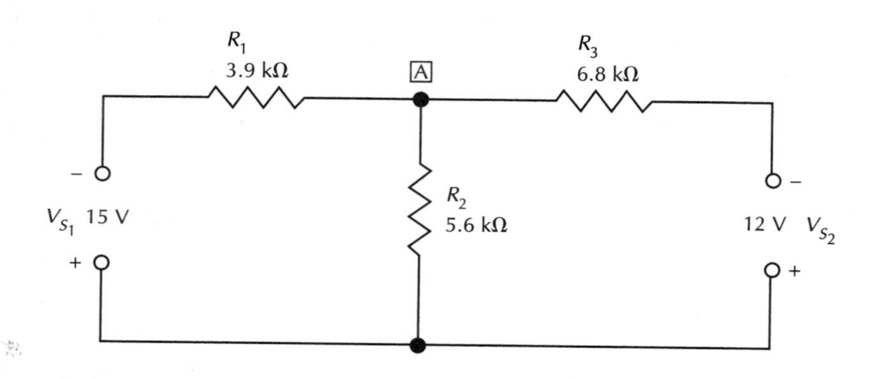

FIGURE 35

PROJECT PURPOSE

To provide practice in using the loop/mesh analysis technique, and to confirm the accuracy of the technique through actual circuit measurements.

PARTS NEEDED

- ☐ DMM/VOM (2)
- ☐ VVPS (dc) (2)
- ☐ CIS
- ☐ Resistors:
  - 3.9 kΩ
  - 5.6 kΩ
  - 6.8 kΩ

Steps to use the loop/mesh analysis approach are summarized as:

1. Label the assumed mesh currents on the diagram. (Use the arbitrary clockwise direction convention.)
2. Assign mathematical polarities to sources. If assumed mesh current enters positive terminal of source, consider source a positive polarity. If mesh current enters negative terminal of source, consider source a negative polarity.
3. Write Kirchhoff's equations for each loop's voltage drops. Assume voltages across resistors in a given loop positive when the voltage drop is caused by its "own mesh" current and negative when caused by an adjacent mesh's assumed current.
4. Solve the resulting equations, using simultaneous equation methods.
5. Verify results using Ohm's law and Kirchhoff's law.

ACTIVITY	OBSERVATION	CONCLUSION
1. Draw appropriate arrows and label the meshes on the diagram of Figure 35.	Number of meshes labeled were ________	The convention used to draw the arrows for assumed mesh currents was to use a (*CW*, *CCW*) ________ direction.
2. Assign appropriate polarities to use for sources when performing loop/mesh calculations.	V_{S_1} = (+, –) ________ V_{S_2} = (+, –) ________	The reason V_{S_1} was assigned (+, –) ________ polarity is because mesh current enters its (+, –) ________ terminal. V_{S_2} was assigned a (+, –) ________ polarity because mesh current enters its (+, –) ________ terminal.
3. Write Kirchhoff's equations for each loop's voltage drops.	Loop A: ________ Loop B: ________	—
4. Solve equations from step 3 by using simultaneous equation techniques.	Equation 1: ________ Equation 2: ________ I_A = ________ I_B = ________	—
5. Use mesh currents and solve for the voltage drops across each resistor.	V_1 = ________ V V_2 = ________ V V_3 = ________ V	Does the sum of $V_1 + V_2$ equal source V_{S_1}'s value? ________ Does the sum of $V_3 + V_2$ equal source V_{S_2}'s value? ________

PROJECT
30
CONTINUED

NETWORK ANALYSIS TECHNIQUES
Loop/Mesh Analysis *(Continued)*

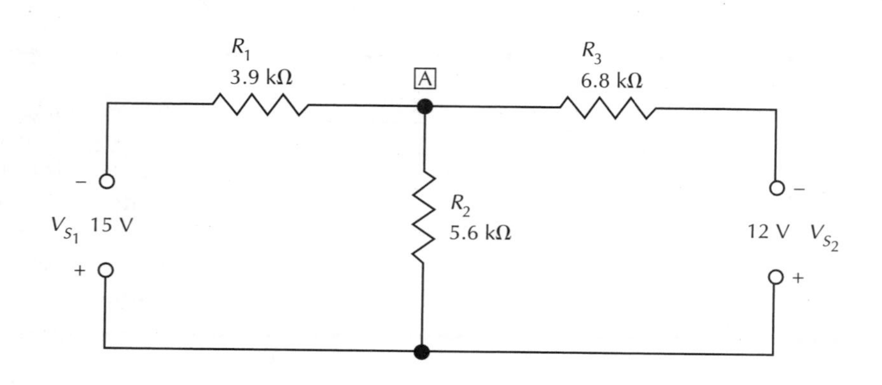

FIGURE 35

PROJECT PURPOSE

To provide practice in using the loop/mesh analysis technique, and to confirm the accuracy of the technique through actual circuit measurements.

PARTS NEEDED

- ☐ DMM/VOM (2)
- ☐ VVPS (dc) (2)
- ☐ CIS
- ☐ Resistors:
  - 3.9 kΩ
  - 5.6 kΩ
  - 6.8 kΩ

ACTIVITY	OBSERVATION	CONCLUSION
6. Connect the circuit of Figure 35. (Be sure to observe source polarities.)	—	—
7. Adjust VVPS #1 so that $V_{S_1} = 15$ V.	—	—
8. Adjust VVPS #2 so that $V_{S_2} = 12$ V.	—	—
9. Measure voltage drops across each resistor and indicate polarity of each of their voltage drops with respect to Point A on the diagram.	$V_1 = (+, -)$ ________ V $V_2 = (+, -)$ ________ V $V_3 = (+, -)$ ________ V	Which resistor's voltage drop represents the result of both meshs' currents? ________.
10. Compare the voltages measured in step 9 with your earlier calculations in step 5.	Step 5 $V_1 =$ ________ V Measured $V_1 =$ ________ V Step 5 $V_2 =$ ________ V Measured $V_2 =$ ________ V Step 5 $V_3 =$ ________ V Measured $V_3 =$ ________ V	Actual circuit operating parameters (*do, do not*) ________ verify the theoretical approach.
11. Calculate the current through each resistor and indicate the direction of current through each *R* by drawing an appropriate arrow on the diagram next to each *R*.	$I_{R_1} =$ ________ mA $I_{R_2} =$ ________ mA $I_{R_3} =$ ________ mA	(*Ohm's, Kirchhoff's*) ________ law was used to find current values. Are the calculated currents the same in every case as the "assumed mesh" currents? ______. Why not? ________.

PROJECT

31

NETWORK ANALYSIS TECHNIQUES
Nodal Analysis

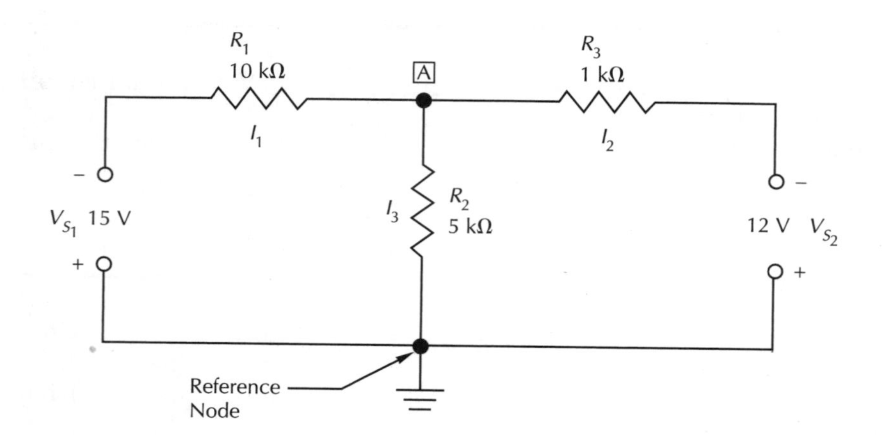

FIGURE 36

PROJECT PURPOSE

To provide practice in using the nodal analysis technique, and to confirm the accuracy of the technique through actual circuit measurements.

PARTS NEEDED

- ☐ DMM/VOM (2)
- ☐ VVPS (dc) (2)
- ☐ CIS
- ☐ Resistors:
  - 1.0 kΩ
  - 5.0 kΩ
  - 10 kΩ

The nodal analysis approach steps are summarized as follows:

1. Find the major nodes and select the reference node.
2. Designate currents into and out of major nodes.
3. Write Kirchhoff's current law equations for currents into and out of major nodes. (Use V/R statements to represent each current.)
4. Develop equations to solve for voltage across resistor(s) associated with the reference node, and solve for V and I for reference node component(s).
5. Using data accumulated, solve for voltages and currents through other components in the circuit.
6. Verify results using Kirchhoff's voltage law.

ACTIVITY	OBSERVATION	CONCLUSION
1. Use arrows and designate currents into and out of the major node on Figure 36.	On the diagram: I_1 is current through R ________. I_2 is current through R ________. I_3 is current through R ________.	—
2. Write Kirchhoff's current equation for the loop with the 15-volt source.	$I_1 + I$ ________ $= I$ ________	Converting these currents into V/R statements, we can say that $V_{R_1}/R_1 + V_{R\_\_}/R\_\_$ equals $V_{R\_\_}/R\_\_$. The voltage and current of special interest, due to being associated with the reference node, is V_R____ and current I___.
3. Write equations for V across R_2 using loop voltage and current statements.	For 15-V loop: $V_{R_1} + V_{R_2} = 15$ V $V_{R_1} = 15 -$ ________ For 12-V loop: $V_{R_3} + V_{R_2} = 12$ V $V_{R_3} = 12 -$ ________	Voltage across R_2 is result of current I ________. Translated into V/R current statements: 15 V − V_{R_2}/10 kΩ PLUS 12 V − V_{R_2}/1 kΩ EQUALS V_{R_2}/5 kΩ
4. Perform appropriate math to solve for V_{R_2}.	$V_{R_2} =$ ________ V	—
5. Use Ohm's law and solve for I_{R_2}.	$I_{R_2} =$ ________ mA	—
6. Solve for V_{R_3}, I_{R_3}, V_{R_1}, and I_{R_1}.	$V_{R_3} = 12 -$ ________ = ________ V $I_{R_3} =$ ________ V ÷ ________ kΩ equals ________ mA $V_{R_1} = 15 -$ ________ = ________ V $I_{R_1} =$ ________ V ÷ ________ kΩ equals ________ mA	

PROJECT

31

CONTINUED

NETWORK ANALYSIS TECHNIQUES
Nodal Analysis *(Continued)*

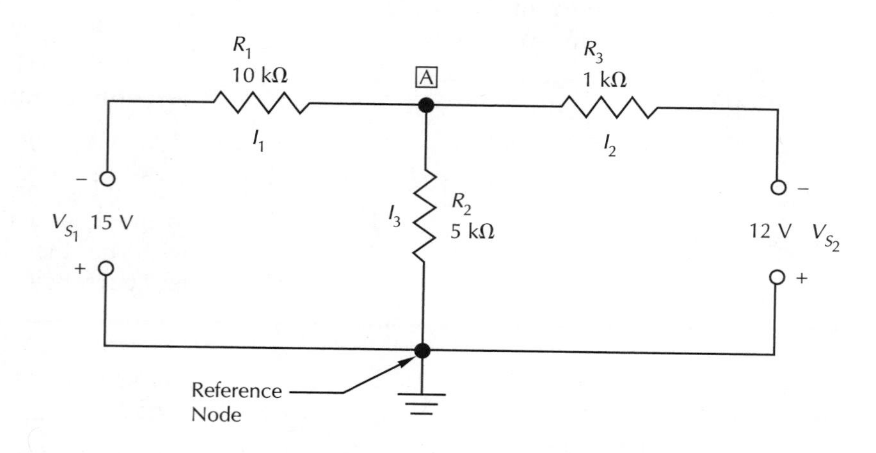

FIGURE 36

PROJECT PURPOSE

To provide practice in using the nodal analysis technique, and to confirm the accuracy of the technique through actual circuit measurements.

PARTS NEEDED

- ☐ DMM/VOM (2)
- ☐ VVPS (dc) (2)
- ☐ CIS
- ☐ Resistors:
  - 1.0 kΩ
  - 5.0 kΩ
  - 10 kΩ

ACTIVITY	OBSERVATION	CONCLUSION
7. Verify results of calculations.	$V_{R_1} + V_{R_2} = V_{S_1}$ _______ + _______ = 15 V $V_{R_2} + V_{R_3} = V_{S_2}$ _______ + _______ = 12 V	Did the calculated values using the nodal approach appropriately check out with Kirchhoff's law(s)? _______.
8. Connect the circuit shown in Figure 36. (Be sure to observe source polarities.)	—	—
9. Adjust VVPS #1 so that V_{S_1} equals 15 volts. Adjust VVPS #2 so that V_{S_2} equals 12 volts.	—	—
10. Measure voltage drops across the resistors.	V_{R_1} measures _______ V V_{R_2} measures _______ V V_{R_3} measures _______ V	Do the measured values of *V*s reasonably agree with your earlier calculations? _______.
11. Using the measured values of *V* and the color-coded values of resistors, calculate current through each resistor.	I_{R_1} = _______ mA I_{R_2} = _______ mA I_{R_3} = _______ mA	Do the current values calculated on the basis of actual circuit measurements reasonably agree with your previous theoretical calculations? _______.

PROJECT

32

NETWORK ANALYSIS TECHNIQUES

Wye-Delta and Delta-Wye Conversions

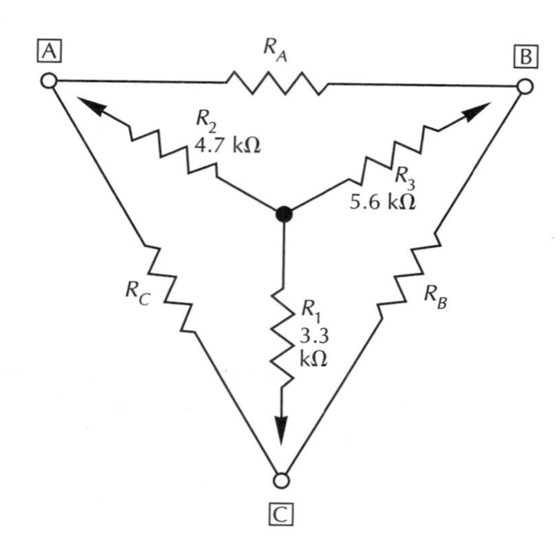

FIGURE 37

PROJECT PURPOSE

To initially connect a wye network and measure its resistive parameters. Next, to determine the equivalent delta network resistance values through the conversion technique. Then, to confirm the conversion technique by actual circuit measurements.

PARTS NEEDED

- ☐ DMM/VOM (2)
- ☐ CIS
- ☐ Resistors:
  - 270 Ω
  - 780 Ω
  - 820 Ω
  - 3.3 kΩ
  - 4.7 kΩ
  - 5.6 kΩ
  - 10 kΩ
  - 12 kΩ
  - 18 kΩ

SAFETY HINTS

Do not connect circuits in this project to any power sources!

The process of converting wye-to-delta and delta-to-wye networks enables selection of appropriate circuit component values that will provide the same electrical load to sources, when converting from one of these circuit configurations to the other. For our purposes, we will use the nomenclatures R_1, R_2, and R_3 to designate wye configuration resistive components and R_A, R_B, and R_C to designate delta configuration resistive components. See Figure 37.

To find values of delta components when the wye component values are known, use:

$$R_A = \frac{(R_1R_2) + (R_2R_3) + (R_3R_1)}{R_1}$$

$$R_B = \frac{(R_1R_2) + (R_2R_3) + (R_3R_1)}{R_2}$$

$$R_C = \frac{(R_1R_2) + (R_2R_3) + (R_3R_1)}{R_3}$$

To find wye component values when the delta component values are known, use:

$$R_1 = \frac{R_BR_C}{R_A + R_B + R_C}$$

$$R_2 = \frac{R_CR_A}{R_A + R_B + R_C}$$

$$R_3 = \frac{R_AR_B}{R_A + R_B + R_C}$$

ACTIVITY	OBSERVATION	CONCLUSION
1. Connect the wye portion of the circuit shown in Figure 37, using the component values shown.	R_1 = ________ kΩ R_2 = ________ kΩ R_3 = ________ kΩ	—
2. Using an ohmmeter, measure and record the resistance readings from Point A to B, from Point B to C, and from Point A to C.	A to B = ________ kΩ B to C = ________ kΩ A to C = ________ kΩ	When measuring from A to B, you are actually measuring the series resistance of R ______ and R ______. The resistance from point B to point C equals the series resistance of R ______ and R ______. The resistance from point A to point C equals the series resistance of R ______ and R ______.

PROJECT **32** CONTINUED

NETWORK ANALYSIS TECHNIQUES
Wye-Delta and Delta-Wye Conversions *(Continued)*

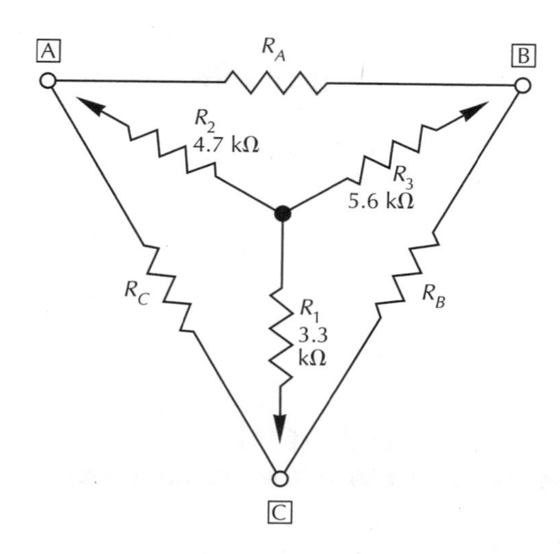

FIGURE 37

PROJECT PURPOSE

To initially connect a wye network and measure its resistive parameters. Next, to determine the equivalent delta network resistance values through the conversion technique. Then, to confirm the conversion technique by actual circuit measurements.

PARTS NEEDED

- ☐ DMM/VOM (2)
- ☐ CIS
- ☐ Resistors:
  270 Ω
  780 Ω
  820 Ω
  3.3 kΩ
  4.7 kΩ
  5.6 kΩ
  10 kΩ
  12 kΩ
  18 kΩ

SAFETY HINTS

Do not connect circuits in this project to any power sources!

ACTIVITY	OBSERVATION	CONCLUSION
3. Use the appropriate formulas and solve for the values of resistors that would form a delta network that is equivalent to the wye network used in steps 1 and 2.	R_A = ______ kΩ R_B = ______ kΩ R_C = ______ kΩ	The equivalent delta resistor values are (*greater*, *smaller*) ______ in value than the wye network resistor values.
4. Connect a delta circuit using the approximate (available) resistance values calculated in step 3. **NOTE:** You may have to use more than one resistor for each delta leg in order to use standard, available resistor values.	—	Is it necessary to use more than one standard resistor for each leg, in this case? ______.
5. Use the space in the Observation column and draw a schematic showing your circuit setup, which shows the *R*s used for each leg. Label the *R*s making up R_A, R_B, R_C, as appropriate. Also label test points A, B, and C on your diagram.	—	—
6. Use an ohmmeter and measure the *R* values found between points A to B, B to C, and A to C.	A to B measures ______ kΩ B to C measures ______ kΩ A to C measures ______ kΩ	Do the measured values for this equivalent delta network come close to the values measured between the same test points in the wye network? (*yes*, *no*) ______. If not, what could cause some of the differences? ______. When measuring *R* between points A and B, you are measuring the equivalent total *R* of R_A in parallel with what components? ______. When measuring *R* between points B and C, you are measuring the equivalent total *R* of R_B in parallel with what components? ______. When measuring *R* between points A and C, you are measuring the equivalent total *R* of R_C in parallel with what components? ______.

NETWORK ANALYSIS TECHNIQUES

Complete the following review questions, indicating the appropriate response by placing a check in the box next to the correct answer.

1. When using the loop/mesh analysis technique, assumed mesh currents:

   ☐ should branch at circuit junction points
   ☐ should not branch at circuit junction points

2. The arbitrary convention regarding directions of assumed mesh currents is that:

   ☐ they should be shown circulating in a clockwise direction through each loop
   ☐ they should be shown circulating in a counterclockwise direction through each loop

3. Voltage drops across resistors caused by their own assumed mesh currents are considered:

   ☐ negative, when used in mesh equations
   ☐ positive, when used in mesh equations
   ☐ polarity assignment is not necessary

4. The convention for source polarity assignment(s) indicates that:

   ☐ if the mesh current returns to the source at its positive terminal, the source will be considered negative in equations
   ☐ if the mesh current returns to the source at its negative terminal, the source will be considered negative in equations
   ☐ polarity assignments are not necessary for sources

5. A major node, for nodal analysis purposes, is a junction where:

   ☐ two components join
   ☐ three components join
   ☐ four components join
   ☐ none of the above

6. To qualify as a "reference node" in the nodal analysis approach:

   ☐ any node or junction point can be called the reference node
   ☐ any major node may be designated as the reference node
   ☐ only the major node with the most junctions can be designated as the reference node

7. The nodal approach is based on:

   ☐ Kirchhoff's voltage law
   ☐ Kirchhoff's current law
   ☐ Ohm's law only

8. Delta networks may also sometimes be called:

   - ☐ Tee networks
   - ☐ Vee networks
   - ☐ Pi networks
   - ☐ L networks

9. When converting from a wye network to a delta network:

   - ☐ the numerator in the mathematical equation stays the same for finding all three equivalent delta resistors
   - ☐ the denominator in the mathematical equation stays the same for finding all three equivalent delta resistors
   - ☐ neither the numerator nor the denominator is the same for the three equations used to find the equivalent delta resistors

10. When converting from a delta network to a wye network:

   - ☐ the numerator in the mathematical equation stays the same for finding all three equivalent delta resistors
   - ☐ the denominator in the mathematical equation stays the same for finding all three equivalent delta resistors
   - ☐ neither the numerator nor the denominator is the same for the three equations used to find the equivalent delta resistors

11. When measuring resistance between outside terminal points, which network configuration results in measuring the series resistance of two resistors?

   - ☐ wye network
   - ☐ delta network

12. When measuring resistance between outside terminal points, which network configuration results in measuring the equivalent resistance of parallel resistors?

   - ☐ wye network
   - ☐ delta network

DC MEASURING INSTRUMENTS (ANALOG)

Objectives

You will connect several dc metering circuits and make measurements and observations about shunting an ammeter to increase its full-scale range; adding a series multiplier resistor to a basic current movement to make it into a voltmeter; and the basic series ohmmeter circuit.

In completing these projects, you will connect circuits, make measurements, perform calculations, draw conclusions, and be able to answer questions about the following items related to dc meter circuits.

- Shunt resistors — application and calculation
- Multiplier resistors — application and calculation
- Operation and characteristics of a series-type ohmmeter circuit
- Loading effects of a voltmeter

PROJECT/TOPIC CORRELATION INFORMATION

PROJECT	TEXT CHAPTER	SECTION	RELATED TEXT TOPIC(S)
33 The Basic Ammeter Shunt	11	11-4	Ammeter Shunts
34 The Basic Voltmeter Circuit	11	11-5	Voltmeters
35 The Series Ohmmeter Circuit	11	11-7	Ohmmeters

PROJECT

33

DC MEASURING INSTRUMENTS (ANALOG)

The Basic Ammeter Shunt

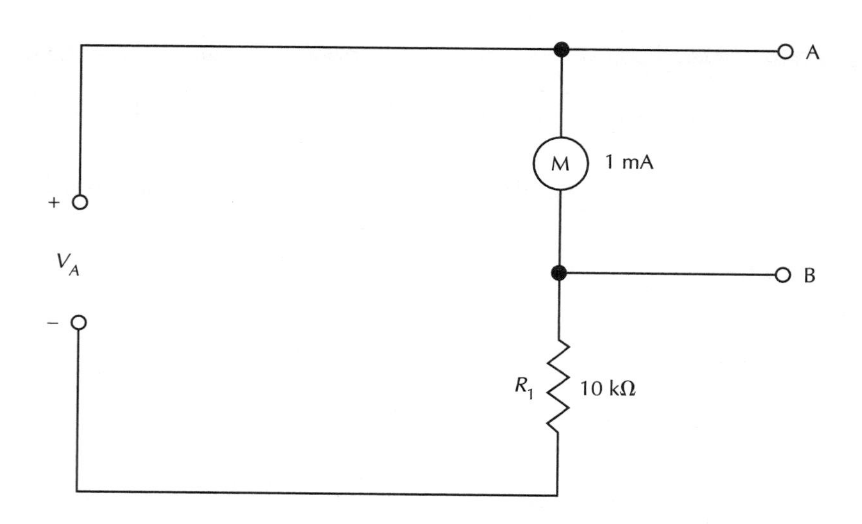

FIGURE 38

PROJECT PURPOSE

To determine the appropriate value of shunt resistance to use to increase the current measuring range of a meter movement. To test the operation of the shunted meter by circuit measurements.

PARTS NEEDED

- ☐ DC Panel Meter, 1 mA (with approximately 360 Ω R_M, if available)
- ☐ DMM/VOM
- ☐ VVPS (dc)
- ☐ CIS
- ☐ Resistors:
  Resistor Decade Box or assorted resistors
  10 kΩ

ACTIVITY	OBSERVATION	CONCLUSION
1. Connect the initial circuit as shown in Figure 38.	—	—
2. Apply just enough V_A to obtain full-scale deflection on the ammeter (1 mA). **NOTE:** DO NOT CHANGE V_A FROM THIS VALUE WHILE PERFORMING THE FOLLOWING STEPS.	V_A = ______ volts	—
3. Assume the meter resistance to be about *360 ohms and calculate what value of shunt resistance is needed to create a 2-mA full-scale meter circuit. **NOTE:** $R_S = \frac{I_M \times R_M}{I_S}$	I_M = ______ mA R_M = ______ ohms I_S = ______ mA R_S = ______ ohms	According to Kirchhoff's current law, if there is 2-mA total current coming to the junction of the ammeter and the shunt R, and we want 1 mA through the meter, then the current through the shunt R must be ______ mA. Since the shunt is in parallel with the meter, then the voltage across the shunt must be equal to the voltage across the meter, which is equal to $I_M \times$ ______.
4. Combine resistors in parallel or series, or series-parallel as required to obtain about *360 ohms in shunt with the meter and note the reading on the ammeter (with V_A the same as above). (**NOTE:** If a "resistance decade" box is available, simply switch in the desired value and connect as appropriate.)	Scale reading equals ______ mA	With the basic 1-mA meter shunted with about *360 ohms, a scale reading of 0.5 mA equals an actual circuit current of ______ mA. Thus, if the current meter showed full-scale deflection, the actual circuit current would be ______ mA.
5. With the shunt still connected, CHANGE V_A to obtain full-scale deflection. Insert the DMM/VOM as required to verify that the circuit current is 2 mA.	V_A = ______ volts	Does the DMM/VOM reading closely correlate with the shunted current meter reading? ______.

*Use appropriate value for ammeter being used. (See manufacturer's specifications for meter model.)

PROJECT 34

DC MEASURING INSTRUMENTS (ANALOG)
The Basic Voltmeter Circuit

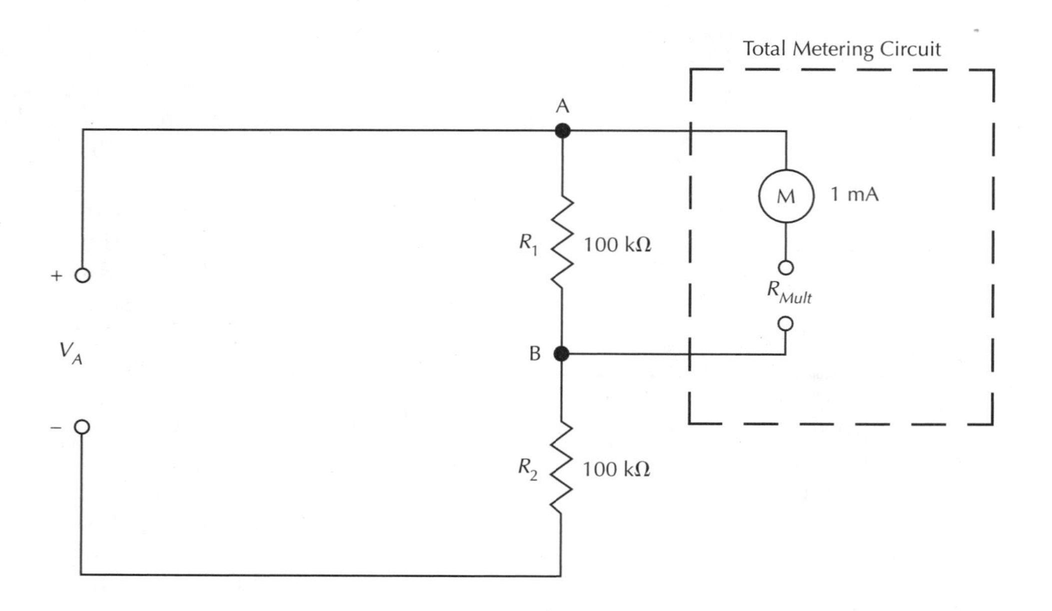

FIGURE 39

PROJECT PURPOSE

To determine the appropriate value of multiplier resistor to use with a meter movement to create a desired voltmeter range. To test the operation of the newly created voltmeter circuit. To observe the effects of "meter loading" on circuit measurements.

PARTS NEEDED

- ☐ DC Panel Meter, 1 mA
- ☐ DMM/VOM
- ☐ VVPS (dc)
- ☐ CIS
- ☐ Resistors:
  - Resistor Decade Box or assorted resistors
  - 100 Ω (2)
  - 1 kΩ (2)
  - 10 kΩ
  - 100 kΩ (2)

ACTIVITY	OBSERVATION	CONCLUSION
1. Connect the initial circuit as shown in Figure 39.	—	—
2. Neglect the small meter resistance of the current meter and calculate the resistance that would be required in series with the meter to limit the current through the meter to 1 mA if 10 volts difference of potential existed between points A and B.	R_{series} = ________ ohms	Since there would theoretically be full-scale deflection of the current meter with 10 volts between points A and B, we might say that the meter in combination with the series current-limiting resistor, called a multiplier resistor, is now a 0-______-volt voltmeter. This means that half-scale deflection would indicate ________ volts, one-fourth scale deflection would indicate ________ volts, and so on.
3. Connect a 10-kΩ resistor in series with the meter; apply 10 volts V_A to the circuit from the VVPS. Measure and record the current indication and translate that into its meaning when measuring V_1.	Current reading = ________ mA V_1 = ________ volts	According to series circuit theory, V_1 should equal what fraction of V applied? ________. Since V_A is 10 volts, then V_1 should equal ________ volts. Some factors that might cause the actual measurement to vary slightly are: resistor (*size, tolerance*) ________, meter (*size, accuracy*) ________, and the fact that we did not consider the (*lead, meter*) ________ resistance in determining R_{Mult}.
4. Use the DMM/VOM and measure V_1. After doing this, vary V_A up and down and observe if the current-meter/multiplier-resistor circuit measures essentially the same as the DMM/VOM at several voltage settings.	V_1 = ________ volts	In observing the comparison of the DMM/VOM readings with the "demonstration voltmeter," it appears that our current meter with "multiplier" (*can, cannot*) ________ be used effectively as a voltmeter.
5. Calculate the value of R_{Mult} needed to make our 1-mA movement into a 0–100-volt voltmeter circuit. **CAUTION:** Remove R_1 and R_2 for this step! Insert proper R_{Mult} as calculated. Rewire meter circuit to measure V_A. Set V_A to 25 volts using the DMM/VOM. Note current.	R_{Mult} = ________ I = ________ mA, interpreted into a voltage reading = ________ volts.	We can conclude that when the basic meter movement is 1 mA, R_{Mult} must limit the current to 1-mA maximum for any given voltage range. This means to have a 0–50-volt range, R_{Mult} must be (*close to*) ______ ohms. In other words, for every volt full-scale range we desire, a value of ______-ohm resistance must be used to limit I to 1 mA.

PROJECT

34

CONTINUED

DC MEASURING INSTRUMENTS (ANALOG)

The Basic Voltmeter Circuit *(Continued)*

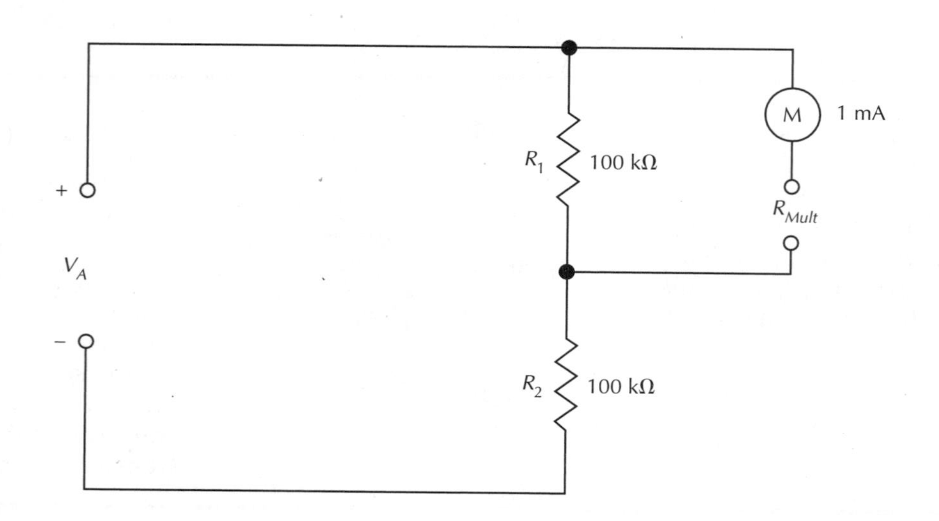

FIGURE 40

PROJECT PURPOSE

To determine the appropriate value of multiplier resistor to use with a meter movement to create a desired voltmeter range. To test the operation of the newly created voltmeter circuit. To observe the effects of "meter loading" on circuit measurements.

PARTS NEEDED

- ☐ DC Panel Meter, 1 mA
- ☐ DMM/VOM
- ☐ VVPS (dc)
- ☐ CIS
- ☐ Resistors:
  - Resistor Decade Box or assorted resistors
  - 100 Ω (2)
  - 1 kΩ (2)
  - 10 kΩ
  - 100 kΩ (2)

ACTIVITY	OBSERVATION	CONCLUSION
6. Connect the circuit as shown in Figure 40.	—	—
7. Use a 10-kΩ R_{Mult} in conjunction with the 1-mA meter and measure V_1 with a V applied of 15 volts (set V_A with the DMM/VOM).	V_1 = ____________ volts	Without the voltmeter circuit being connected across R_1, V_1 should read ____________ volts drop. Instead of reading this value, it measured ____________ volts. From this we conclude a voltmeter can change circuit conditions if the meter circuit resistance is low compared to the resistance of the circuit across which it is measuring. One way of rating voltmeter "sensitivity" is by the "ohms-per-volt" required to limit the meter circuit current to the basic movement's full-scale current limit. A 50-microamp basic movement would require 20,000 ohms R_{Mult} if 1 volt were applied to the meter and R_{Mult}, in order to limit I to 50 microamps. If we wanted a 0–10-volt range, then 200,000 ohms would be the value for R_{Mult}, and so on. What would be the "ohms-per-volt" rating of this type voltmeter? ____________ ohms per volt.
8. Change R_1 and R_2 in the circuit to 1-kΩ resistors and using the 1-mA meter with R_{Mult} of 10 kΩ, measure V_1 with a circuit V_A of 15 volts.	V_1 = ____________ volts	Comparing these results with those of step 2 above, we conclude that a low "sensitivity" meter circuit affects the circuit being measured less when the measured circuit's resistances are of (*high, low*) ______ value. It should be noted that the higher a voltmeter's "ohms-per-volt" sensitivity rating the less the voltmeter will alter the circuit under test.

PROJECT

35

DC MEASURING INSTRUMENTS (ANALOG)

The Series Ohmmeter Circuit

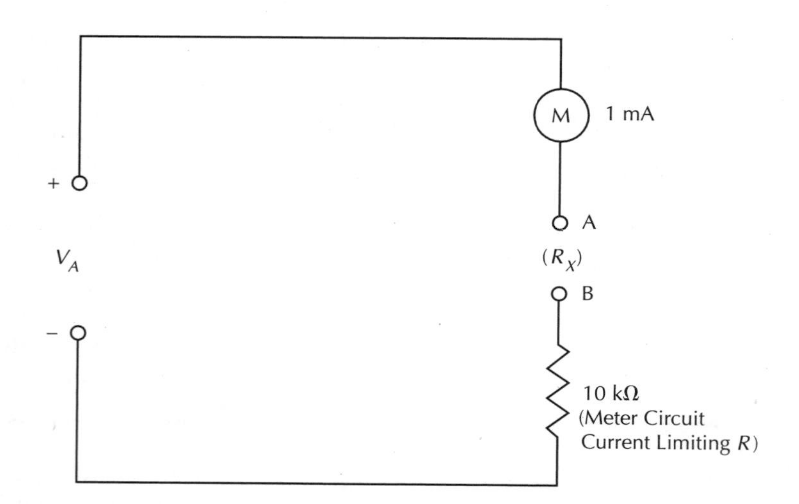

FIGURE 41

PROJECT PURPOSE

To demonstrate a simple series-type ohmmeter circuit and observe the "backoff-type" ohmmeter scale operation for this type circuit. To predict the value of R_x that, when measured, would cause a specified amount of meter deflection.

PARTS NEEDED

- ☐ DC Panel Meter, 1 mA
- ☐ VVPS (dc)
- ☐ CIS
- ☐ Resistors:
  10 kΩ (2)
  27 kΩ (2)

ACTIVITY	OBSERVATION	CONCLUSION
1. Connect the initial circuit as shown in Figure 41.	—	—
2. Jumper points A and B with a short and adjust V_A from VVPS for exactly full-scale deflection on the current meter. **NOTE:** DO NOT CHANGE V_A FROM THIS SETTING FOR THE REST OF THIS DEMONSTRATION OR PROJECT.	V_A = ______ volts	For this project or demonstration, points A and B will represent the "test prods" of our demonstration ohmmeter. When there is full-scale deflection of the current meter, what resistance is present between points A and B? ______ ohms. How much deflection would there be if points A and B were "open" (infinite ohms)? ______ deflection. In other words, a reading of 0 mA represents a resistance of ______ ohms; a reading of 1 mA represents a resistance of ______ ohms; and a reading in between 0 and 1 mA represents some value of resistance.
3. Calculate what value of resistance would be between points A and B if there were 1/2-scale deflection (0.5 mA).	R_{A-B} = ______ ohms	If the current is 0.5 mA with 10 volts applied, then the formula for solving R_T is: ($R_T = V_T/I_T$). Once R_T is known, we subtract the known value of resistance from it, and we then know the value of R between points A and B.
4. Insert a resistor of the value calculated in step 3 between points A and B, and note the meter reading.	Reading = ______ mA	This current reading represents a resistance between points A and B of ______ ohms.

PROJECT 35 CONTINUED

DC MEASURING INSTRUMENTS (ANALOG)

The Series Ohmmeter Circuit

(Continued)

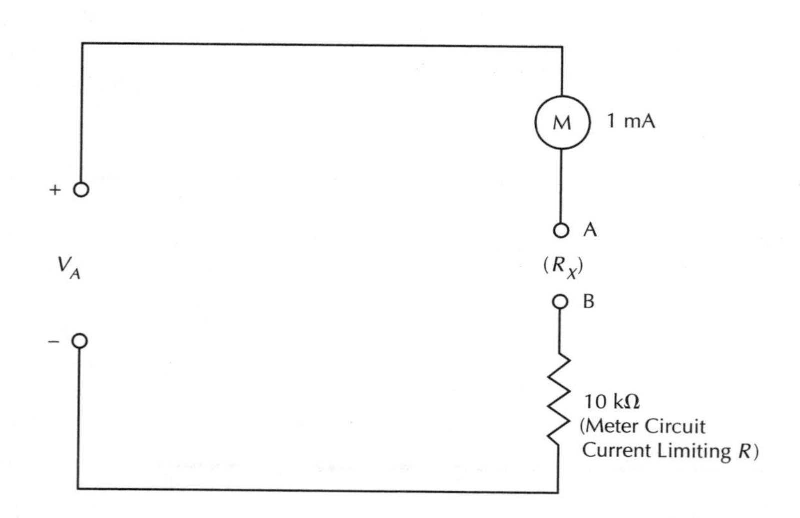

FIGURE 41

PROJECT PURPOSE

To demonstrate a simple series-type ohmmeter circuit and observe the "backoff-type" ohmmeter scale operation for this type circuit. To predict the value of R_x that, when measured, would cause a specified amount of meter deflection.

PARTS NEEDED

- ☐ DC Panel Meter, 1 mA
- ☐ VVPS (dc)
- ☐ CIS
- ☐ Resistors:
  10 kΩ (2)
  27 kΩ (2)

ACTIVITY	OBSERVATION	CONCLUSION
5. Calculate what value of resistance (R_X) would limit the current reading to 0.25 mA. Then, insert an R that is the closest available to this value and note the meter reading.	R_X = ____________ ohms	The closest available resistor was ________ ohms. Was the current reading close to 0.25 mA? ______. Was it higher or lower than 0.25 mA? __________. The higher the current reading, the (*higher*, *lower*) ______________ the value of R_X. If you were making calibration marks on the meter for an ohmmeter scale, would the scale be linear or nonlinear? __________. We may conclude that, in essence, the full-scale rating of the meter and "internal" ohmmeter battery determine the scale divisions on the ohmmeter.

DC MEASURING INSTRUMENTS (ANALOG)

Complete the following review questions, indicating the appropriate response by placing a check in the box next to the correct answer.

1. To extend the range of a current meter, a resistor is used:
   - ☐ in series with the meter called a multiplier
   - ☐ in parallel with the meter called a shunt
   - ☐ neither of these

2. To use a basic current meter movement as a voltmeter, __________ is used.
   - ☐ a multiplier
   - ☐ a shunt
   - ☐ a divider
   - ☐ none of these

3. For a series-type ohmmeter, the unknown resistance (R_x) is connected:
   - ☐ in series with the ohmmeter circuit
   - ☐ in parallel with the ohmmeter circuit

4. Because a current meter is connected in series with the circuit under test, the current meter's resistance should be:
   - ☐ as high as possible
   - ☐ as low as possible
   - ☐ makes no difference

5. Because a voltmeter is connected in parallel with the circuit under test, the voltmeter and associated meter circuit resistance should be:
   - ☐ as high as possible
   - ☐ as low as possible
   - ☐ makes no difference

6. On a series-type ohmmeter, zero ohms is on the:
   - ☐ right end of the meter scale
   - ☐ left end of the meter scale

7. If a 100-microamp movement is to be used as a voltmeter, what would be its ohms-per-volt sensitivity rating?
   - ☐ 1,000
   - ☐ 10,000
   - ☐ 20,000
   - ☐ 50,000
   - ☐ none of these

8. If a 50-microamp meter has an internal resistance of 5,000 ohms, what precise value of R_{Mult} is needed to make this meter a 0–10-volt meter?

   - ☐ 100 kΩ
   - ☐ 150 kΩ
   - ☐ 195 kΩ
   - ☐ 205 kΩ
   - ☐ none of these

9. If the ohmmeter "internal source" voltage for a series-type ohmmeter is doubled and the zero-adjust control is reset to properly zero the meter, the value of R represented by a half-scale deflection of the ohmmeter would be ________________________________ that of the original.

   - ☐ double
   - ☐ one-half
   - ☐ three times
   - ☐ four times

10. Voltmeter "loading" is more of a problem when the voltmeter is measuring across a:

    - ☐ high R value
    - ☐ low R value
    - ☐ makes no difference

THE OSCILLOSCOPE

Objectives

You will adjust a variety of controls on an oscilloscope and connect voltage or signal sources to the scope in order to demonstrate several of the various functions for which an oscilloscope may be used by technicians.

In completing these projects, you will set up and connect an oscilloscope and various voltage/signal source(s), manipulate controls on the equipment, make observations and measurements, draw conclusions, and be able to answer questions about the following items related to the oscilloscope.

- Use of various input/output terminals
- Use of various controls on the front panel
- How to measure voltage(s)
- How to observe waveforms
- How to make phase comparisons
- How to determine frequency

PROJECT/TOPIC CORRELATION INFORMATION

PROJECT	TEXT CHAPTER	SECTION	RELATED TEXT TOPIC(S)
36 Basic Operation	13	13-2 13-4	Key Parts of the Scope Combining Horizontal and Vertical Signals to View a Waveform
37 Voltage Measurements	13	13-5	Measuring Voltage and Determining Current with a Scope
38 Phase Comparisons	13	13-6	Using the Scope for Phase Comparisons
39 Determining Frequency	13	13-7	Determining Frequency with a Scope

PROJECT

36

THE OSCILLOSCOPE
Basic Operation: Part A

FRONT PANEL/CONTROLS SKETCH
(To be drawn by student)

PROJECT PURPOSE

To familiarize you with the most-used controls on an oscilloscope via observation of an actual oscilloscope, and through reading of its operating manual.

PARTS NEEDED

- ☐ Oscilloscope with cable(s) (preferably a triggered sweep-single or dual-trace oscilloscope)
- ☐ Equipment Manual for oscilloscope used
- ☐ Audio or Function Generator with cable(s)
- ☐ 1.5 V Dry Cell
- ☐ CIS

The oscilloscope is a versatile test instrument that can visually display the relationship between: 1) two electrical quantities, 2) an electrical quantity and time. In this first project, you will become acquainted with the basic "jacks" and "controls" used in obtaining a visual waveform with the scope. (**CAUTION!** Do not leave a bright "spot" on the screen.)

ACTIVITY	OBSERVATION	CONCLUSION
1. Obtain the available scope, audio or function generator, dc battery or cell, appropriate test leads, and equipment manuals.	—	—
2. Referring to the oscilloscope and appropriate operating manual, locate and point out the following jacks and controls to your instructor. **NOTE:** Different brands and models may use different names for some of these items. However, you should be able to interpret these differences quite easily.	Locate and point out the following: – Ground jack – AC-GND-DC switch – Vertical input jack(s) – Horizontal input jack – Ext. sync. signal jack – Off/On switch – Intensity (brilliance) control – Focus control(s) – Horizontal position control(s) – Vertical position control(s) – Horizontal (sweep) frequency control(s) – Horizontal gain control(s) – Vertical gain control(s) – Synchronization control(s) (Instructor initial ______________)	The intensity control adjusts the (*brightness, size*) ______________ of the display. Fuzziness of display can be changed with the ________ control. To center waveforms on the scope, the ______________ and ______________ position controls are used. To change the size of a display, the ____________ and ____________ gain controls are used. To get the display to "just fill" the scope face from left to right, the __________ gain control is used. To adjust the amount of "up-and-down" deflection caused by the waveform, the ________ ________ control(s) is/are used.
3. Draw a sketch of the scope front panel and controls in the space provided on the previous page (page 152).	—	—

PROJECT 36

THE OSCILLOSCOPE
Basic Operation: Part B

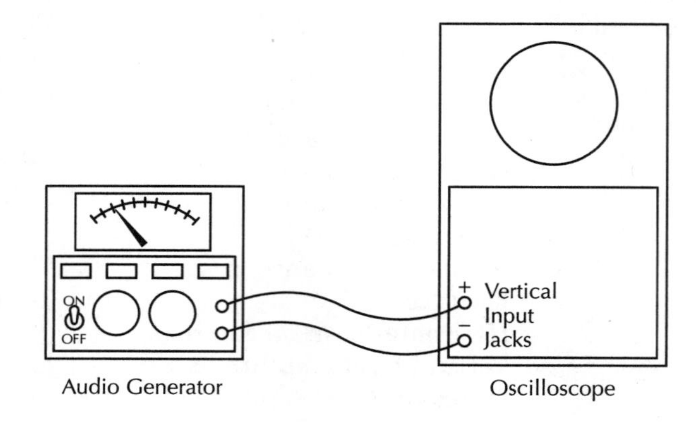

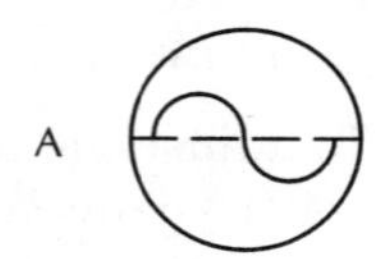

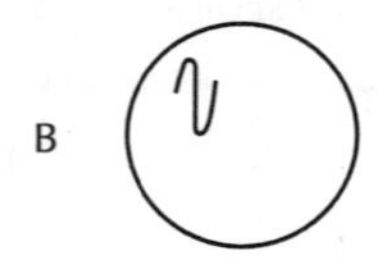

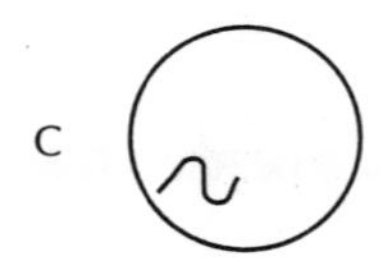

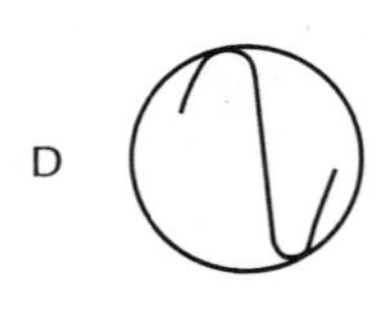

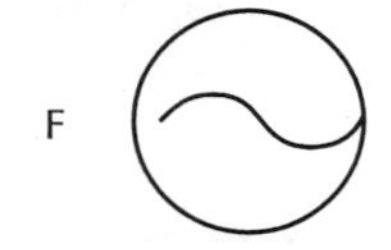

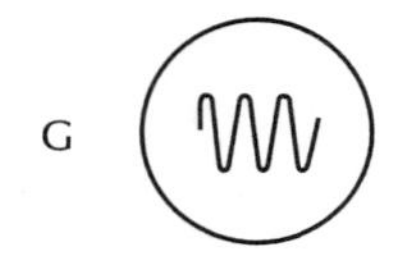

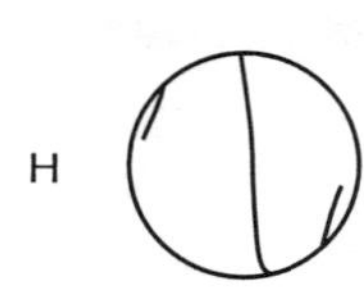

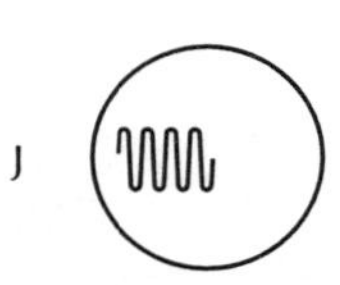

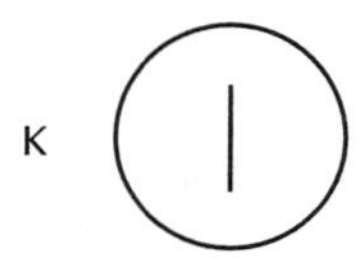

FIGURE 42

PROJECT PURPOSE

To provide practice in manipulating oscilloscope controls to achieve desired waveforms.

ACTIVITY	OBSERVATION	CONCLUSION
1. Connect the function generator or audio generator to the oscilloscope as shown in Figure 42. Adjust the generator frequency to 100 Hz.	—	—
2. Using the various controls, obtain each of the waveforms shown on the opposite page. Demonstrate waveforms A, G, and H for your instructor. (**NOTE:** If you have trouble achieving stable waveforms (that is, "stopping the waveform(s)"), ask the instructor to help you adjust the synchronization/trigger control(s), as appropriate.)	Demonstration of waveforms: A ____________ G ____________ H ____________ (Instructor initial ________)	The main controls manipulated to achieve waveforms A,B,C,D,E,G,H, and I were the ________ controls and the ________ controls. To obtain the waveform shown in G required changing the ________ ________ controls.
3. Disconnect the audio generator from the circuit. Make sure the scope is in the dc input mode. Adjust scope controls to obtain a straight horizontal line display that just fills the screen from left to right. (H frequency controls, V and H position controls, and H gain control(s).)	—	—

PROJECT

36

THE OSCILLOSCOPE
Basic Operation: Part C

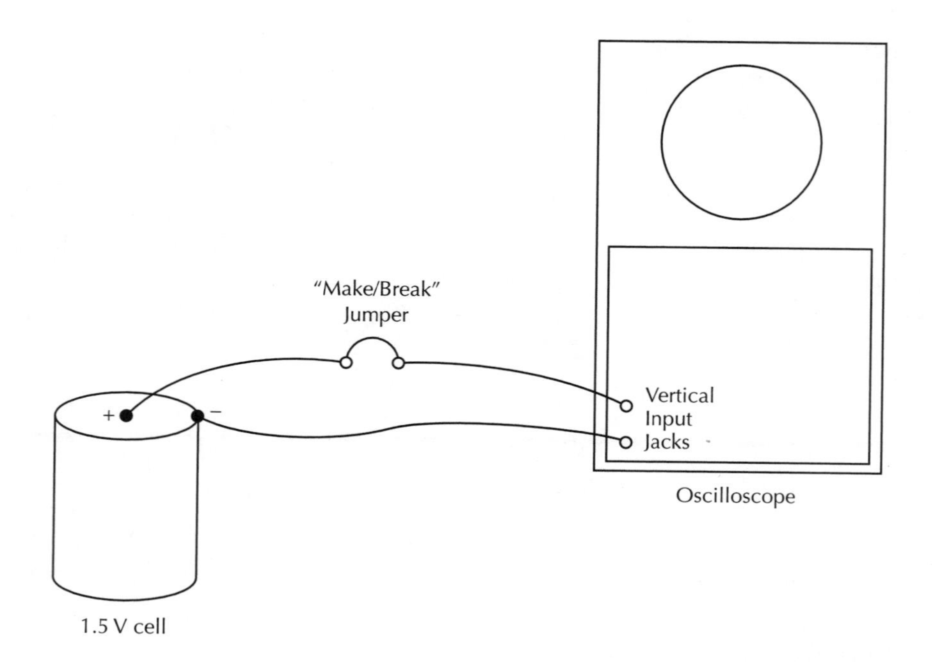

FIGURE 43

PROJECT PURPOSE

To observe the "on-screen" results of connecting dc voltage(s) to the vertical input of an oscilloscope and to see the effect of changing, and to practice using, vertical gain control(s).

ACTIVITY	OBSERVATION	CONCLUSION
1. Adjust the scope to obtain a straight horizontal line display that just fills the screen from left to right. Use the horizontal frequency controls, vertical and horizontal position controls, and horizontal gain controls as appropriate.	—	—
2. Connect a 1.5-volt dc source (such as a dry cell) to the vertical input of the scope, (– terminal to ground) (+ terminal to vertical input), Figure 43.	—	—
3. "Make" and "break" the input connection from the dc source to the vertical input jack and observe the display.	Does the horizontal trace (line) move when the dc voltage is applied? ________________. Which direction? ____________.	The dc voltage applied to the vertical input causes deflection of the scope trace. If the polarity of the input were reversed, would the trace react differently? __________ Explain. ____________________________.
4. Reverse the polarity of the connections from the dc source to the vertical input of the scope.	Does the direction of the display deflection reverse from that shown in step 3?	—
5. Make appropriate vertical gain control adjustments to cause the display to "jump" three vertical calibration squares on the scope face "graticule" when the input is connected and disconnected from the scope. (**NOTE:** When NOT connected, the trace should be in the middle of the screen.) When you have made the adjustments, demonstrate this action for your instructor.	Demonstrate that gain controls have been adjusted to achieve specified results. (Instructor initial __________)	Was more than one gain control adjusted to achieve the desired results? ________. What were the names of the controls? __________________________.

PROJECT

36

THE OSCILLOSCOPE
Basic Operation: Part D

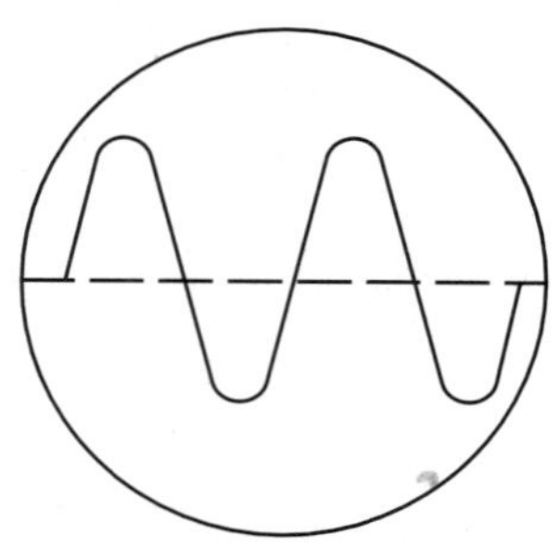

Sine Waves

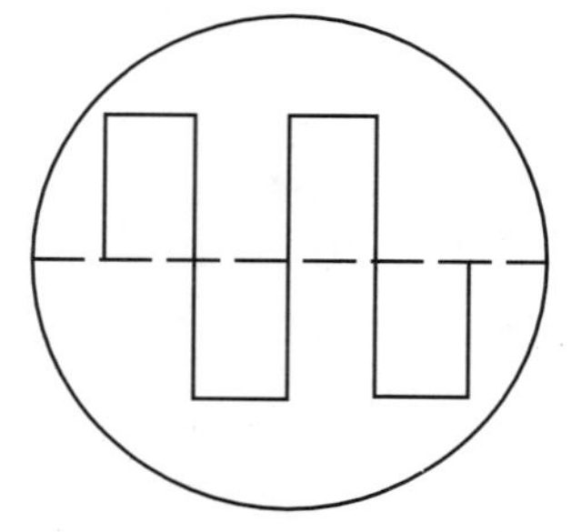

Square Waves

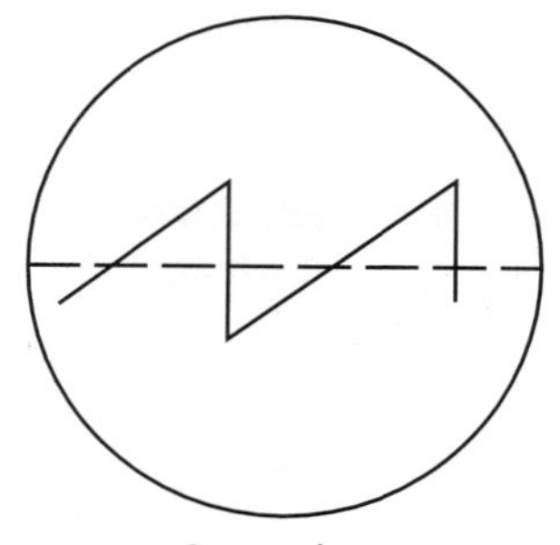

Sawtooth or
Triangular Waves

FIGURE 44

PROJECT PURPOSE

To observe various types of waveforms, using the oscilloscope.

ACTIVITY	OBSERVATION	CONCLUSION
1. If appropriate signal/function generator(s) are available, demonstrate to your instructor that you can obtain the waveforms shown in Figure 44.	Sine wave; Square wave; Triangular wave. (Instructor initial ______)	—

PROJECT 37

THE OSCILLOSCOPE Voltage Measurements

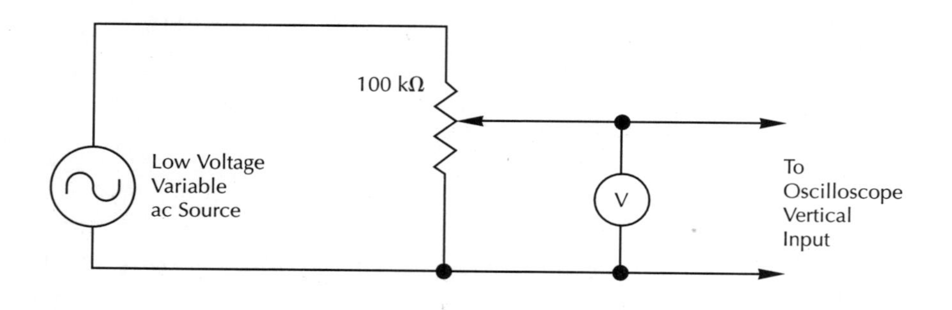

FIGURE 45

PROJECT PURPOSE

To provide familiarization with the techniques for measuring voltages with an oscilloscope. To learn to calibrate the scope and interpret the measurement results.

PARTS NEEDED

- ☐ Low voltage variable ac source
- ☐ Oscilloscope
- ☐ DMM/VOM
- ☐ CIS
- ☐ 100-kΩ potentiometer
- ☐ Appropriate cables for equipment

The following facts and considerations are important when using the oscilloscope to make voltage measurements.

1. Waveform deflection on the scope indicates the *peak-to-peak* value of the voltage under test.
2. For any SINE WAVE, the peak-to-peak deflection on the screen is *directly proportional* to the peak and rms values of the ac applied voltage that produce the peak-to-peak waveform. For example, if the deflection on the CRO screen is 1" (peak-to-peak deflection) when 1-volt rms is applied, it will be 2" of total deflection when a 2-volt rms ac signal is applied. Therefore, it is easy to measure ac voltage values with the scope even though the deflection is peak-to-peak in nature.
3. It is easier to read voltages on the CRO screen when the presentation is a vertical line. To achieve this type display, set the horizontal gain control fully counterclockwise (CCW).

ACTIVITY	OBSERVATION	CONCLUSION
1. Connect the circuit as shown in Figure 45. 2. Monitor input voltage to the scope with the meter and adjust the ac input voltage to 2 volts rms (as indicated by the meter).	— —	— —
3. Adjust the scope positioning and V and H gain controls to obtain a vertical deflection of ONE SQUARE (peak to peak). **CAUTION!** DO NOT MOVE THE VERTICAL V/DIV OR GAIN CONTROLS ONCE THIS IS DONE UNLESS DIRECTED TO DO SO. HOWEVER, YOU CAN CHANGE THE VERTICAL POSITION CONTROL FOR EASE OF VIEWING.	—	One large square of deflection equals ______ V rms; ______ V peak; and ______ V peak-peak. Each small division on the screen equals ______ V rms.
4. REMOVE the meter. Adjust the source voltage to obtain two squares of deflection.	Screen deflection is ______ squares.	What rms value does this deflection represent? ______ V.
5. Now measure the voltage with the meter.	Voltage measures ______ volts.	Does this answer agree with the scope measurement? ______.
6. Use the meter and adjust the source voltage to 6 volts rms.	Amount of deflection equals ______ squares.	Is the change in deflection essentially linear in nature? ______.

PROJECT
37
CONTINUED

THE OSCILLOSCOPE
Voltage Measurements *(Continued)*

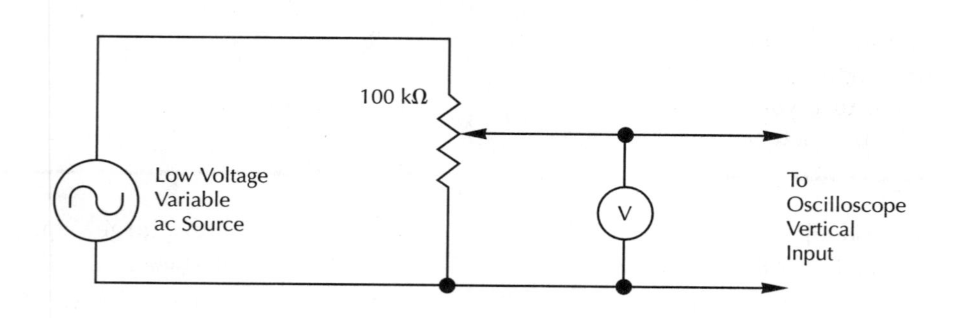

FIGURE 45

PROJECT PURPOSE

To provide familiarization with the techniques for measuring voltages with an oscilloscope. To learn to calibrate the scope and interpret the measurement results.

PARTS NEEDED

- ☐ Low voltage variable ac source
- ☐ Oscilloscope
- ☐ DMM/VOM
- ☐ CIS
- ☐ 100-kΩ potentiometer
- ☐ Appropriate cables for equipment

ACTIVITY	OBSERVATION	CONCLUSION
7. If a low-voltage transformer is available, CALIBRATE the scope and measure an unknown transformer secondary voltage using the oscilloscope as the measuring device.	Scope is calibrated so that each large square = ________ V_{P-P} or ________ V_{rms}. The unknown voltage is causing ________ large squares of deflection.	The value of the unknown voltage must equal approximately ________ volts rms.
8. Use the meter and verify the measurement taken in step 7 above.	Meter measures ________ V rms.	The scope and meter voltage measurements (*are, are not*) ________ close to equal.

PROJECT

38

THE OSCILLOSCOPE
Phase Comparisons: Part A

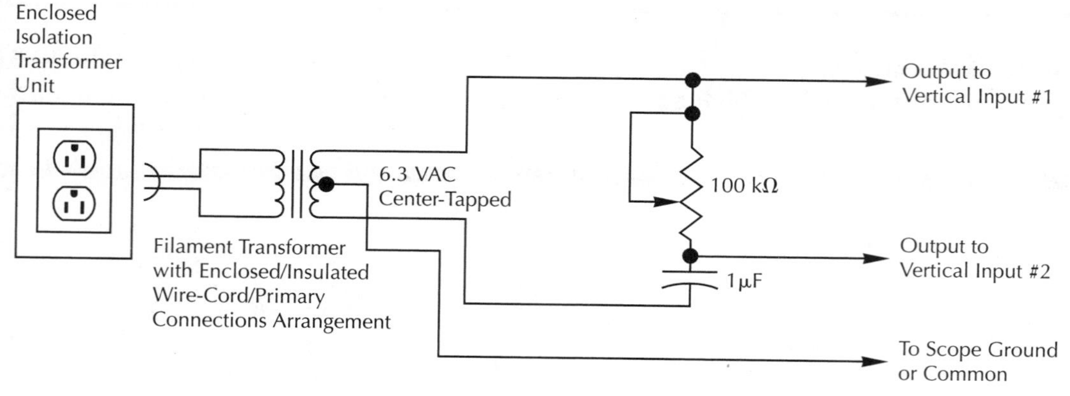

FIGURE 46
Variable phase-shift network

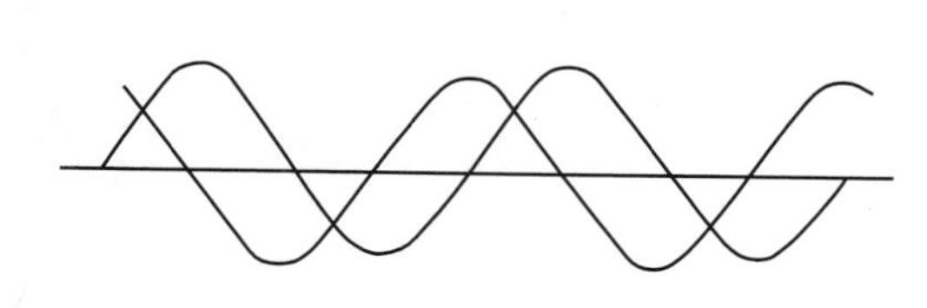

FIGURE 47
Superimposed waveforms

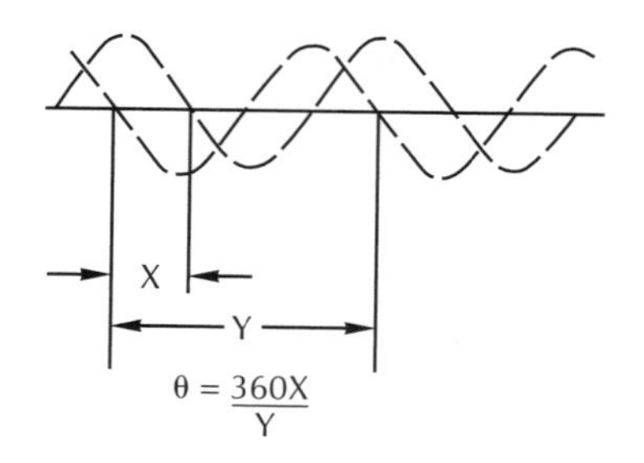

FIGURE 48
Calculating phase difference

PROJECT PURPOSE

To provide hands-on practice in using the scope with the direct ("overlay") phase comparison technique for determining phase difference between signals.

PARTS NEEDED

- ☐ Dual-trace oscilloscope
- ☐ Isolation transformer with outlet sockets
- ☐ *6.3-V filament transformer with insulated connections to power plug
- ☐ 100-kΩ potentiometer
- ☐ 1.0-μF paper/mylar capacitor

***NOTE:** Use a low-voltage ac power supply with center-tapped output, if available.

SAFETY HINTS

CAUTION: DO NOT USE EXPOSED 120 VAC CONNECTIONS ANYWHERE IN THE CIRCUIT SETUP! BE CAREFUL! 120 VAC CAN BE LETHAL!

In this project, you will briefly look at two approaches that may be used to compare the phase of two sine-wave signals (of the same frequency) using an oscilloscope. These two methods include:

1. Direct comparison using a dual-trace scope or a single-trace scope with the aid of an electronic switch device.
2. A Lissajous pattern technique. (**NOTE:** This technique is used infrequently; however, this optional project is provided for those who would like to see the technique for familiarization.)

ACTIVITY	OBSERVATION	CONCLUSION
1. Connect the variable phase shift circuit shown in Figure 46 on the adjacent page.	—	—
2. Connect the outputs of the network to the dual-trace scope vertical inputs #1 and #2, as appropriate. Adjust the scope H FREQUENCY, GAIN CONTROLS, AND CENTERING CONTROLS to achieve superimposed waveforms, centered on the face of the scope, similar to that shown in Figure 47.	*V* input #1 gain control set at: ______ V/div. *V* input #2 gain control set at: ______ V/div.	Are the *V* levels fed to the two vertical inputs equal? ______.
3. Refer to Figure 48. Using the technique shown, determine the phase difference between the two signals you are displaying on your scope.	Distance X = ______ calibration marks on the scope screen. Distance Y = ______ calibration marks on the scope screen.	The number of degrees difference between the two signals is: ______ degrees.
4. Change the setting on the 100-kΩ potentiometer enough to see a noticeable change in the phase and repeat steps 2 and 3 from the previous diagram in Figure 46.	Distance X = ______ calibration marks on the scope screen. Distance Y = ______ calibration marks on the scope screen.	The number of degrees difference between the two signals is: ______ degrees.

PROJECT 38

THE OSCILLOSCOPE
Phase Comparisons: Part B (Optional Project)

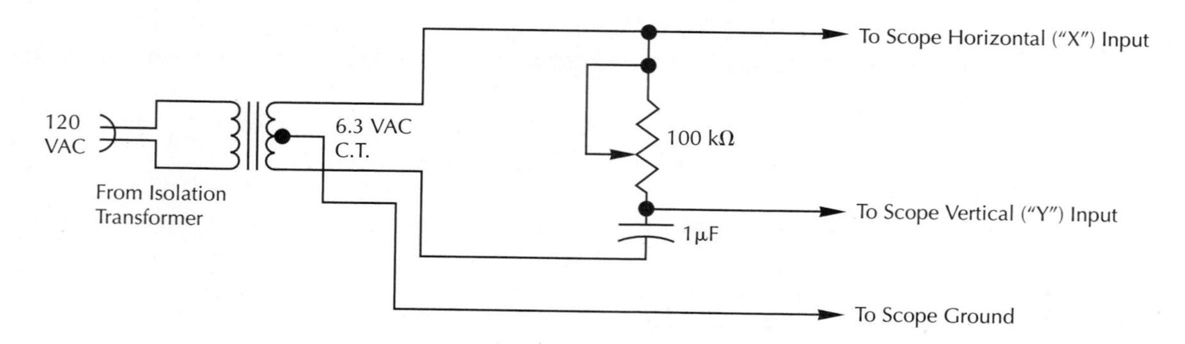

FIGURE 49
Phase-shift network

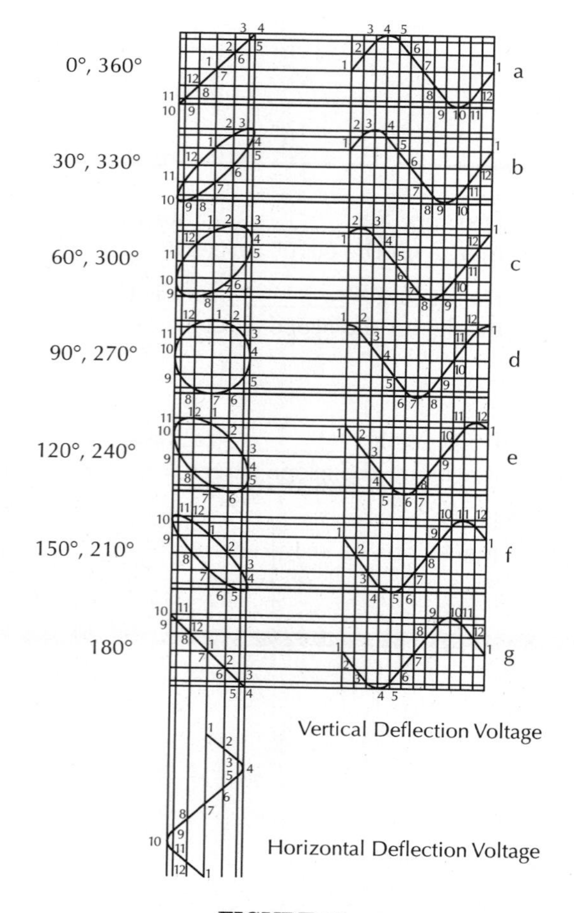

FIGURE 50
Lissajous patterns

PROJECT PURPOSE

To (optionally) enable practice in interpreting Lissajous patterns to determine phase comparisons between signals.

ACTIVITY	OBSERVATION	CONCLUSION
1. Connect the circuit shown in Figure 49.	—	—
2. Adjust the potentiometer through its range slowly and see how many of the Lissajous patterns shown in Figure 50 you can obtain.	Check the patterns you were able to approximately match: 0 degrees ____________ 30 degrees ____________ 60 degrees ____________ 90 degrees ____________ 120 degrees ____________ 150 degrees ____________ 180 degrees ____________	The range of phase shift obtainable with the given circuit is approximately from ________ degrees to ____________ degrees. What shape Lissajous pattern indicates a 90-degree difference in phase? ____________. What shape Lissajous pattern indicates a 0-degree difference in phase? ____________.
3. Adjust the potentiometer to obtain the pattern of a 90-degree phase difference and show your instructor the pattern.	Pattern O.K. (Instructor initial ____________)	—

PROJECT
39
THE OSCILLOSCOPE
Determining Frequency

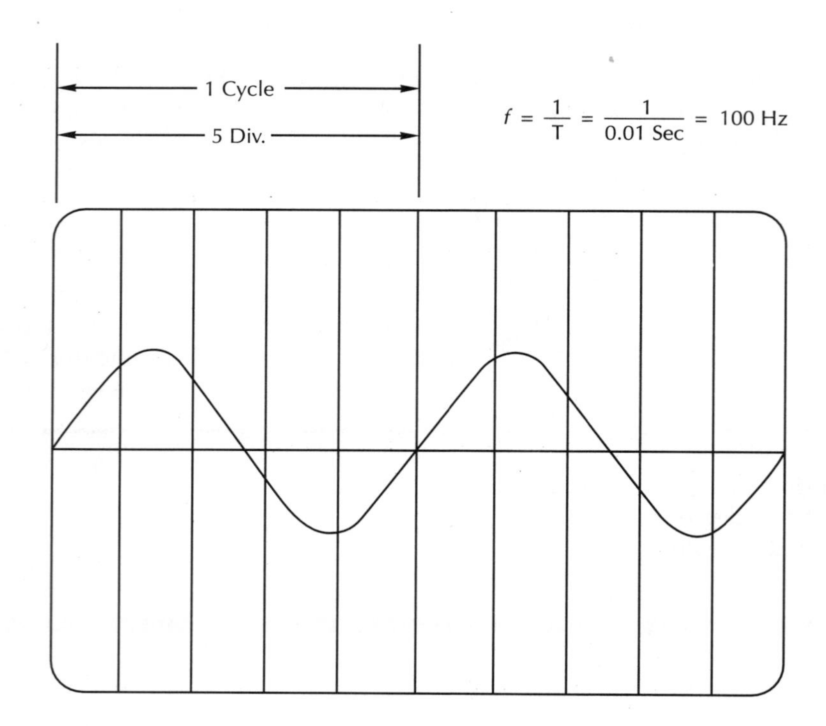

FIGURE 51
Scope sweep time setting at 2 ms/div.

PROJECT PURPOSE

To provide hands-on practice in determining frequency using the direct time measurement technique with an oscilloscope.

One method used to determine frequency with an oscilloscope is the direct method. This method uses a scope having a triggered sweep with calibrated sweep times. Since the horizontal sweep is linear and the calibrated sweep provides information about how many milliseconds or microseconds are required for the sweep to travel 1 div. horizontally on the screen, we can then easily determine the time for one cycle of the waveform being viewed, Figure 51. Once we know the time it takes for one cycle of the waveform (its period), we can then use the $f = 1/T$ formula to find the frequency of the signal causing the pattern. For example, if the calibrated sweep time is set at 2 milliseconds per division and the scope display shows one cycle of the signal's waveform starts and finishes in 5 horizontal divisions, the signal must have a period of 5×2 ms, or 10 ms. The frequency of the signal equals $1/T$, or $1/0.01$ sec = 100 Hertz.

ACTIVITY	OBSERVATION	CONCLUSION
1. Connect the output of an audio oscillator or function generator to a vertical input on a scope having a calibrated sweep system.	—	—
2. Set the sweep for a time of 1 ms per division.	The number of ms for the sweep to travel all the way across the screen is ________ ms.	A signal having a frequency of ________ Hz would cause a waveform display wherein one cycle would take 10 div. on the scope display.
3. Adjust the signal generator frequency to obtain a one-cycle display across 10 divisions.	Generator frequency dial calibration reads ________ Hertz.	The measured period for this signal is ________ ms. This indicates that the frequency is ________ Hz. Does the generator calibration approximately agree with the measured signal frequency? ________ What could cause any differences? ________ ________
4. Double the signal input frequency with the scope sweep setting still at 1 ms per division.	How many cycles of the signal are now displayed? ________ cycle(s).	The time for one cycle of this frequency is ______ ms.; f = ______ Hz. Increasing the frequency of an input signal while keeping the sweep speed the same causes (*more, less*) ________ cycles to be displayed.

PROJECT
39
CONTINUED

THE OSCILLOSCOPE
Determining Frequency *(Continued)*

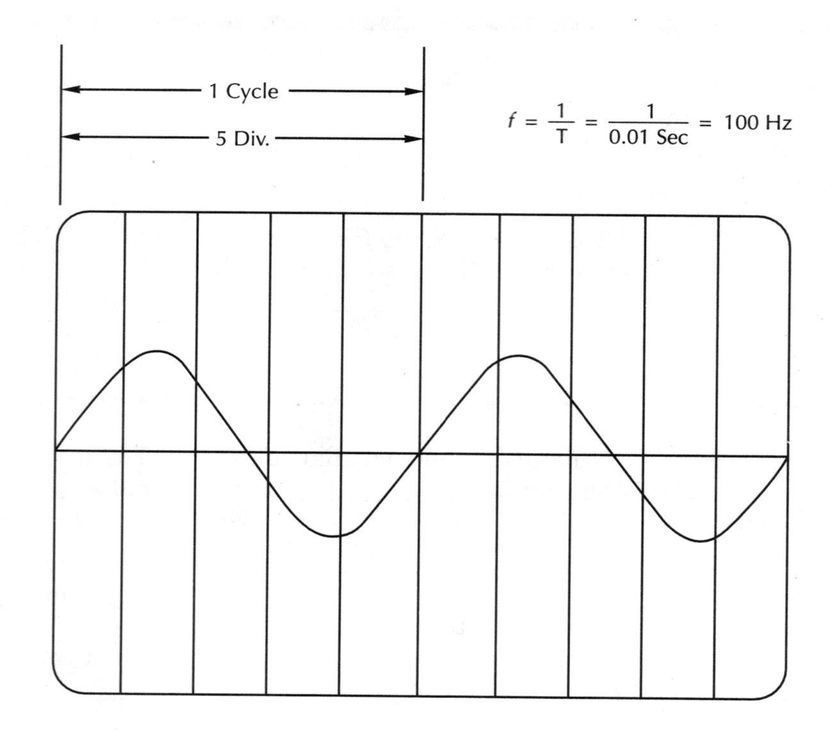

FIGURE 51
Scope sweep time setting at 2 ms/div.

PROJECT PURPOSE

To provide hands-on practice in using the direct time measurement technique for determining frequency with an oscilloscope.

ACTIVITY	OBSERVATION	CONCLUSION
5. Have the instructor set the frequency of the signal source to a completely different setting. Preferably, the frequency should be set to one that will cause you to change sweep speeds in order to determine the frequency of the signal.	To get a readable display, the sweep speed had to be changed to ______ per div.	The period of one cycle is ______. The frequency determined by the scope's direct measurement system is ______ kHz.
6. Have the instructor check your results in step 5.	Instructor: Indicate whether the measurement is correct. Yes ______ No ______.	—
7. Practice determining other frequencies, as time permits.	—	—

THE OSCILLOSCOPE

Complete the multiple-choice questions by placing a check in the box next to the best answer option. Respond to the other types of questions by filling in the blanks or responding as appropriate.

1. Four important uses of the oscilloscope are:

   ☐ Waveform display, voltage measurement, current measurement, and power measurement.
   ☐ Waveform display, voltage measurement, determining frequency, and power measurement.
   ☐ Waveform display, voltage measurement, determining phase, and power measurement.
   ☐ Waveform display, voltage measurement, determining phase, and determining frequency.

2. Why should the intensity or brightness control be set at the lowest point that makes the display readable?

   ____

3. What oscilloscope terminal is used as the input terminal for the signal whose waveform is to be viewed?

   ____

4. If an ac voltage of 125 volts *rms* causes 2 div. deflection on the scope screen, what is the peak-to-peak voltage of a signal that causes a deflection of 5 div. with the scope controls unchanged? Show your calculations below.

   Voltage is ____________ volts peak-to-peak.

5. If a frequency of 150 Hz gives a display on the scope screen of three cycles, what is the frequency of a signal that gives a display of four cycles with the scope controls unchanged? Show your calculations below.

   Frequency is ______________ Hz.

6. The most common way in which the scope might be used to show phase comparisons between two signals is:

   ____.

7. A scope having a calibrated sweep system can be used to directly determine frequency because the time per division of the ________________ sweep allows us to determine the ______________________ of the signal being observed.

8. What is the frequency of a signal whose display indicates a period of 0.1 millisecond?

   ____.

9. What is the period of a 25-kHz signal? ______________________.

10. By which of the following two methods can you more accurately measure a given signal's frequency?

    ☐ Setting the sweep speed so that two cycles of the signal's waveform covers the entire distance of the horizontal sweep.
    ☐ Setting the sweep speed so that four cycles of the signal's waveform covers the entire x-axis display.

INDUCTANCE

Objectives

In these projects you will connect circuits that illustrate the property of an inductor to oppose a change in current and the resultant total inductance of two inductors connected either in series or in parallel.

In completing these projects, you will connect circuits, make measurements, perform calculations, draw conclusions, and be able to answer questions about the following items related to inductance.

- The property of inductance
- Effect of changing value of L
- Total inductance of inductors in series and in parallel
- Significance of the L/R ratio

PROJECT/TOPIC CORRELATION INFORMATION

PROJECT	TEXT CHAPTER	SECTION	RELATED TEXT TOPIC(S)
40 L Opposing Change of Current	14	14-1	Background Information
		14-2	Review of Faraday's and Lenz's Laws
		14-3	Self-Inductance
41 Total Inductance in Series and Parallel	14	14-5	Inductance in Series
		14-5	Inductance in Parallel

PROJECT 40

INDUCTANCE
L Opposing Change of Current

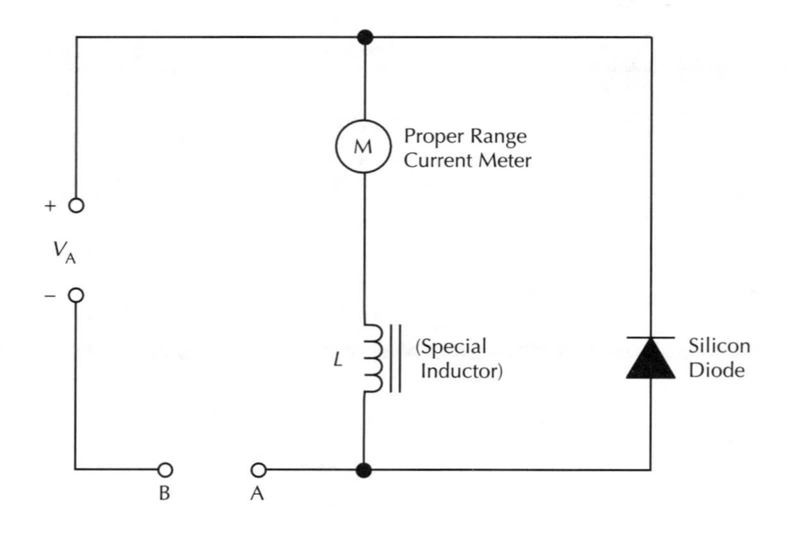

FIGURE 52

PROJECT PURPOSE

To demonstrate, through observation, that an inductor opposes changes in current, whether they be increasing or decreasing in nature. (**NOTE:** This takes a special inductor that has very high inductance and very low dc wire resistance.)

PARTS NEEDED

- ☐ DMM/VOM (2)
- ☐ VVPS (dc)
- ☐ CIS
- ☐ Special inductor (with high L, low R)
- ☐ Silicon diode (with ratings of at least 1A, 600V)

SAFETY HINTS

DO NOT EXCEED CURRENT RANGE OF METER. BE AWARE OF HIGH TRANSIENT VOLTAGE POSSIBILITIES WHEN INDUCTOR CIRCUIT IS "OPENED."

SPECIAL NOTE: Before attempting to perform this project, students should check with the instructor! In many cases, the instructor may elect to use this project as a class "demonstration" due to the special inductor needed, and the special precautions to be used in performing these procedures.

In order for this project to be most effective, a special large iron-core choke is needed that has a very high inductance and very low dc resistance winding. With a large *L*/*R* ratio, it is possible to visually illustrate the action of the inductor opposing a change in current. (**NOTE**: The choke used for this demonstration/project should have a time constant (*L*/*R* ratio) of AT LEAST 0.2 seconds. This means that five time constants would be at least 1 second . . . enabling the student to observe that the rise time of current from zero to its final value is not instantaneous; thus, illustrating the choke's opposition to a change in current.) The silicon diode is put in the circuit to protect the meter. You will learn about this device later in the manual. **CAUTION**: The value of V_A and the current range of the current meter to be used should be determined by the value of the dc resistance of the choke. Be sure not to exceed the current limitations of the meter by applying too much V_A.

ACTIVITY	OBSERVATION	CONCLUSION
1. Connect the initial circuit as shown in Figure 52.	—	—
2. Measure the dc resistance of the choke. Determine what value of V_A and meter current range should be used ($I_m = V_A/R_{dc}$).	V_A = (__________) volts Current range equals (0–__________) mA	The purpose of this calculation is for protection of the (*inductor, meter*) __________.
3. Apply amount of V_A calculated as safe and then insert jumper at points A and B to close the circuit and observe the action of the current meter.	The current (*did, did not*) ______ rise instantly to full value.	Approximately how much time did it take current to rise to full value? __________. Did the inductor apparently oppose the current change in current from zero to full value? __________.
4. Remove the jumper from points A and B and note the action of the current meter.	The current (*did, did not*) ______ fall instantly from full value to zero.	The inductor (*did, did not*) ______ oppose the change of current from maximum value to zero. Did the inductor tend to keep current flowing through the circuit in the same direction as it was flowing previous to removing the source? __________. This was evidenced by the fact that the polarity of the current meter (*did, did not*) __________ have to be reversed. We can conclude then that inductance will oppose a change in current either increasing or __________.

PROJECT 41

INDUCTANCE
Total Inductance in Series and Parallel

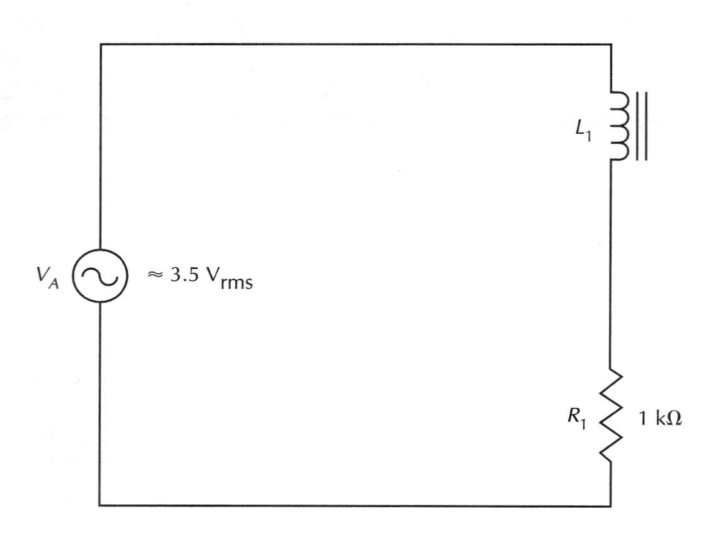

FIGURE 53

PROJECT PURPOSE

To demonstrate, through circuit connections and measurements, that inductances in series add like resistances in series, and that inductances in parallel are analyzed in the same fashion that resistances in parallel are analyzed.

PARTS NEEDED

- ☐ DMM/VOM (2)
- ☐ Low voltage ac source
- ☐ CIS
- ☐ Inductors, 1.5 H, 95 Ω, or approximate (2)
- ☐ Resistor(s):
  1 kΩ

For this project, we will observe the property of inductance to oppose a change in current by applying a continuously changing ac voltage to the circuit and noting the current limiting effects. The higher the inductance, or L, the higher the opposition to ac current. By noting the ac circuit current with a single inductor, then two inductors in series, then two inductors in parallel, we will illustrate the effects on total inductance of connecting two coils in series and in parallel. **NOTE:** For convenience, the voltage drop across a 1-kΩ resistor will be used as a current indicator. Since $I = V/R$, the number of volts divided by 1 kΩ automatically yields I in mA, e.g., 10 volts across a 1-kΩ resistor indicates 10 mA through the resistor, and so on.

ACTIVITY	OBSERVATION	CONCLUSION
1. Measure the dc resistance of the two inductors that will be used for this demonstration. Also, measure the ac source voltage that will be used.	L_1 = ________ ohms L_2 = ________ ohms (approximately) V = ________ volts (no load)	—
2. Connect the initial circuit as shown in Figure 53.	—	If V_A were dc, what would be the current through this circuit? ________ mA.
3. Measure V_1 (voltage drop across R_1) and calculate the ac current.	V_1 = ________ volts I = ________ mA	The back emf produced by the ________ is limiting the current to a lower value than it would be if dc were applied to the circuit.
4. Insert the second L (L_2) in series with the circuit. Measure V_1 and calculate the ac current.	V_1 = ________ volts I = ________ mA	Since the current was lower with the two inductors in series, the L total is obviously (*more*, *less*) than with one inductor. ________. We conclude that inductors in series add like resistors in (*series*, *parallel*) ________.
5. Change the circuit so L_2 is in parallel with L_1. Measure V_1 and calculate the ac current.	V_1 = ________ volts I = ________ mA	Since the current is higher with the two inductors in parallel, then L total must be (*more*, *less*) ________ than with one inductor. We conclude that inductors in parallel add like resistors in (*series*, *parallel*) ________.

INDUCTANCE

Complete the following review questions, indicating the appropriate response by placing a check in the box next to the correct answer.

1. Inductance is that property in an electrical circuit that opposes:

   ☐ a change in voltage
   ☐ a change in resistance
   ☐ a change in current

2. In an ac circuit if L is increased, the circuit current will:

   ☐ increase
   ☐ decrease
   ☐ remain the same

3. If two equal inductances are connected in series, total inductance will be:

   ☐ two times that of one
   ☐ one-half that of one
   ☐ neither of these

4. If two equal inductances are connected in parallel, total inductance will be:

   ☐ two times that of one
   ☐ one-half that of one
   ☐ neither of these

5. If an inductor has an L/R ratio of 2, how much time will be required to obtain a complete change of current level from one static level to another?

   ☐ 1 second
   ☐ 2 seconds
   ☐ 5 seconds
   ☐ 10 seconds
   ☐ none of these

INDUCTIVE REACTANCE IN AC

Objectives

You will connect several ac inductive circuits and make measurements and observations regarding their important electrical characteristics.

In completing these projects, you will connect circuits, make measurements, perform calculations, draw conclusions, and be able to answer questions about the following items related to inductive reactance.

- Relationship of L, induced voltage, and inductive reactance
- Relationship of frequency to inductive reactance
- The X_L formula
- Solving for L when X_L and frequency are known

PROJECT/TOPIC CORRELATION INFORMATION

PROJECT	TEXT CHAPTER	SECTION	RELATED TEXT TOPIC(S)
42 Induced Voltage	14	14-2 14-3	Review of Faraday's and Lenz's Laws Self-Inductance
43 Relationship of X_L to L and Frequency	15	15-4 15-5	Relationship of X_L to Inductance Value Relationship of X_L to Frequency of AC
44 The X_L Formula	15	15-6	Methods to Calculate X_L

SPECIAL NOTE: For this project, and all the following projects that require *setting* or *measuring* ac voltages (or currents) — assume that the rms values are desired, unless specified otherwise!

PROJECT 42

INDUCTIVE REACTANCE IN AC

Induced Voltage

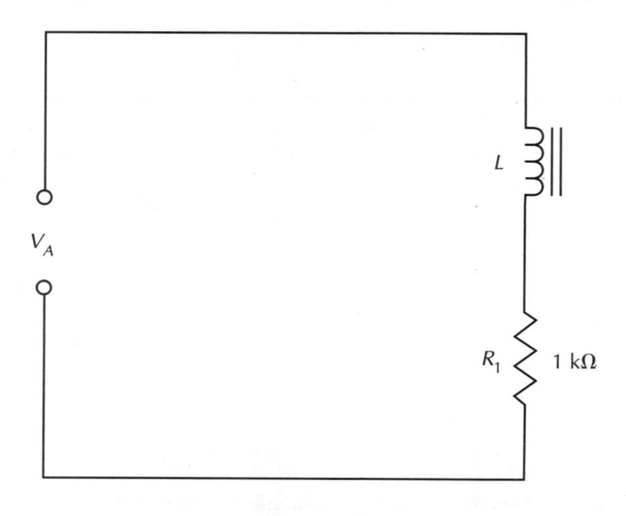

FIGURE 54

PROJECT PURPOSE

To illustrate the circuit effects of inductor counter-emf by noting the difference in circuit current when dc is applied to the inductor circuit, and when ac is applied.

PARTS NEEDED

- ☐ DMM/VOM
- ☐ VVPS (dc)
- ☐ Low voltage ac source (approximately 7 VAC)
- ☐ CIS
- ☐ Inductor, 1.5 H, 95 Ω (or approximate)
- ☐ Resistor(s): 1 kΩ

The methods used during this project are used to illustrate the concept of back emf and are not a precise scientific method of measuring or calculating exact values of back emf. Also remember that the ac equivalent of a given dc value is the "effective" (rms) value of ac.

ACTIVITY	OBSERVATION	CONCLUSION
1. Connect the initial circuit as shown in Figure 54.	—	—
2. Measure the ac voltage source that will be used for this demonstration. Next, connect the VVPS to the circuit and adjust the dc input V to a value that matches the ac voltage you will be using in a later step.	ac source equals ______volts (rms) dc source connected to circuit set to ______________________ volts	—
3. Measure V_1 (voltage across R_1) and calculate circuit current.	I (dc) = ______________________ mA	The circuit current is limited only by the dc resistance of R_1 and ______________________.
4. Remove the dc source, connect the ac source and measure V_1. Now calculate the circuit current.	I (ac) = ______________________ mA	The inductor developed a back emf that opposes the changing current. Discount the small dc resistance of the coil. The current that flowed with dc applied was equivalent to having a V_A of ______________ volts across a circuit resistance of 1 kΩ. When the same value of ac voltage was applied to the circuit, a current flowed that would be equivalent to applying only ______________ volts across the circuit resistance. The effect simulated was as if there must be a back emf of approximately ____________ volts. **NOTE:** This disregards any R in the circuit and assumes that the total limiting effect on the current is due to back emf.

PROJECT 43

INDUCTIVE REACTANCE IN AC

Relationship of X_L to L and Frequency

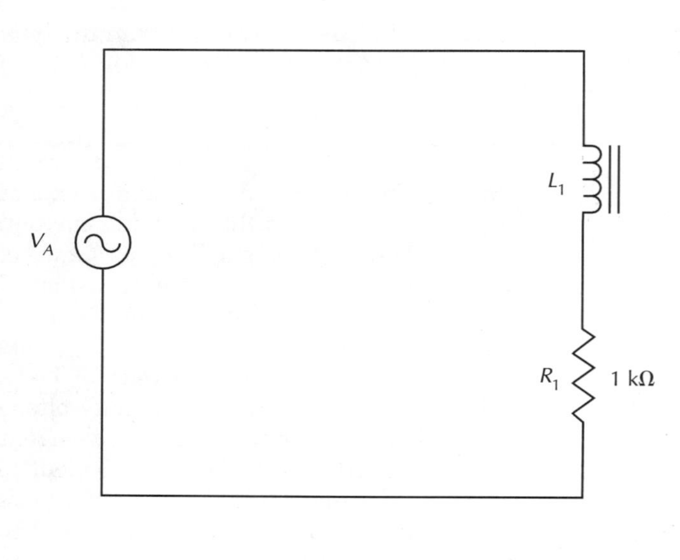

FIGURE 55

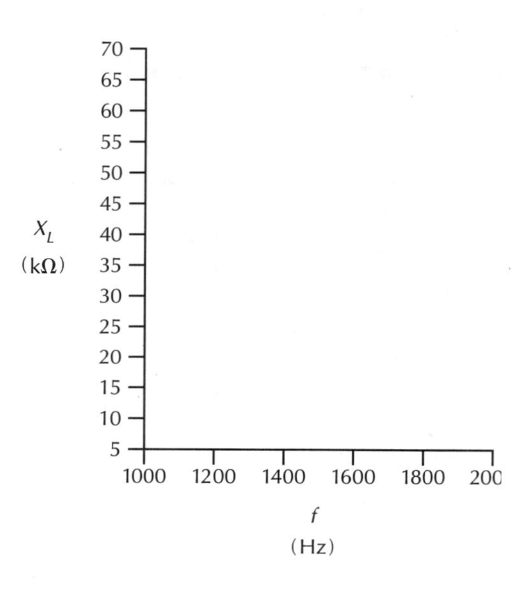

FIGURE 56
X_L versus f

NOTE: If using a different value of *L* than shown, renumber the *y*-axis of the graph, as appropriate.

PROJECT PURPOSE

To observe the relationship of X_L to L by changing L_T, and to frequency by applying various frequencies to an inductor circuit, and by measuring the changing circuit parameters.

PARTS NEEDED

- ☐ DMM/VOM (2)
- ☐ Function generator or audio oscillator
- ☐ CIS
- ☐ Inductor, 1.5 H, 95 Ω (2) (or approximate)
- ☐ Resistor(s): 1 kΩ

SPECIAL NOTE: The inductors used in this project, *and a number of projects following this one*, are iron-core, filter-choke-type inductors. Under normal operating conditions, this inductor has dc current as well as ac signal components present. The manufacturer has rated the inductance value based on the normal operating environment for this type inductor. *Students and instructors* should be aware that this inductor will exhibit "apparent" inductance values quite different from the manufacturer's rating under the varying operating conditions in our projects. Due to the inductor being used under different conditions than those specified by the manufacturer, such factors as "incremental permeability" enter into the results in terms of how much "acting" inductance the inductor "appears" to have. Also, the voltage dropped by the inductor is due to the inductor's impedance — not just its reactance. Add into this scenario the tolerances of component values, variances in calibration of signal sources, test equipment, and so on, and it is obvious that results will vary from those that would result if the inductor were acting at the "rated" 1.5 H value.

ACTIVITY	OBSERVATION	CONCLUSION
1. Connect the initial circuit as shown in Figure 55.	—	—
2. Set the function generator to a frequency of 100 Hz and V_A from the source to 3 volts. Measure V_1 and calculate the circuit current. Next, measure V_L and calculate X_L by Ohm's law ($X_L = V_L/I$).	V_A = ________ volts V_1 = ________ volts I = ________ mA V_L = ________ volts X_L = ________ ohms	Since this is a simple series circuit, the current through the inductor is the same as the current through the ________.
3. Insert a second inductor (same type as L_1). You should now have a series circuit of L_1, L_2, and R_1. With the same frequency and V_A as step 2 above, measure V_1; calculate I; measure the voltage across the total inductance of L_1 and L_2; then calculate X_L total.	V_A = ________ volts V_1 = ________ volts I = ________ mA V_{L_T} = ________ volts X_{L_T} = ________ ohms	Connecting a second inductor of nearly equal value as L_1 in series with L_2 caused the total inductance to approximately (*double, halve*) ________. In analyzing the results of step 2 and comparing with this step, we conclude that doubling total inductance caused the total inductive reactance (X_L) to approximately (*double, halve*) ________. We therefore conclude that inductive reactance (X_L) is (*directly, inversely*) ________ proportional to inductance (L). Increasing L causes X_L to (*increase, decrease*) ________ at any given frequency.

PROJECT 43 CONTINUED

INDUCTIVE REACTANCE IN AC
Relationship of X_L to L and Frequency
(Continued)

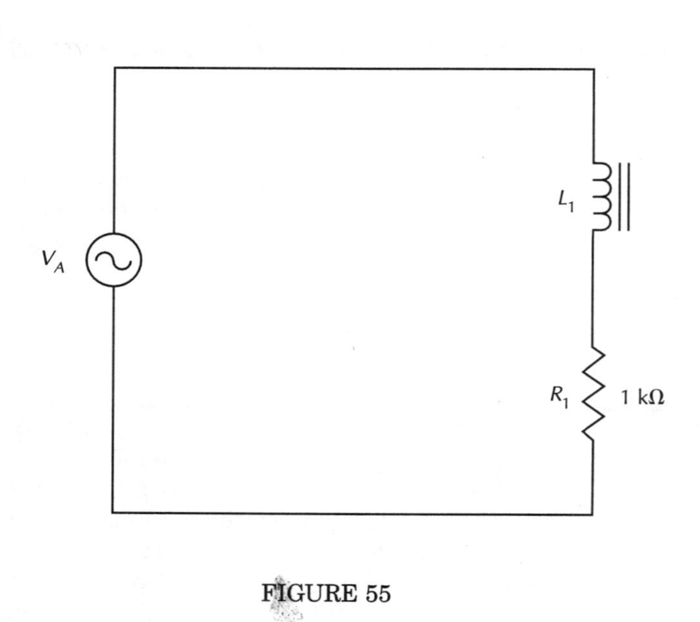

FIGURE 55

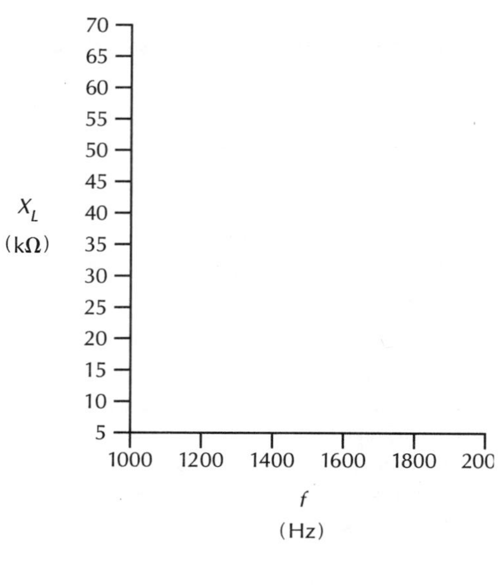

FIGURE 56
X_L versus f

NOTE: If using a different value of L than shown, renumber the y-axis of the graph, as appropriate.

PROJECT PURPOSE

To observe the relationship of X_L to L by changing L_T, and to frequency by applying various frequencies to an inductor circuit, and by measuring the changing circuit parameters.

PARTS NEEDED

- ☐ DMM/VOM (2)
- ☐ Function generator or audio oscillator
- ☐ CIS
- ☐ Inductor, 1.5 H, 95 Ω (2) (or approximate)
- ☐ Resistor(s):
  1 kΩ

ACTIVITY	OBSERVATION	CONCLUSION
4. Remove L_2 and replace with a jumper. Change the input frequency to 1,000 Hz and keep V_A at 3 volts. Measure V_1; calculate I; measure V_L; and calculate X_L.	V_A = ______ volts V_1 = ______ volts I = ______ mA V_{L_1} = ______ volts X_L = ______ ohms	Increasing the frequency from 100 Hz to 1,000 Hz caused the inductive reactance (X_L) to (*increase, decrease*) ______. If this had been a perfect inductive circuit, would X_L have increased 10 times? ______. We conclude that inductive reactance is (*inversely, directly*) ______ proportional to f (frequency). If the frequency were decreased to one-half its original value, then theoretically the X_L would (*increase, decrease*) ______ to (*double, one-half*) ______ its original value.
5. Make a graphic plot of X_L versus f from 1,000 Hz to 2,000 Hz on the graph coordinates shown in Figure 56. (**NOTE:** Calculate X_L for each 200 Hz change, as appropriate.)	X_L at 1,000 Hz = ______ X_L at 1,200 Hz = ______ X_L at 1,400 Hz = ______ X_L at 1,600 Hz = ______ X_L at 1,800 Hz = ______ X_L at 2,000 Hz = ______	Did X_L act like it is directly related to f? ______.

EXTRA CREDIT STEP(S)

6. Use the appropriate version of the $X_L = 2\pi fL$ formula to find the "apparent" L of the inductor for the operating conditions used in step 2. **NOTE:** (Use the X_L value found by Ohm's law in step 2 when solving for L.)	Calculated "apparent" L = ______ H	Is the calculated (apparent) inductance different from the manufacturer's rated value for this inductor? ______. What do you think the inductance value would appear to be if the inductor were operated under the conditions used by the manufacturer when rating the inductor? ______ H
7. Use the appropriate version of the $X_L = 2\pi fL$ formula to find the "apparent" L_T of the series inductors for the operating conditions used in step 3. **NOTE:** (Use the X_L value found by Ohm's law in step 3 when solving for L_T.)	Calculated "apparent" L_T = ______ H	Is the calculated (apparent) total inductance different from the manufacturer's rated value for these inductors? ______. What do you think the total inductance value would appear to be if the inductors were operated under the conditions used by the manufacturer when rating these inductors? ______ H

PROJECT 43 CONTINUED

INDUCTIVE REACTANCE IN AC

Relationship of X_L to L and Frequency *(Continued)*

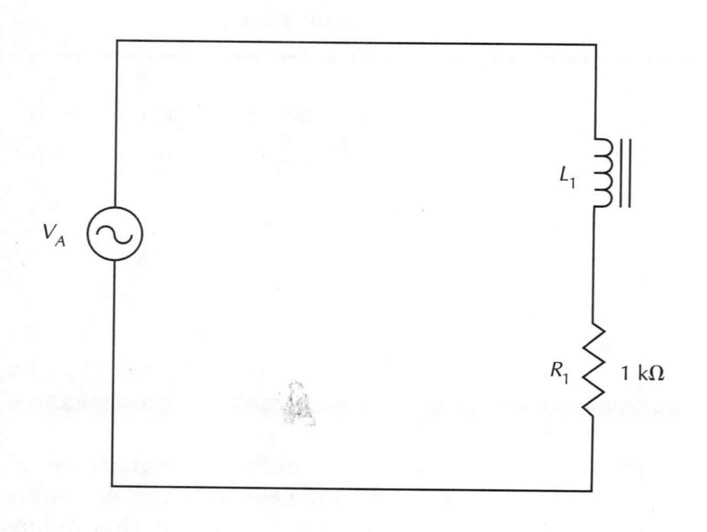

FIGURE 55

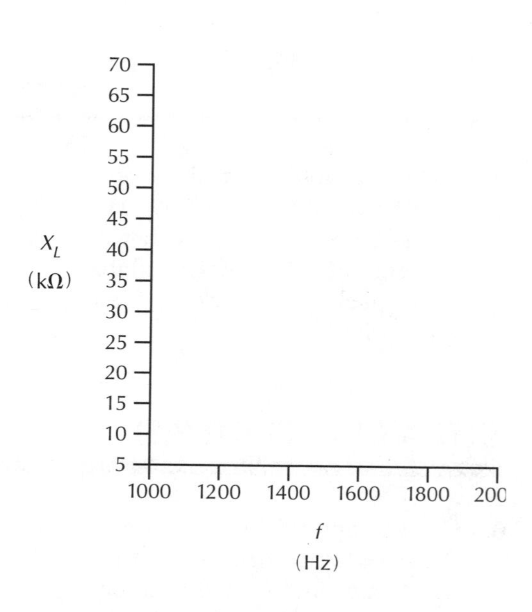

FIGURE 56
X_L versus f

NOTE: If using a different value of L than shown, renumber the *y*-axis of the graph, as appropriate.

PROJECT PURPOSE

To observe the relationship of X_L to L by changing L_T, and to frequency by applying various frequencies to an inductor circuit, and by measuring the changing circuit parameters.

PARTS NEEDED

- ☐ DMM/VOM (2)
- ☐ Function generator or audio oscillator
- ☐ CIS
- ☐ Inductor, 1.5 H, 95 Ω (2) (or approximate)
- ☐ Resistor(s): 1 kΩ

EXTRA CREDIT STEP(S)

8. Use the appropriate version of the $X_L = 2\pi fL$ formula to find the "apparent" L of the inductor for the operating conditions used in step 4. **NOTE:** (Use the X_L value found by Ohm's law in step 4 when solving for L.)

Calculated "apparent" L = ________ H

Is the calculated (apparent) inductance different from the manufacturer's rated value for this inductor, in this case? __________. Is it different under step 4's operating conditions than it was for step 2? __________. If so, what might account for the difference? _______
____

PROJECT 44

INDUCTIVE REACTANCE IN AC

The X_L Formula: Part A

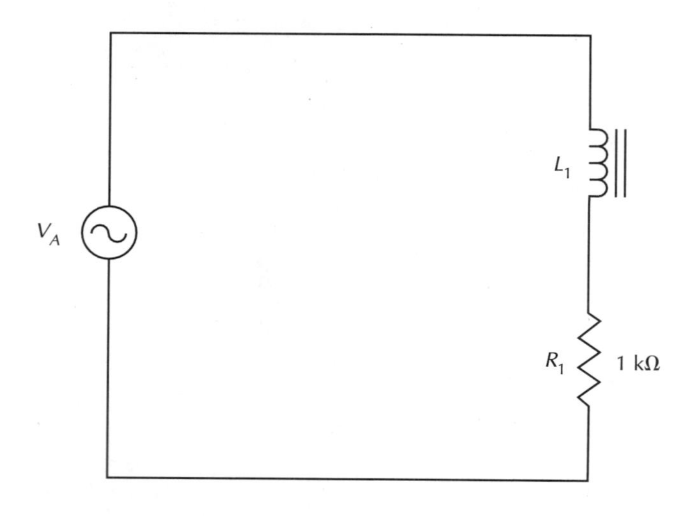

FIGURE 57

PROJECT PURPOSE

To use and verify the X_L formula by making measurements and calculating X_L by Ohm's law and comparing the results with the values calculated using the X_L formula.

PARTS NEEDED

- ☐ DMM/VOM (2)
- ☐ Function generator or audio oscillator
- ☐ CIS
- ☐ Inductor, 1.5 H, 95 Ω (2) (or approximate)
- ☐ Resistor(s):
  1 kΩ

Since the X_L of an inductor is directly related to the amount of induced voltage generated, we might relate this to the X_L formula as follows:

$$X_L = 2\pi fL$$

A higher frequency means a greater rate of change of current, hence a higher induced voltage. A higher inductance means a higher induced back emf because more flux lines will be cut per unit time. (Relate back to the induced voltage formula, $V = N \times \theta/\text{Time} \times 10^8$.)

In this project, we will illustrate that the X_L formula can be used to predict X_L if L is known, and also to calculate L if X_L is known but L is not known.

NOTE: Be aware that since the inductors used for this project were rated with a given value of dc current through them, you will find that their "apparent" inductance is different than the rated value, since there is no dc current through them. Also, they were designed to operate at low frequency, thus at higher frequency their "stray capacitance" alters results.

ACTIVITY	OBSERVATION	CONCLUSION
1. Connect the initial circuit as shown in Figure 57.	—	—
2. Assume L_1 has an inductance of 3 H and calculate the X_L if the input frequency were to be 500 Hz. (Use the X_L formula.)	X_L calculated = ________ ohms	—
3. Set the function generator frequency to 500 Hz and the circuit input voltage to 3 volts. Measure V_1; calculate I; measure V_L; and calculate X_L using Ohm's law.	V_A = ________ volts V_1 = ________ volts I = ________ mA V_L = ________ volts X_L = ________ ohms	Did the X_L calculated by the X_L formula and the X_L calculated using Ohm's law approximately agree? _____. Any difference might be due to L not being exactly the assumed value. Also, when we calculated X_L by Ohm's law, we did not take into account the resistance of the (*resistor*, *inductor*) ________.
4. Change the function generator frequency to 1 kHz and keep the circuit input voltage at 3 volts. Measure V_1; calculate I; measure V_L; and calculate X_L using Ohm's law. Then calculate X_L by the X_L formula.	V_A = ________ volts V_1 = ________ volts I = ________ mA V_L = ________ volts X_L = ________ ohms (by Ohm's law) X_L = ________ ohms (by X_L formula)	Doubling the frequency caused X_L to approximately (*double*, *halve*) ________. According to the measured results and the X_L formula, X_L is (*inversely*, *directly*) ________ proportional to frequency.

PROJECT 44

INDUCTIVE REACTANCE IN AC

The X_L Formula: Part B

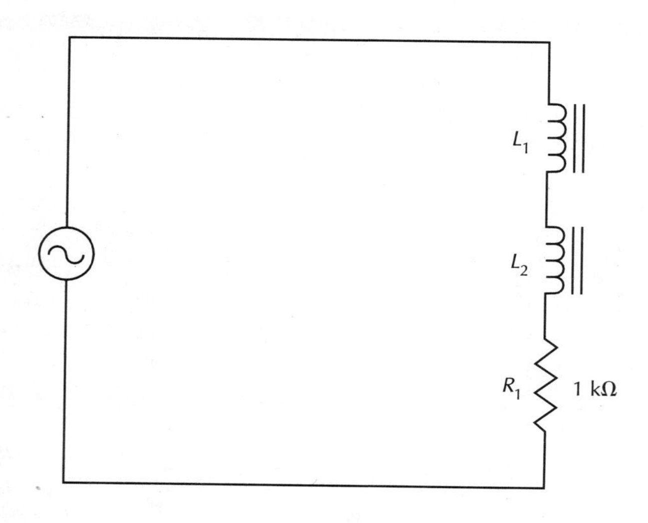

FIGURE 58

PROJECT PURPOSE

To observe the effect of increased inductance on total inductive reactance. To determine the effective L in a circuit, using the X_L formula.

PARTS NEEDED

- ☐ DMM/VOM (2)
- ☐ Function generator or audio oscillator
- ☐ CIS
- ☐ Inductor, 1.5 H, 95 Ω (or approximate) (2)
- ☐ Resistor(s):
  1 kΩ

ACTIVITY	OBSERVATION	CONCLUSION
1. Connect the initial circuit as shown in Figure 58.	—	—
2. Set the function generator frequency to 500 Hz and the circuit input voltage to 3 volts. Measure V_1 and calculate the circuit current. Measure the voltage across both coils (V_L total or V_{L_T}) and calculate X_{L_T} by Ohm's law. Use the formula $L = X_{L_T}/2\pi \times f$ and calculate L total.	V_A = ________ volts V_1 = ________ volts I = ________ mA V_{L_T} = ________ volts X_{L_T} = ________ ohms L_T calculated = ________ H	Adding L_2 to the circuit caused the circuit total inductance to approximately (*double, halve*) ________ from the assumed 2.3 H of Project 44, Part A, step 2. Did the rearrangement of the X_L formula to solve for L, when X_L was known, closely agree with the anticipated results? ________. We can also conclude inductive reactance is (*inversely, directly*) ________ proportional to inductance (L), for when we approximately doubled the circuit total inductance by adding L_2, we found the circuit total inductive reactance approximately (*doubled, halved*) ________ compared to the circuit with only L_1 used in Part A, step 2. Finally, we can conclude the X_L formula is very useful in analyzing circuits with inductance.

EXTRA CREDIT STEP(S)

3. Rearrange the circuit so that the two inductors (L_1 and L_2) are in parallel with each other. With the function generator set at 500 Hz and the circuit input voltage set at 3 volts, make appropriate measurements and calculations to enable filling the blanks in the Observation column.	V_A = ________ volts V_1 = ________ volts $I_{calc.}$ = ________ mA V_{L_T} = ________ volts X_{L_T} = ________ ohms L_T ________ H	Connecting the inductors in parallel caused the total circuit inductance to ________. The circuit inductance changed to a value which is approximately (*double, half, one-quarter*) ________ the value found when the two inductors were in series. (Step 2) Even though the results are not precise, due to the variables, the data tends to prove that as L decreases, X_L will (*increase, decrease*) ________ proportionately.

INDUCTIVE REACTANCE IN AC

Complete the following review questions, indicating the appropriate response by placing a check in the box next to the correct answer.

1. The induced voltage and X_L of a coil are directly proportional to:

   - ☐ L and R
   - ☐ L/R
   - ☐ L and f
   - ☐ R and f
   - ☐ none of these

2. If frequency is doubled and L is halved, the net resultant X_L will:

   - ☐ double
   - ☐ halve
   - ☐ quadruple
   - ☐ remain the same
   - ☐ none of these

3. If two equal inductors that were in series are now parallel connected, the resulting total X_L compared to the original circuit will be:

   - ☐ two times greater
   - ☐ one-half as great
   - ☐ four times greater
   - ☐ one-fourth as great

4. As f increases, the rate of change of current:

   - ☐ increases
   - ☐ decreases
   - ☐ remains the same

5. The X_L formula shows that inductive reactance is:

   - ☐ directly proportional to L and inversely proportional to f
   - ☐ inversely proportional to L and directly proportional to f
   - ☐ neither of these

6. To solve for L when X_L and frequency are known, use the formula:

   - ☐ $2\pi f/X_L$
   - ☐ $2\pi X_L/f$
   - ☐ $X_L/2\pi f$
   - ☐ none of these

INDUCTIVE REACTANCE IN AC

7. The unit of X_L is the:

   - ☐ back emf
   - ☐ ohm
   - ☐ ampere
   - ☐ volt
   - ☐ none of these

8. The opposition that an inductor shows to ac is:

   - ☐ purely inductive
   - ☐ purely resistive
   - ☐ a combination of resistance and inductive reactance
   - ☐ none of these

9. The amount of inductive reactance that a given coil exhibits is directly related to:

   - ☐ the amount of current
   - ☐ the applied voltage
   - ☐ neither of these

10. If frequency is tripled and inductance halved, the resultant X_L will be:

    - ☐ two-thirds the original
    - ☐ three-halves the original
    - ☐ six times the original
    - ☐ one-sixth the original

RL CIRCUITS IN AC

Objectives

You will connect several ac *RL* circuits and make measurements and observations regarding their important electrical characteristics.

In completing these projects, you will connect circuits, make measurements, perform calculations, draw conclusions, and be able to answer questions about the following items related to *RL* circuits.

- Relationship of R and X_L to circuit phase angle
- Relationships of current and voltage for an inductor and for a resistor
- Circuit impedance
- Simple vector diagram(s)
- Reference vector(s)

PROJECT/TOPIC CORRELATION INFORMATION

PROJECT		TEXT CHAPTER	SECTION	RELATED TEXT TOPIC(S)
45	Review V, I, R, Z, and θ Relationships in a Series R Circuit	16	16-1	Review of Simple R and L Circuits
46	V, I, R, Z, and θ Relationships in a Series RL Circuit	16	16-4	Fundamental Analysis of Series RL Circuits
47	V, I, R, Z, and θ Relationships in a a Parallel R Circuit	16	16-1	Review of Simple R and L Circuits
48	V, I, R, Z, and θ Relationships in a Parallel RL Circuit	16	16-5	Fundamental Analysis of Parallel RL Circuits

PROJECT

45

RL CIRCUITS IN AC
Review *V, I, R, Z,* and *θ* Relationships in a Series *R* Circuit

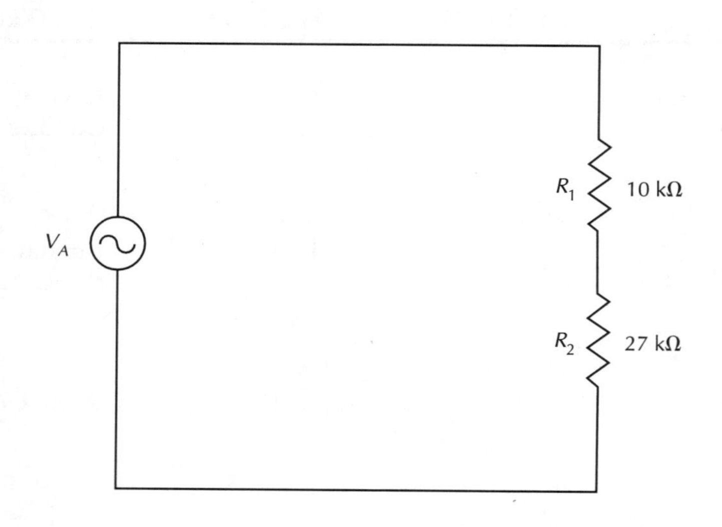

FIGURE 59

PROJECT PURPOSE

To review and demonstrate the key electrical parameter relationships in a purely resistive ac circuit via measurement and observation.

PARTS NEEDED

- ☐ DMM/VOM (2)
- ☐ Function generator or audio oscillator
- ☐ CIS
- ☐ Resistor(s):
  10 kΩ
  27 kΩ

ACTIVITY	OBSERVATION	CONCLUSION
1. Connect the initial circuit as shown in Figure 59.	—	—
2. Set the frequency of the function generator at 1000 Hz and the circuit input voltage to 3 volts. Measure the circuit voltages.	V_A = __________ volts V_1 = __________ volts V_2 = __________ volts	Is the voltage distribution of a purely resistive series circuit at a frequency of 1000 Hz the same as it would be with dc applied? __________
3. Calculate the circuit current using Ohm's law (V_1/R_1); calculate R_T also using Ohm's law (V_A/I_T).	I_T = __________ mA R_T = __________ ohms	From the observations of steps 2 and 3, we may conclude that *V* total can be calculated as the arithmetic __________ of the individual voltage drops. *R* total is the simple (*arithmetic, vector*) __________ sum of the individual resistances. Since V_T is the sum of the individual voltage drops, the voltages across the components must be (*in phase, out of phase*) __________ with each other. Further, we may note that since the *IR* drop across a resistor is maximum when *I* is maximum, then the current and voltage are in phase for a resistive component. The total opposition of a circuit to ac current flow is called impedance (*Z*). In the purely resistive circuit, the circuit total impedance or *Z* is equal to the total __________. The phase angle between *V* total and *I* total in a purely resistive ac circuit may be assumed to be __________ degrees.

PROJECT **46**

RL CIRCUITS IN AC
V, I, R, Z, and θ Relationships in a Series *RL* Circuit

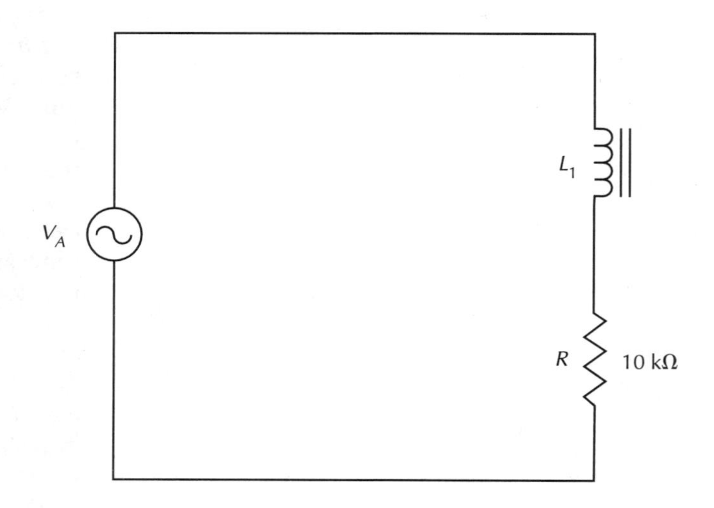

FIGURE 60

PROJECT PURPOSE

To demonstrate the key electrical parameter relationships in a series *RL* circuit. To observe that due to out-of-phase elements, simple dc analysis techniques cannot be used to determine circuit parameters in ac circuits containing reactive components. Furthermore, to provide practice in using simple ac analysis techniques and in drawing ac circuit vector diagrams.

PARTS NEEDED

- ☐ DMM/VOM (2)
- ☐ Function generator or audio oscillator
- ☐ Dual-trace oscilloscope
- ☐ CIS
- ☐ Inductor, 1.5 H, 95 Ω
- ☐ Resistor(s):
  10 kΩ

ACTIVITY	OBSERVATION	CONCLUSION
1. Connect the initial circuit as shown in Figure 60.	—	—
2. Set the frequency of the function generator to 500 Hz and the circuit input voltage to 3 volts. Measure V_A, V_R, and V_L.	V_A = ________ volts V_R = ________ volts V_L = ________ volts	Does V_A equal the arithmetic sum of V_R and V_L? ________. We conclude that to find V_A we must vectorially (*add, subtract*) ________ V_R and V_L. We can also use the ________ theorem.
3. Calculate I_T from V_R/R. Calculate X_L from V_L/I. Calculate circuit total impedance from $Z = V_T/I_T$. 4. Determine the apparent value of L from the known frequency and X_L parameters. 5. Determine Z using the Pythagorean theorem (using the R and X_L parameters). 6. Draw an impedance diagram in the Observation column space. 7. Use trigonometry and determine the phase angle.	I_T = ________ mA X_L = ________ ohms Z = ________ ohms L = ________ H Z = ________ ohms θ = ________ degrees	Does Z equal the arithmetic sum of R and X_L? ________. We again conclude that we must use the vector sum or the (Pythagorean) theorem used in analysis of right ________. From the observations of steps 2 and 3 we may conclude that V_R and V_L are (*in phase, not in phase*) ________. Since inductance opposes a change in current, we may assume that V_L (*leads, lags*) ________ I_L by some angle. If L were a perfect inductor, ________ would lead ________ by 90 degrees. Since the circuit is not composed of purely resistance in which the phase angle between V and I = ________ degrees, nor purely inductance in which the phase angle between V and I equals ________ degrees, but rather a composite of both, we might expect the phase angle between V_T and I_T to be between ________ and ________ degrees. Further, the larger the X_L is compared to the circuit R, the more like a purely inductive circuit the results will be; thus the (*greater, lesser*) ________ will be the circuit phase angle. The converse is also true.

PROJECT **46** CONTINUED

RL CIRCUITS IN AC
V, I, R, Z, and θ Relationships in a Series *RL* Circuit *(Continued)*

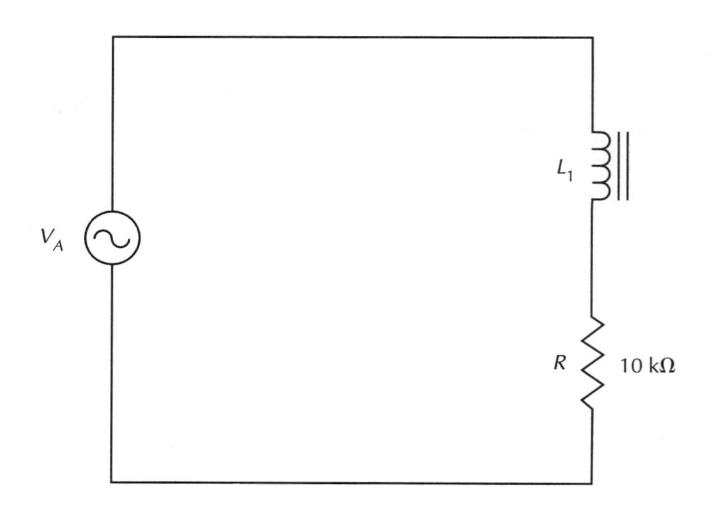

FIGURE 60

PROJECT PURPOSE

To demonstrate the key electrical parameter relationships in a series *RL* circuit. To observe that due to out-of-phase elements, simple dc analysis techniques cannot be used to determine circuit parameters in ac circuits containing reactive components. Furthermore, to provide practice in using simple ac analysis techniques and in drawing ac circuit vector diagrams.

PARTS NEEDED

- ☐ DMM/VOM (2)
- ☐ Function generator or audio oscillator
- ☐ Dual-trace oscilloscope
- ☐ CIS
- ☐ Inductor, 1.5 H, 95 Ω
- ☐ Resistor(s):
  10 kΩ

EXTRA CREDIT STEP(S)

8. Use the measured and calculated data in previous steps 2 and 3, and draw a *V-I* vector diagram in the Observation column.

9. Use trigonometry and determine the phase angle.

θ = ________ degrees

Does the phase angle from the *V-I* vector diagram agree reasonably with the phase angle you determined from the *Z* diagram? ________.

10. Use a dual-trace oscilloscope and perform a phase comparison of V_A and circuit current, (represented by the voltage across the resistor). **CAUTION:** Be sure the signal source ground and the scope ground(s) are connected to the same end of the resistor when making the measurements to prevent the two grounds from shorting out a portion of the circuit! Determine the phase difference between the two signals. (If possible, demonstrate your scope waveforms and calculations to your instructor.)

θ determined by the scope phase comparison:
θ = ________ degrees

Do the scope phase measurements and the phase angle calculations agree reasonably with your earlier findings? (Considering tolerances in components, source and scope frequency calibration tolerances, etc.) ________.

PROJECT
47

RL CIRCUITS IN AC
Review *V*, *I*, *R*, *Z*, and θ Relationships in a Parallel *R* Circuit

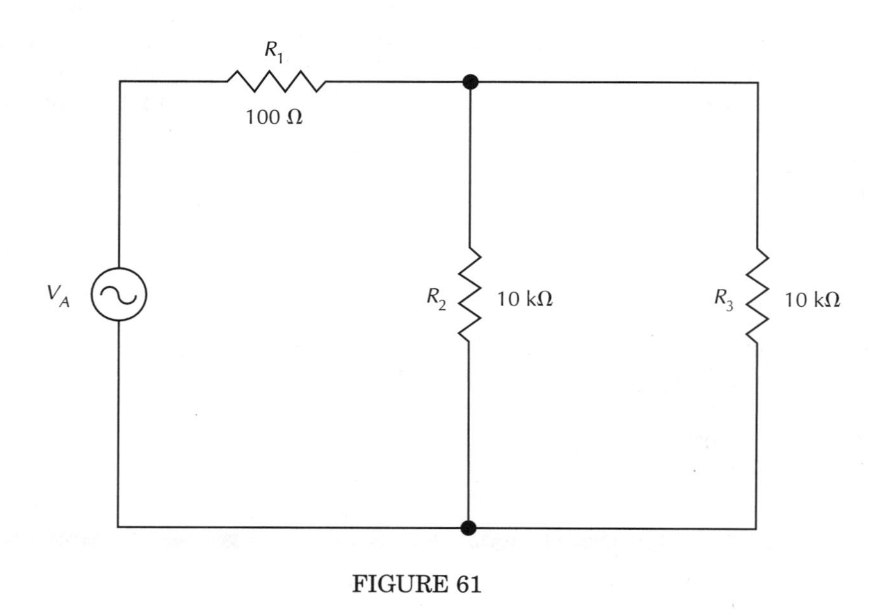

FIGURE 61

PROJECT PURPOSE

To review and demonstrate the key electrical parameter relationships in a purely resistive parallel ac circuit via measurement and observation.

PARTS NEEDED

- ☐ DMM/VOM (2)
- ☐ Function generator or audio oscillator
- ☐ CIS
- ☐ Resistor(s):
  100 Ω
  10 kΩ (2)

It should be noted that for this project and several other later ones involving parallel ac circuitry, we will use a 100-Ω resistor in series with the "main line" as a method of indicating total circuit current. You will recall that the voltage dropped across a 1-kΩ resistor is the same value as the amount of milliamperes through it (as shown by Ohm's law where $I = V/R$); however, when a 100-Ω resistor is used, the current is ten times the voltage dropped by the 0.1-kΩ (100-Ω) resistor.

ACTIVITY	OBSERVATION	CONCLUSION
1. Connect the initial circuit as shown in Figure 61.	—	—
2. Set the function generator to a frequency of 1,000 Hz and V_A to 3 volts. Measure the circuit voltages and calculate the circuit currents.	V_A = ________ volts V_1 = ________ volts V_2 = ________ volts V_3 = ________ volts I_T = ________ mA I_2 = ________ mA I_3 = ________ mA	Does I_T equal the arithmetic sum of $I_2 + I_3$? ________. Does V_T equal the summation of the voltage drops around each closed loop? ________. From this we conclude all the voltage drops (*are, are not*) ________ in phase with each other, and the branch currents (*are, are not*) ________ in phase with each other. Also, we conclude that since this is a purely resistive circuit, V_T and I_T are ________ phase. Is Z equal to R_T? ________. In summary, we might state that a purely resistive circuit with ac applied (*can, cannot*) ________ be analyzed the same as a dc resistive circuit.

PROJECT
48

RL CIRCUITS IN AC
V, I, R, Z, and θ Relationships in a Parallel *RL* Circuit

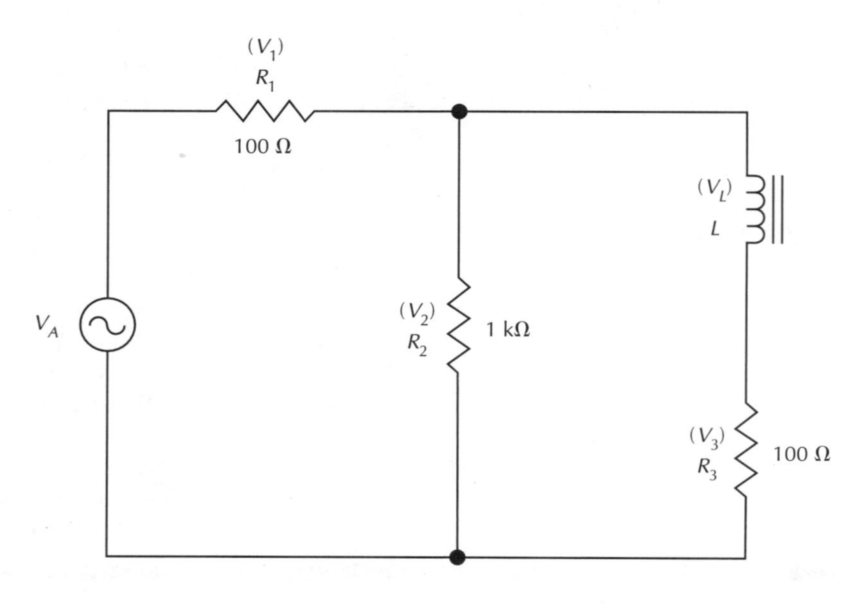

FIGURE 62

PROJECT PURPOSE

To demonstrate the key electrical parameter relationships in a parallel *RL* circuit. To observe that due to out-of-phase elements, simple dc analysis techniques cannot be used to determine circuit parameters in ac circuits containing reactive components. Also, to provide practice in using simple ac analysis techniques and in drawing ac circuit vector diagrams.

PARTS NEEDED

- ☐ DMM/VOM (2)
- ☐ Function generator or audio oscillator
- ☐ CIS
- ☐ Inductor, 1.5 H, 95 Ω (or approximate)
- ☐ Resistor(s):
  - 100 Ω (2)
  - 1 kΩ (1)

It should be noted that once again, we will be using the 100-Ω resistor in series with the "main line" as a circuit current indicator. The current equals 10 times the voltage drop.

ACTIVITY	OBSERVATION	CONCLUSION
1. Connect the initial circuit as shown in Figure 62.	—	—
2. Set the 60-Hz source so that V_A = 6.8 V. Measure the circuit voltages.	V_A = ____ volts V_1 = ____ volts V_2 = ____ volts V_3 = ____ volts V_L = ____ volts	Do the voltages around any given closed loop add up by addition to V_A? ____. From this we conclude that the circuit current(s) and voltage(s) are (*in phase, out of phase*) ____.
3. Calculate the total circuit current from V_1. Calculate the current through R_2 by Ohm's law. Calculate the current through L by using the voltage drop across R_3. Use Ohm's law to solve for I_L ($I_L = V_3/R_3$) and X_L ($X_L = V_L/I_L$). 4. Use measured and calculated values of I_2 and I_L. Apply the Pythagorean theorem formula and calculate I_T. 5. Draw the appropriate *V-I* vector diagram in the Observation column. 6. Use trigonometry and determine the phase angle. (Neglect R_1 parameters and use only R_2 and L parameters.)	I_T = ____ mA I_2 = ____ mA I_L = ____ mA X_L = ____ ohms I_T calculated = ____ mA θ = ____ degrees	Does total current equal the arithmetic sum of the branch currents? ____. This is because the branch currents are ____ ____ ____. The current through R_2 is in phase with V_2 (*True* or *False*) ____. The current through the coil (*leads, lags*) ____ the voltage across the coil by close to ____ degrees. If the inductor were perfect, it would be exactly ____ degrees. Since the total circuit current is the vector resultant of the two branch currents, it would seem logical to assume the circuit total current would be (*leading, lagging*) ____ V_A by some angle between ____ and ____ degrees. Also note from our measurements and calculations that the total circuit impedance (Z) cannot be found by the product-over-the-sum method but is most easily solved by Ohm's law, where $Z = V_T/I_T$.

PROJECT 48 CONTINUED

RL CIRCUITS IN AC
V, I, R, Z, and θ Relationships in a Parallel *RL* Circuit *(Continued)*

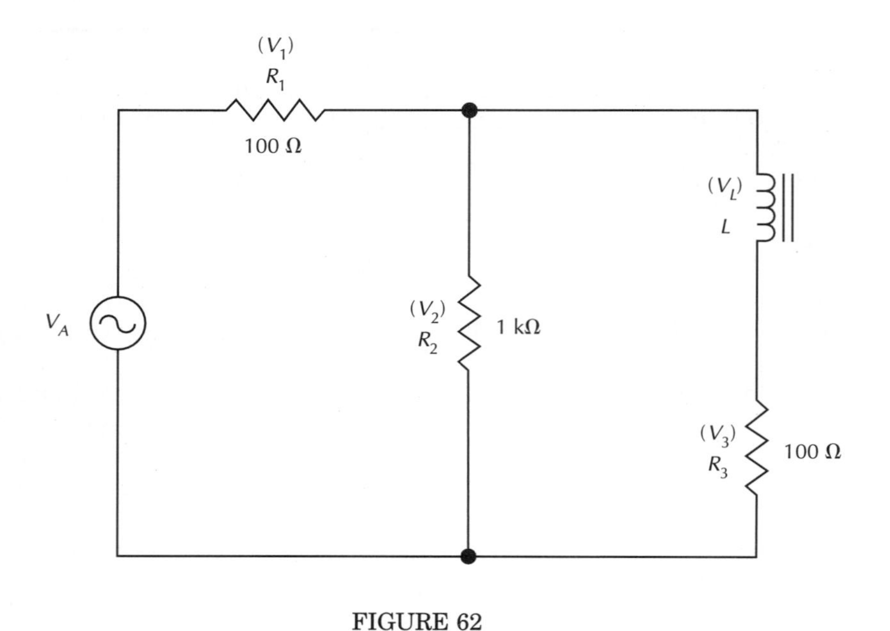

FIGURE 62

PROJECT PURPOSE

To demonstrate the key electrical parameter relationships in a parallel *RL* circuit. To observe that due to out-of-phase elements, simple dc analysis techniques cannot be used to determine circuit parameters in ac circuits containing reactive components. Also, to provide practice in using simple ac analysis techniques and in drawing ac circuit vector diagrams.

PARTS NEEDED

- ☐ DMM/VOM (2)
- ☐ Function generator or audio oscillator
- ☐ CIS
- ☐ Inductor, 1.5 H, 95 Ω (or approximate)
- ☐ Resistor(s):
  100 Ω (2)
  1 kΩ (1)

EXTRA CREDIT STEP(S)

7. Use a dual-trace oscilloscope and perform a phase comparison of the current through the R_2 branch and the current through the inductor branch. Do this by letting the voltage across R_2 represent the current through R_2, and the voltage across R_3 represent the current through the inductor branch.
   CAUTION: Be sure the signal source ground and the scope ground(s) are connected to the same end (the bottom end) of the circuit network when making the measurements to prevent the grounds from shorting out a portion of the circuit! Determine the phase difference between the two signals. (If possible, demonstrate your scope waveforms and calculations to your instructor.)

Measured θ between the resistor branch and the inductor branch as determined by the scope phase comparison:
θ = ______ degrees

Was the phase difference between the two branches reasonably close to 90°? ________. If not, what variables and factors might account for the difference? ______________________________
____
____
____
____
____
______________________________.

RL CIRCUITS IN AC

Complete the review questions, indicating the appropriate response by placing a check in the box next to the correct answer.

1. An ac circuit whose phase angle is 45 degrees is composed of:

   - ☐ an equal amount of resistance and reactance
   - ☐ an unequal amount of resistance and reactance
   - ☐ purely resistance
   - ☐ purely reactance

2. The current through an inductor:

   - ☐ leads V_L
   - ☐ lags V_L
   - ☐ is in phase with V_L

3. To find the Z of a series RL circuit:

   - ☐ simply add X_L total and R total
   - ☐ subtract X_L from R
   - ☐ neither of these

4. When making a V-I vector diagram of a series RL circuit, the reference vector is:

   - ☐ I_T
   - ☐ V_T
   - ☐ Z_T
   - ☐ R_T
   - ☐ none of these

5. When making a V-I vector diagram of a parallel RL circuit, the reference vector is:

   - ☐ I_T
   - ☐ V_T
   - ☐ Z_T
   - ☐ R_T
   - ☐ none of these

6. In a series RL circuit, if L is increased while R remains the same, the circuit phase angle (angle between V_A and I_T) will:

   - ☐ increase
   - ☐ decrease
   - ☐ remain the same

7. In a series *RL* circuit, if *R* is increased while *L* remains the same, the circuit phase angle will:

   ☐ increase
   ☐ decrease
   ☐ remain the same

8. In a parallel *RL* circuit, if *L* is increased while *R* remains the same, the circuit phase angle will:

   ☐ increase
   ☐ decrease
   ☐ remain the same

9. In a parallel *RL* circuit, if *R* is increased while *L* remains the same, the circuit phase angle will:

   ☐ increase
   ☐ decrease
   ☐ remain the same

10. The value of total circuit impedance (*Z*) for both series and parallel ac circuits can be solved by Ohm's law.

    ☐ True
    ☐ False

BASIC TRANSFORMER CHARACTERISTICS

Objectives

In these projects you will connect circuits that will illustrate the concepts of step-up and step-down transformer actions.

In completing these projects, you will connect circuits, make measurements, perform calculations, draw conclusions, and be able to answer questions about the following items relative to transformers.

- Transformer turns ratios
- Transformer voltage ratios
- Transformer current ratios
- The relationship of transformer impedances to turns and/or voltage ratios

PROJECT/TOPIC CORRELATION INFORMATION

PROJECT	TEXT CHAPTER	SECTION	RELATED TEXT TOPIC(S)
49 Turns, Voltage, and Current Ratios	17	17-5	Important Transformer Ratios
50 Turns Ratios Versus Impedance Ratios	17	17-5	Impedance Ratio

PROJECT 49

BASIC TRANSFORMER CHARACTERISTICS

Turns, Voltage, and Current Ratios

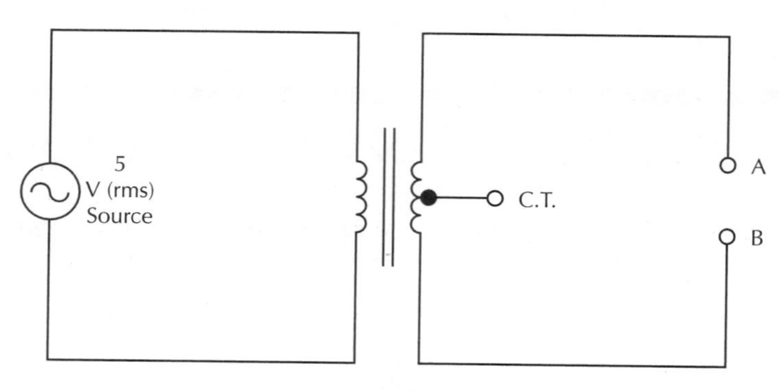

FIGURE 63
12.6 V center-tapped step-down transformer

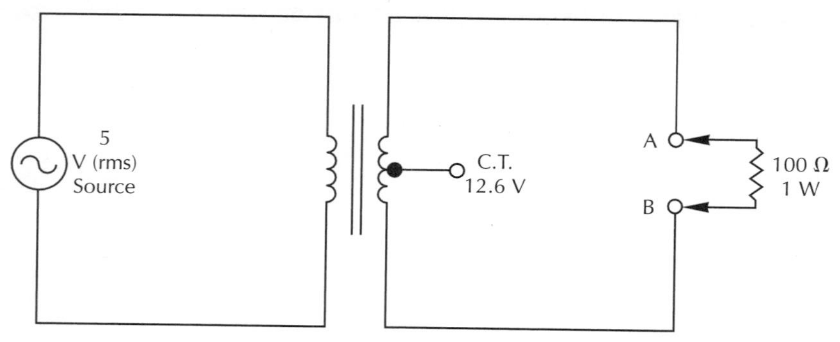

FIGURE 64
Loaded transformer

PROJECT PURPOSE

To demonstrate the relationships of turns ratios, voltage ratios, and current ratios in transformers by circuit measurements and observations.

PARTS NEEDED

- ☐ DMM/VOM (2)
- ☐ Low voltage source
- ☐ 12.6 V C.T. step-down transformer
- ☐ Typical power transformer
- ☐ CIS
- ☐ Resistors:
  100 Ω, 1 W

Transformers may be used to "step-up" or "step-down" voltages. If it were possible to have 100% efficiency, all of the power in the primary would be transferred to the secondary. If this were true, then the product of V and I in the secondary would equal the product of V and I in the primary, and the current step-up or step-down ratio would be inverse to the voltage step-up or step-down ratio in order for the $V \times I$ products to be the same.

To illustrate the concepts of basic transformer action, we want you to assume 100% efficiency of the transformers used for any calculations you are required to perform.

CAUTION! You will be working with some HIGH VOLTAGE windings in some cases. **USE ALL THE SAFETY PROCEDURES RELATIVE TO WORKING WITH HIGH VOLTAGE** while performing the project(s).

ACTIVITY	OBSERVATION	CONCLUSION
1. Obtain a 12.6-volt transformer and connect the circuit shown in Figure 63.	—	—
2. Use a voltmeter and measure the voltage present on the primary and on the secondary.	Primary V = ______ V Secondary V = ______ V	The primary-to-secondary voltage ratio is ______. The secondary-to-primary voltage ratio is ______. What is the N_S/N_P turns ratio? ______. Are the voltage and turns ratios the same? ______.
3. Connect a 100 Ω, 1 watt resistor between points A and B as shown in Figure 64. Use Ohm's law to determine the secondary current.	V_R = ______ V	$I = V/R$ I = ______ A Is the secondary current the same as the current through the load R? ______.
4. Use the previously determined voltage ratio from above and calculate the primary-to-secondary current ratio.	V_P:V_S = ______ I_P:I_S = ______	The current ratio is the ______ of the voltage ratio. Using the secondary current previously determined in Figure 64, step 3 and the current ratio just calculated, what is the approximate primary current value? ______ A.

PROJECT 49 CONTINUED

BASIC TRANSFORMER CHARACTERISTICS
Turns, Voltage, and Current Ratios *(Continued)*

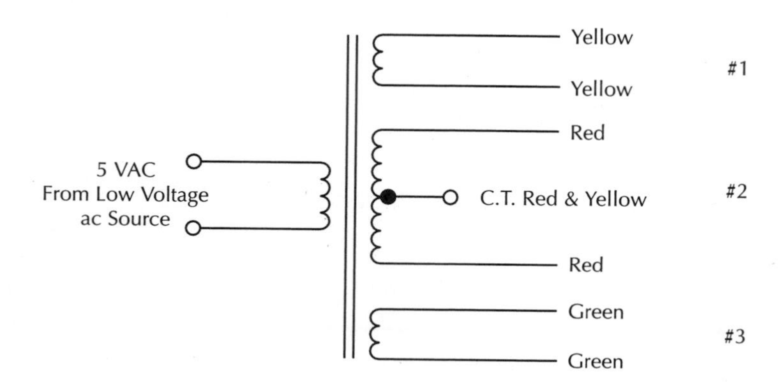

FIGURE 65
Power transformer

PROJECT PURPOSE

To demonstrate the relationships of turns ratios, voltage ratios, and current ratios in transformers by circuit measurements and observations.

PARTS NEEDED

- ☐ DMM/VOM (2)
- ☐ Low voltage source
- ☐ 12.6 V C.T. step-down transformer
- ☐ Typical power transformer
- ☐ CIS
- ☐ Resistors:
  100 Ω, 1 W

ACTIVITY	OBSERVATION	CONCLUSION
5. Obtain a power transformer similar to the one shown in Figure 65. Apply 5 V ac to the primary and measure and record all the voltages on the transformer.	Primary = ____________ V Secondary #1 = ____________ V Secondary #2 = ____________ V 1/2 Secondary #2 = ____________ V Secondary #3 = ____________ V	The voltage ratio Primary-to-Secondary #1 = ____________. The voltage ratio Primary-to-Secondary #2 = ____________. The voltage ratio Primary-to-Secondary #3 = ____________. The turns ratios theoretically should be the same as the voltage ratios. (*True* or *False*) ____________.

PROJECT 50

BASIC TRANSFORMER CHARACTERISTICS

Turns Ratios Versus Impedance Ratios

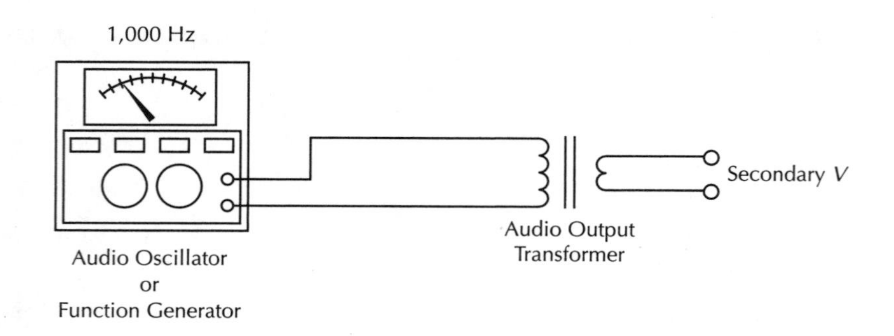

FIGURE 66

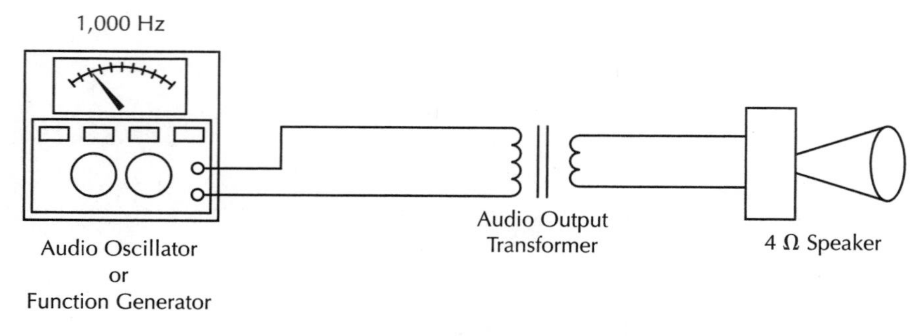

FIGURE 67
Optional circuit

PROJECT PURPOSE

To demonstrate that the impedance ratio of a transformer is related to the square of the turns ratio.

PARTS NEEDED

- ☐ DMM/VOM (2)
- ☐ Function generator or audio oscillator
- ☐ CIS
- ☐ Audio output transformer
- ☐ 4 Ω speaker

The impedance ratio of a transformer is related to the square of the turns ratio. This project will let you determine the turns ratio of a transformer, then theoretically determine the impedance ratio. Once this has been determined, you or your instructor will compare your results with the catalog specifications for the transformer you used. You can then see how closely your measurements and calculations correlated with the transformer specifications.

ACTIVITY	OBSERVATION	CONCLUSION
1. Obtain an audio output transformer and connect a circuit similar to that shown in Figure 66.	—	—
2. Apply a 1,000 Hz signal at a 3–5 V voltage level to the transformer primary. Measure the secondary voltage with an appropriate measuring instrument.	Voltage applied to the primary is ______ volts. Measured voltage on secondary is ______ volts.	The N_P/N_S ratio must be about ______ :1. Using the formula $Z_P/Z_S = N_P^2/N_S^2$, the primary-to-secondary impedance ratio is ______ :1.
3. Assume that the secondary load is going to be 4 ohms. Using the impedance ratio previously calculated, compute the nominal impedance of the circuit that should be connected to the primary of the output transformer.	Nominal Z calculated = ______ ohms.	Using the catalog parameters for the transformer, the primary-to-secondary Z ratio is ______ :1. Does the impedance ratio you measured and calculated via voltage measurements reasonably agree with the catalog specifications for the transformer? ______. What might cause any differences?
4. OPTIONAL STEP: Connect a 4-ohm speaker to the secondary, Figure 67. Measure the voltages and determine the Z ratio.	Primary V = ______ V Secondary V = ______ V Calculated turns ratio = ______ Calculated Z ratio = ______	Does this measured and calculated Z ratio more closely match the transformer specifications? ______.

BASIC TRANSFORMER CHARACTERISTICS

Answer the following questions with "T" for true and "F" for false. (Put your answer in the appropriate blank.)

1. A transformer's voltage and current ratios are inverse. ____

2. The impedance ratio of a transformer is related to the square root of the turns ratio. ____

3. If a transformer has a 3:1 turns ratio (secondary-to-primary) and a primary impedance of 5,000 ohms, the load impedance on the secondary should be 45,000 ohms to affect a proper impedance match. ____

4. If a transformer's turns ratio is doubled, the related impedance ratio will quadruple. ____

5. A transformer with a secondary-to-primary turns ratio of 10:1 will have a secondary voltage of 50 volts with 0.5 volts applied to its primary. ____

6. If a transformer's current ratio is 1:3, then the related voltage ratio will be 1:3. ____

7. If a transformer's voltage ratio is doubled, its related current ratio will halve. ____

8. If the turns on a transformer secondary are doubled, the impedance of the secondary will be four times as great as the original impedance. ____

9. Transformer efficiency is typically 100%. ____

10. The voltage across a transformer secondary will be higher when it is "unloaded" as compared to when it is supplying current to a load. ____

CAPACITANCE (DC CHARACTERISTICS)

Objectives

You will connect circuits illustrating the characteristics of capacitor(s) in dc circuits.

In completing these projects, you will connect circuits, make measurements, perform calculations, draw conclusions, and be able to answer questions about the following items related to capacitance in dc circuits.

- Charge and discharge action
- *RC* time
- Total capacitance of series capacitors
- Total capacitance of parallel capacitors

PROJECT/TOPIC CORRELATION INFORMATION

PROJECT		TEXT CHAPTER	SECTION	RELATED TEXT TOPIC(S)
51	Charge and Discharge Action and *RC* Time	18	18-3 18-10	Charging and Discharging Action The *RC* Time Constant
52	Total Capacitance in Series and Parallel	18	18-8	Total Capacitance in Series and Parallel

PROJECT

51

CAPACITANCE (DC CHARACTERISTICS)

Charge and Discharge Action and *RC* Time

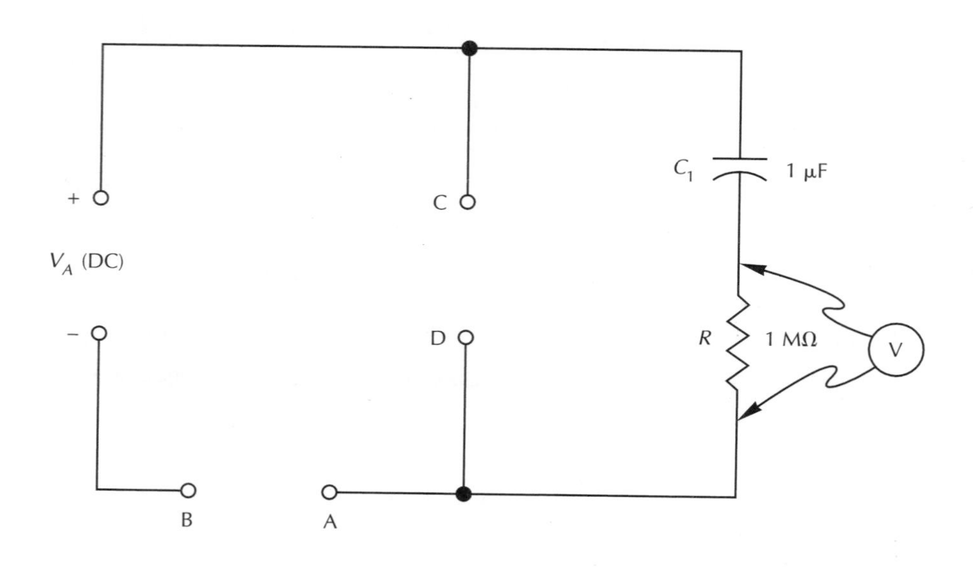

FIGURE 68

PROJECT PURPOSE

To demonstrate the charging and discharging action of a capacitor, and to observe that a capacitor takes five *RC* time constants to change from one set voltage level to another.

PARTS NEEDED

- ☐ DMM/VOM (2)
- ☐ VVPS (dc)
- ☐ CIS
- ☐ Capacitor(s), 0.1 μF, 1.0 μF
- ☐ Resistor(s):
  100 kΩ
  1 MΩ
  10 MΩ

SAFETY HINTS

Remember that charged capacitors remain charged and can shock you (until they are discharged).

For this and the next project, if a VOM is used for monitoring voltage(s), the resistance of the VOM will enter into the determination of the times for charge and discharge. For example, if a 20,000 ohms-per-volt voltmeter is used on the 50-volt range, the meter resistance is 1 MΩ. If we use this setup to measure across a 1-MΩ resistor, the actual circuit conditions will be 500 kΩ. On the 250-volt range, this same meter has a resistance of 5 MΩ, so, *take into account the meter resistance*, as appropriate (even if you use a DMM with 10 MΩ input *R*).

ACTIVITY	OBSERVATION	CONCLUSION
1. Connect the initial circuit as shown in Figure 68.	—	—
2. Set V_A at 20 volts. Insert a jumper wire between points A and B and observe the meter action.	Current flowed for approximately __________ seconds as evidenced by the voltage measured across *R*. The *rate* of charge was (*linear, non-linear*) __________.	The current that flowed was the charging current that charged the capacitor to a voltage equal to (V_R, V_A) __________. Was charging current maximum at the beginning of the charge time or near the end? __________. At the beginning of the charge time, the voltage across *R* was equal to (V_C, V_A) __________ (the first instant). At the end of the charge time, the voltage across the resistor is ______ volts; the voltage across the capacitor is equal to V_A. (**NOTE:** The charged capacitor voltage is equal to V_A and series opposing the source. Hence, no current can flow once the capacitor is charged.)
3. Remove the jumper from points A and B. Reverse the polarity of the voltmeter. Insert a jumper between points C and D and observe the meter action during discharge of the capacitor.	Discharge time was approximately __________ seconds. The *rate* of discharge was (*linear, nonlinear*) __________.	Did the capacitor take the same time to discharge through *R* as it did to charge? __________. At the end of the discharge time V_C = ______ volts; V_R = ______ volts.
4. Remove the jumper from points C and D. Change the VOM or DMM polarity from the original setup. Change *R* to a 10-MΩ resistor. Repeat the sequence of steps 2 and 3.	Charge time was approximately __________ seconds. Discharge time was __________ seconds.	Increasing *R* increased the charge time because the charging current was limited to a smaller value. Thus, it took longer to obtain a given potential difference or ______ across the *C*.

PROJECT

51

CONTINUED

CAPACITANCE (DC CHARACTERISTICS)

Charge and Discharge Action and *RC* Time *(Continued)*

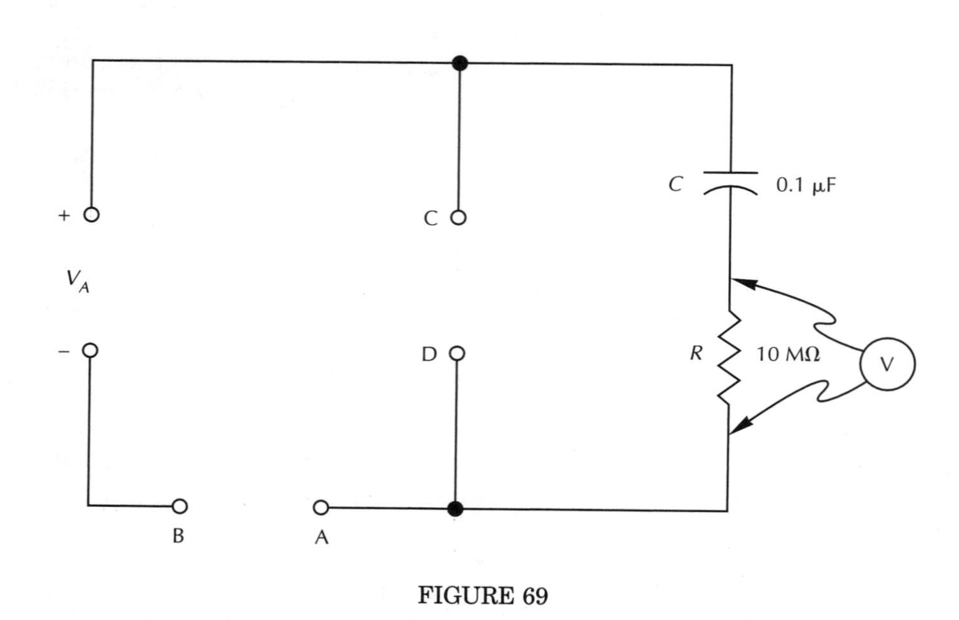

FIGURE 69

PROJECT PURPOSE

To demonstrate the charging and discharging action of a capacitor, and to observe that a capacitor takes five *RC* time constants to change from one set voltage level to another.

PARTS NEEDED

- ☐ DMM/VOM (2)
- ☐ VVPS (dc)
- ☐ CIS
- ☐ Capacitor(s), 0.1 μF, 1.0 μF
- ☐ Resistor(s): 100 kΩ, 1 MΩ, 10 MΩ

SAFETY HINTS

Remember that charged capacitors remain charged and can shock you (until they are discharged).

ACTIVITY	OBSERVATION	CONCLUSION
5. Connect the circuit as shown in Figure 69.	—	—
6. Insert a jumper between points A and B and note the charge time.	Charge time was approximately ____________ seconds.	Changing C from a 1.0 μF to a 0.1 μF caused the charge time to ____________. We conclude that both the value of ________ and of ____________ determine charge and discharge time. It should be noted that if it were not for the effect of the multimeter resistance, the time to charge the capacitor or discharge it would be directly proportional to R and to C. The formula relating to this is called the formula for the *time constant*. This formula states that one RC time constant = R in ohms times C in farads, and the answer is in seconds. Also, it should be observed that it takes 5 time constants to charge or discharge the capacitor. Calculate the R_e of R and the meter circuit's resistance in parallel, and determine if the charge time is about equal to the expected 5 TC. Is it? ________.
7. Use the RC time constant formula and determine how long it would take to charge a 1-μF capacitor in series with a 100-kΩ resistor.	5 RC time constants = ________ seconds.	—
8. Connect the circuit described in step 7 and note the charge time.	Charge time measured approximately ____________ seconds.	A multimeter with a meter circuit R of 5 MΩ or greater does not alter the circuit resistance of 100 kΩ very much. The measured charge time was (*close, not close*) ____________ to the theoretical value.

PROJECT 52

CAPACITANCE (DC CHARACTERISTICS)

Total Capacitance in Series and Parallel

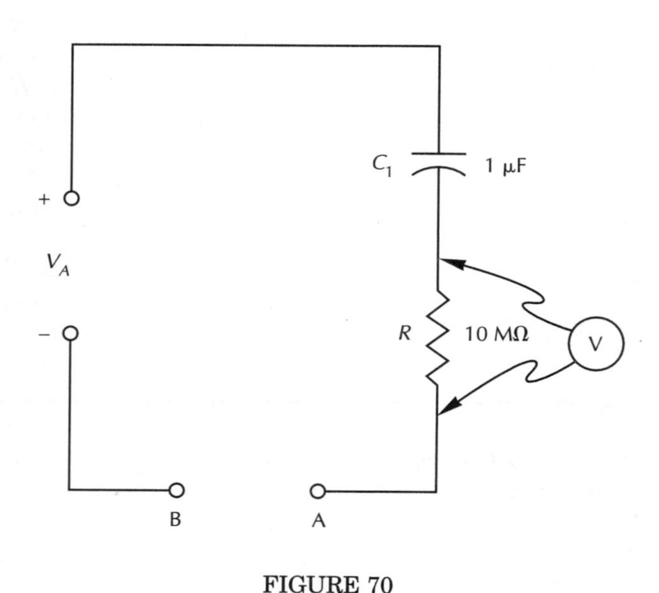

FIGURE 70

PROJECT PURPOSE

To demonstrate that capacitors in series add like resistances in parallel, and that capacitors in parallel add like resistances in series, using circuit observations or *RC* times.

PARTS NEEDED

- ☐ DMM (2)
- ☐ VVPS (dc)
- ☐ CIS
- ☐ Capacitor(s), 1.0 μF (2)
- ☐ Resistor(s): 10 MΩ

SAFETY HINTS

DON'T FORGET THAT CAPACITORS HOLD THEIR CHARGE UNLESS A DISCHARGE PATH IS PROVIDED! DON'T LET THAT DISCHARGE PATH BE YOU!

For this project, we will take advantage of the fact that the charge time of a capacitor is directly proportional to capacitance. By noting the charge time of a single capacitor, then noting the charge time for two capacitors in series, then in parallel, we should be able to conclude the effect on total capacitance of connecting capacitors in series or parallel. (**NOTE:** A DMM is highly preferred over a VOM for voltage measurements here.)

ACTIVITY	OBSERVATION	CONCLUSION
1. Connect the initial circuit as shown in Figure 70.	—	—
2. Set V_A to 20 volts. Insert a jumper between points A and B and note the charge time by observing the voltmeter.	Charge time was approximately ________ seconds.	One *RC* time is approximately ________ seconds. This means that the R_e of the meter and the 10-MΩ resistor is approximately ________ohms.
3. Obtain a second 1-µF capacitor. Remove the jumper from points A and B. Carefully discharge C_1, then insert the second capacitor (C_2) in series with C_1. Insert a jumper between points A and B and note the charge time of C_1 and C_2 in series.	Charge time was approximately ________ seconds.	Since the charge time has decreased to ________ the value it was with only C_1 in the circuit, it may be concluded that the total capacitance has (*increased, decreased*) ________. Since the new *RC* time is (*double, half*) ________ the original and the *R* has not been changed, then we conclude that the new total capacitance is ______ µF. Our observations tell us that capacitors in series add like resistors in (*series, parallel*) ________.
4. Remove the jumper from points A and B. Discharge the capacitors. Change the circuit as required to achieve a circuit with C_1 and C_2 in parallel, and this combination in series with *R*. Insert the jumper again, and note the charge time of C_1 and C_2 in parallel.	Charge time was approximately ________ seconds.	The charge time for this step is approximately (*2, 3, 4*) ______ times the time recorded in step 2. This indicates that the total capacitance of C_1 and C_2 in parallel is (*1/2, 2×, 3×*) ______ the capacitance of C_1 alone. We may conclude that capacitors in parallel add like resistors in (*series, parallel*) ________. If C_2 were not the same value as C_1, would the statements concerning total capacitance of capacitors in series and parallel still hold true? (That is, series *C*s add like parallel *R*s, and parallel *C*s add like series *R*s.) ________.

CAPACITANCE (DC CHARACTERISTICS)

Complete the following review questions, indicating the appropriate response by placing a check in the box next to the correct answer.

1. A charged capacitor has a difference of potential between its plates due to:
   - ☐ an excess of electrons on one plate and a deficiency of electrons on the other
   - ☐ an excess of electrons on both plates
   - ☐ neither of these

2. A capacitor in a given circuit will take the same amount of time to discharge as it does to charge.
   - ☐ True
   - ☐ False

3. If the value of R is doubled and the value of C is halved in a given RC circuit, the time it will take to charge the capacitor will:
   - ☐ increase
   - ☐ decrease
   - ☐ remain the same

4. One RC time constant is equal to:
   - ☐ $R + C$
   - ☐ R/C
   - ☐ $R - C$
   - ☐ $R \times C$
   - ☐ none of these

5. In order for a capacitor to fully charge or discharge, it takes:
   - ☐ one time constant
   - ☐ two time constants
   - ☐ four time constants
   - ☐ five time constants

6. For a given RC circuit, increasing the value of V_A will cause the time needed for the capacitor to fully charge or discharge to:
   - ☐ increase
   - ☐ decrease
   - ☐ remain the same

7. The total capacitance of a 0.05-μF and a 0.1-μF capacitor in parallel is:

   - ☐ 0.05 μF
   - ☐ 0.1 μF
   - ☐ 0.06 μF
   - ☐ 0.15 μF
   - ☐ 0.033 μF
   - ☐ none of these

8. The total capacitance of a 0.05-μF and a 0.1-μF capacitor in series is:

   - ☐ 0.05 μF
   - ☐ 0.1 μF
   - ☐ 0.06 μF
   - ☐ 0.15 μF
   - ☐ 0.033 μF
   - ☐ none of these

9. How long would it take two parallel 0.1-μF capacitors to charge through a 1-MΩ resistance?

   - ☐ 0.2 seconds
   - ☐ 1 second
   - ☐ 0.1 seconds
   - ☐ 5 seconds
   - ☐ none of these

10. How long would it take two series 0.1-μF capacitors to charge through a 1-MΩ resistance?

    - ☐ 1/4 second
    - ☐ 1/2 second
    - ☐ 3/4 second
    - ☐ 1 second
    - ☐ 2.5 seconds
    - ☐ none of these

CAPACITIVE REACTANCE IN AC

Objectives

You will connect several ac circuits illustrating the characteristics of capacitance in ac and the relationship of capacitive reactance to frequency and capacitance.

In completing these projects, you will connect circuits, make measurements, perform calculations, draw conclusions, and be able to answer questions about the following items related to capacitance and capacitive reactance in ac circuits.

- Characteristic(s) of capacitance in ac circuits
- Relationship of X_C to capacitance value
- Relationship of X_C to frequency
- Total reactance of series capacitors
- Total reactance of parallel capacitors
- The X_C Formula

PROJECT/TOPIC CORRELATION INFORMATION

PROJECT	TEXT CHAPTER	SECTION	RELATED TEXT TOPIC(S)
53 Capacitance Opposing a Change in Voltage	19	19-2	*V* and *I* Relationships in a Purely Capacitative ac Circuit
54 X_C Related to Capacitance and Frequency	19	19-4	Relationship of X_C to Capacitance Value
		19-5	Relationship of X_C to Frequency of ac
55 The X_C Formula	19	19-6	Methods to Calculate X_C

PROJECT 53

CAPACITIVE REACTANCE IN AC
Capacitance Opposing a Change in Voltage

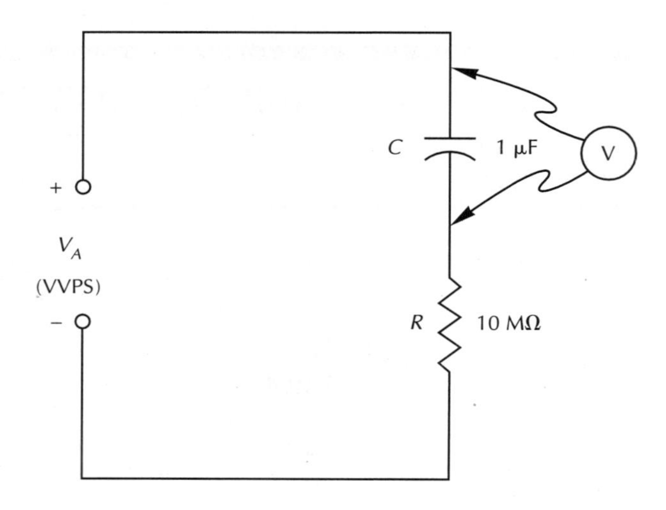

FIGURE 71

PROJECT PURPOSE

To demonstrate the delaying effect a capacitor has on a change in voltage by varying a dc source level and observing the time it takes for the capacitor voltage to track the change in source voltage.

PARTS NEEDED

- ☐ DMM/VOM (2) — use a 20,000 Ω/V VOM
- ☐ VVPS (dc)
- ☐ CIS
- ☐ Capacitor(s): 1.0 μF
- ☐ Resistor(s): 10 MΩ

SAFETY HINTS

DON'T FORGET THE CHARGED CAPACITOR WARNING YOU HAVE SEEN BEFORE!

ACTIVITY	OBSERVATION	CONCLUSION
1. Connect the initial circuit as shown in Figure 71.	—	—
2. Temporarily remove one lead from the VVPS and then set its output voltage at 10 volts. Observe how much time it takes the capacitor to stop charging after the lead is reinserted into the VVPS by watching the voltmeter measuring V_C.	Approximate time to reach a steady state for V_C was ________ seconds.	V_C did not reach 50 volts because the (*wire*, *meter*) ________ resistance formed a voltage divider with the 10-MΩ resistor.
3. Quickly turn the VVPS voltage control knob to a higher voltage setting, and note how much time it takes V_C to reach its new steady-state value. **CAUTION:** DON'T SET V_A HIGHER THAN THE MULTIMETER VOLTAGE RANGE SETTING.	Approximate time to reach the new steady-state value was ________ seconds.	It takes the voltage on the capacitor (*the same*, *a different*) ________ amount of time to change from some given value to a new value, as it does for V_C to change from zero to any steady-state value.
4. Quickly turn the VVPS voltage control knob to a lower voltage setting and observe if V_C changes instantaneously, or takes time.	It (*did*, *didn't*) ________ take time for V_C to decrease to its new steady-state value.	From our observations, we conclude that a capacitor seems to oppose a change in (*current*, *voltage*) ________ in a similar fashion to a coil or inductor opposing a change in (*current*, *voltage*) ________.

PROJECT 54

CAPACITIVE REACTANCE IN AC
X_C Related to Capacitance and Frequency

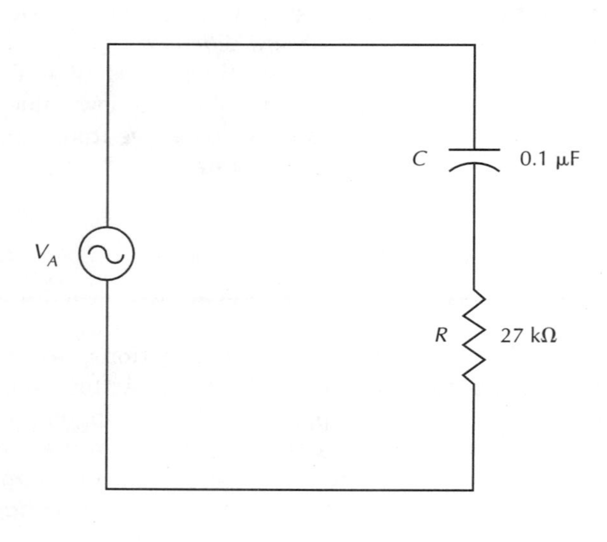

FIGURE 72

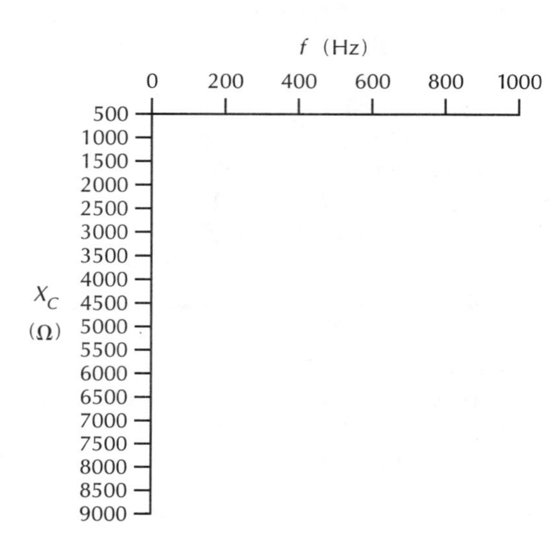

FIGURE 73

PROJECT PURPOSE

To verify the inverse relationship of X_C to both capacitance value and frequency by changing the values of circuit C and f, and measuring and analyzing the circuit parameter changes caused by changes in C and f.

PARTS NEEDED

- ☐ DMM/VOM (2)
- ☐ Function generator or audio oscillator
- ☐ CIS
- ☐ Capacitor(s): 0.1 μF, 1.0 μF
- ☐ Resistor(s): 27 kΩ

SAFETY HINTS

Here's the charged capacitor warning, again! Just a reminder!

ACTIVITY	OBSERVATION	CONCLUSION
1. Connect the initial circuit as shown in Figure 72.	—	—
2. Set the function generator or audio oscillator to a frequency of 100 Hz and V_A to 3 volts. Measure V_R and V_C. Calculate the circuit I by Ohm's law, and also calculate X_C ($X_C = V_C/I$).	V_A = ______ volts V_R = ______ volts V_C = ______ volts I_T = ______ mA X_C = ______ ohms	Since the current is the same through all parts of a series circuit, and the voltage drop across R is approximately 2 times V_C, it appears that X_C must be about (*double, half*) ______ the value of R.
3. Change C to a 1.0-μF capacitor and repeat step 2.	V_A = ______ volts V_R = ______ volts V_C = ______ volts I_T = approximately ______ mA X_C = approximately ______ ohms	Increasing the value of C while maintaining all other parameters the same caused X_C to (*increase, decrease*) ______. The X_C for the 1.0-μF capacitor was approximately (*ten times, one-tenth*) ______ the X_C of the 0.1-μF C. From this we conclude that X_C is (*directly, inversely*) ______ proportional to capacitance. This means as C decreases, X_C (*increases, decreases*) ______ or, if X_C has decreased, then C must have (*increased, decreased*) ______, all other factors being constant.
4. Keeping the 1.0-μF capacitor, change the frequency to 200 Hz and maintain 3 volts V_A. Repeat the measurements and calculations of the previous steps.	V_A = ______ volts V_R = ______ volts V_C = ______ volts I_T = approximately ______ mA X_C = approximately ______ ohms	Increasing the frequency and keeping the C the same caused X_C to (*increase, decrease*) ______. If there were no voltmeter loading effects on the circuit, the indicated X_C at 200 Hz would have been (*two times, one-half*) ______ the X_C value at 100 Hz. From this we see that X_C is (*directly, inversely*) ______ proportional to frequency.

PROJECT 54 CONTINUED

CAPACITIVE REACTANCE IN AC
X_C Related to Capacitance and Frequency *(Continued)*

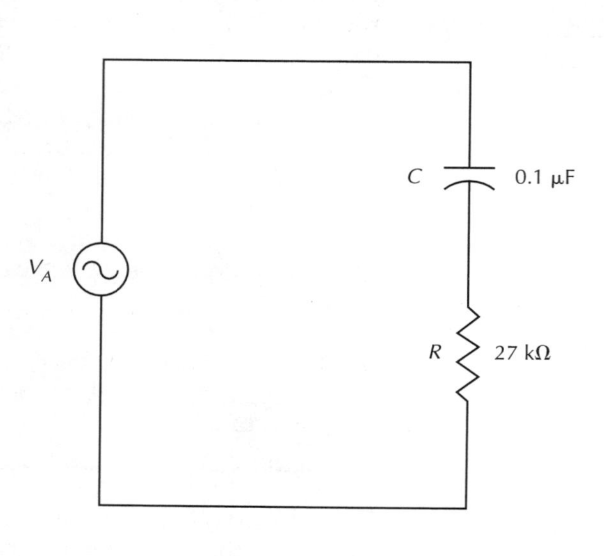

FIGURE 72

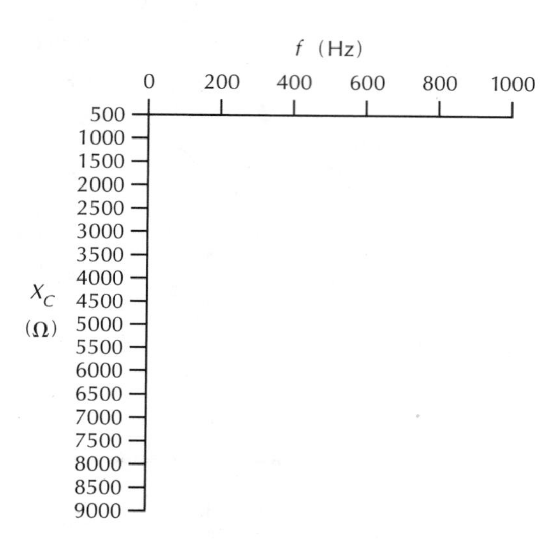

FIGURE 73

PROJECT PURPOSE

To verify the inverse relationship of X_C to both capacitance value and frequency by changing the values of circuit *C* and *f*, and measuring and analyzing the circuit parameter changes caused by changes in *C* and *f*.

PARTS NEEDED

- ☐ DMM/VOM (2)
- ☐ Function generator or audio oscillator
- ☐ CIS
- ☐ Capacitor(s): 0.1 μF, 1.0 μF
- ☐ Resistor(s): 27 kΩ

SAFETY HINTS

Here's the charged capacitor warning, again! Just a reminder!

ACTIVITY	OBSERVATION	CONCLUSION
5. Keep f at 200 Hz and change C back to a 0.1-μF. Repeat the measurements and calculations of the previous steps.	V_A = ______ volts V_R = ______ volts V_C = ______ volts I_T = ______ mA X_C = ______ ohms	Referring back to step 2 observations, does it appear that the X_C of the 0.1 μF C at 200 Hz (this step) is about one-half that at 100 Hz? ______. We conclude from the preceding that X_C is inversely proportional to ______ and ______.
6. Plot a graph of X_C versus f from 200 Hz to 1000 Hz using the 0.1 μF capacitor. **NOTE:** Calculate X_C for each 200 Hz change in frequency and plot in Figure 73.	X_C at 200 Hz = ______ X_C at 400 Hz = ______ X_C at 600 Hz = ______ X_C at 800 Hz = ______ X_C at 1000 Hz = ______	Did X_C act inversely proportional to f? ______

PROJECT 55

CAPACITIVE REACTANCE IN AC
The X_C Formula

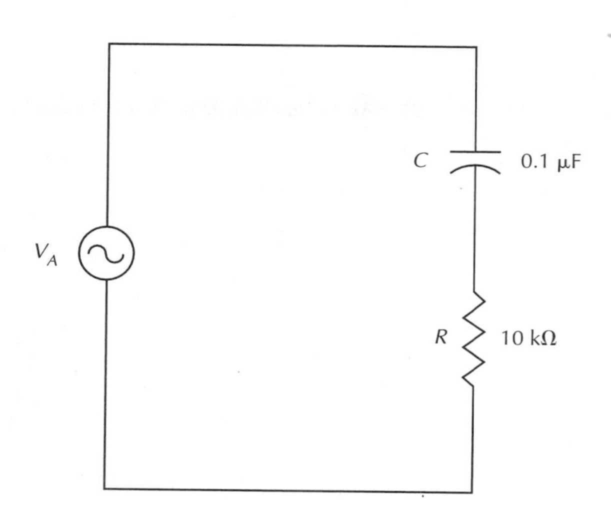

FIGURE 74

PROJECT PURPOSE

To verify the X_C formula by changing C and f values in a simple RC circuit and making measurements and appropriate Ohm's law calculations to see if the formula is valid.

PARTS NEEDED

- ☐ DMM/VOM (2)
- ☐ Function generator or audio oscillator
- ☐ CIS
- ☐ Capacitor(s):
  0.1 μF
  1.0 μF
- ☐ Resistor(s):
  10 kΩ

As shown in the previous project, the X_C of a capacitor in ohms is inversely proportional to the frequency and also to the capacitance value. This relationship is shown by the X_C formula, which may be discussed in the following manner:

$$X_C = \frac{1}{2\pi fC} \text{ or } \frac{0.159}{fC}$$

NOTE: Using the calculator reciprocal function, the $X_C = \frac{1}{2\pi fC}$ is easiest to use.

(Since 1 divided by $2\pi = 0.159$, we can use the 0.159 as a constant, if desired.) The rationale we might use in understanding this formula is as follows. A larger C requires more charge (stored electrons) to arrive at a given difference of potential between its plates. This means that more charging (and discharging) current must flow in the circuit in order for the capacitor to "follow" the ac voltage applied to it at a given frequency than a smaller C would require under the same conditions. More current flowing for a given applied voltage indicates a lower opposition. A higher frequency also causes more charge and discharge current to flow per unit time in order for the voltage across the capacitor to follow the applied voltage.

ACTIVITY	OBSERVATION	CONCLUSION
1. Connect the initial circuit as shown in Figure 74.	—	—
2. Set the function generator at a frequency of 150 Hz and V_A at 3 volts. Measure V_R and then calculate I_T. Measure V_C and use the calculated I to determine X_C by Ohm's law. Calculate X_C by the X_C formula.	V_A = ______ volts V_R = ______ volts I = ______ mA V_C = ______ volts X_C by Ohm's law approximately = ______ ohms X_C by X_C formula = ______ ohms	Since this is a series circuit and V_C is virtually the same as V_R, it is apparent that the X_C must essentially be equal to ______. Was the X_C calculated by the capacitive reactance formula close to the value determined by Ohm's law? ______.
3. Change the frequency to 75 Hz and keep V_A at 3 volts. Make the measurements and calculations described in step 2 above.	V_A = ______ volts V_R = ______ volts I = ______ mA V_C = ______ volts X_C by Ohm's law approximately = ______ ohms X_C by X_C formula = ______ ohms	Lowering the frequency to one-half its previous value caused the X_C to (*increase, decrease*) ______ to a value nearly (*double, half*) ______ the original. Two reasons for the X_C values calculated by Ohm's law method and X_C formula not being the same might be the resistor and capacitor ______ and meter ______ effects.

PROJECT

55

CONTINUED

CAPACITIVE REACTANCE IN AC
The X_C Formula *(Continued)*

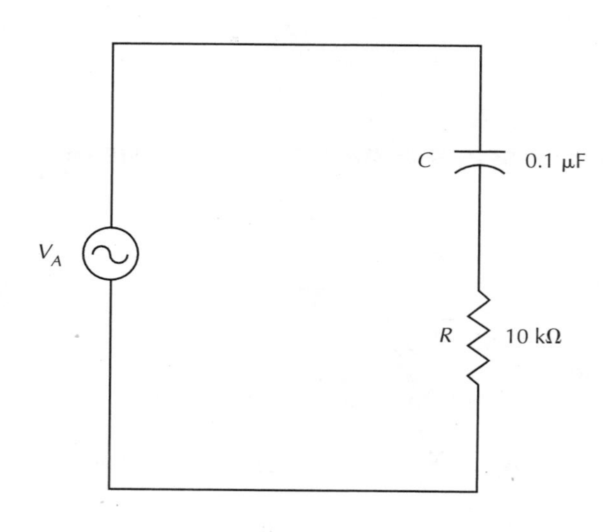

FIGURE 74

PROJECT PURPOSE

To verify the X_C formula by changing C and f values in a simple RC circuit and making measurements and appropriate Ohm's law calculations to see if the formula is valid.

PARTS NEEDED

- ☐ DMM/VOM (2)
- ☐ Function generator or audio oscillator
- ☐ CIS
- ☐ Capacitor(s):
  0.1 μF
  1.0 μF
- ☐ Resistor(s):
  10 kΩ

ACTIVITY	OBSERVATION	CONCLUSION
4. Keep the frequency at 75 Hz, but change C to 1 μF. Measure and calculate as before.	V_A = ______ volts V_R = ______ volts I = ______ mA V_C = ______ volts X_C by Ohm's law approximately = ______ ohms X_C by X_C formula = ______ ohms	Increasing C by a factor of 10 caused the X_C to (*increase, decrease*) ______. Did the X_C approximately change by a factor of 10? ______. From the observations we have made, does it appear that the X_C formula is functional for predicting parameters in practical circuits? ______.

CAPACITIVE REACTANCE IN AC

Complete the following review questions, indicating the appropriate response by placing a check in the box next to the correct answer.

1. X_c is measured in ohms because it limits ac current to a value of:

   ☐ $0.159 \times f \times C$
   ☐ V/X_c
   ☐ $0.159/fC$
   ☐ none of these

2. The formula for X_c is:

   ☐ $X_c = 2\pi/fC$
   ☐ $X_c = 1/2\pi fC$
   ☐ $X_c = fC/0.159$
   ☐ none of these

3. For a given value of C, if f is increased, then X_c will:

   ☐ increase
   ☐ decrease
   ☐ remain the same

4. For a given value of f, if C is decreased, then X_c will:

   ☐ increase
   ☐ decrease
   ☐ remain the same

5. A capacitor appears to oppose a change in voltage because current must flow before a difference of potential can be established across a capacitor.

   ☐ True
   ☐ False

6. Since capacitors in series add like resistors in parallel, the X_c of two series capacitors will be:

   ☐ greater than either one alone
   ☐ less than either one alone

7. Capacitive reactances in parallel add like resistances in:

   ☐ series
   ☐ parallel
   ☐ neither of these

8. If C and f are both doubled in a given circuit, the X_c will:

   - ☐ increase two times
   - ☐ decrease two times
   - ☐ increase four times
   - ☐ decrease four times

9. If C is doubled and f is halved in a given circuit, the X_c will:

   - ☐ increase
   - ☐ decrease
   - ☐ remain the same

10. What is the total capacitive reactance of two series 1-μF capacitors at a frequency of 200 Hz?

    - ☐ 3180 ohms
    - ☐ 15.9 ohms
    - ☐ 318 ohms
    - ☐ 1.59 kΩ
    - ☐ 31.8 kΩ
    - ☐ none of these

RC CIRCUITS IN AC

Objectives

You will connect several ac *RC* circuits and make measurements and observations regarding their important electrical characteristics.

In completing these projects, you will connect circuits, make measurements, perform calculations, draw conclusions, and be able to answer questions about the following items related to *RC* circuits.

- Relationship of circuit phase angle to R and X_C
- Relationships of current and voltage for capacitors and resistors
- Circuit impedance
- Simple vector diagram(s)
- Reference vector(s)

PROJECT/TOPIC CORRELATION INFORMATION

PROJECT	TEXT CHAPTER	SECTION	RELATED TEXT TOPIC(S)
56 V, I, R, Z, and θ Relationships in a Series *RC* Circuit	20	20-2	Series *RC* Circuit Analysis
57 V, I, R, Z, and θ Relationships in a Parallel *RC* Circuit	20	20-4	Parallel *RC* Circuit Analysis

PROJECT

56

RC CIRCUITS IN AC *V, I, R, Z,* and θ Relationships in a Series *RC* Circuit

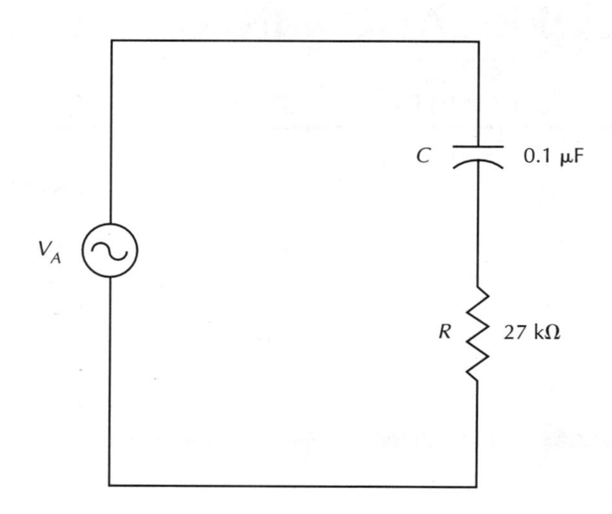

FIGURE 75

PROJECT PURPOSE

To demonstrate the key electrical parameter relationships in a series *RC* circuit. To observe that due to out-of-phase elements, simple dc analysis techniques cannot be used to determine circuit parameters in ac circuits containing reactive components. Furthermore, to provide practice in using simple ac analysis techniques and in drawing ac circuit vector diagrams.

PARTS NEEDED

- ☐ DMM/VOM (2)
- ☐ Function generator or audio oscillator
- ☐ Dual-trace oscilloscope
- ☐ CIS
- ☐ Capacitor(s): 0.1 μF
- ☐ Resistor(s): 27 kΩ

By way of review, recall that in a purely resistive series ac circuit, the circuit current was in phase with the applied voltage. Also, the voltage drops across the individual resistors were in phase with current and with each other. The total opposition to current flow in the purely resistive circuit of this type was the arithmetic sum of the individual resistances. If a vector diagram of voltage and current were drawn, we would use the current as the reference vector, since it is the common factor in a series circuit. Summarizing: $Z = R_T$, $\theta = 0$ degrees, V_T = simple sum of $V_1 + V_2 + \ldots$ (etc.) for purely resistive series ac circuits. In this project, you will examine parameters for a series *RC* circuit.

ACTIVITY	OBSERVATION	CONCLUSION
1. Connect the initial circuit as shown in Figure 75.	—	—
2. Set the frequency of the function generator to 100 Hz and V_A to 3 volts. Measure V_A, V_R, and V_C.	V_A = ________ volts V_R = ________ volts V_C = ________ volts	Does the sum of V_R and V_C equal V_A? ________. This is because V_R and V_C are (*in phase, out of phase*) ________.
3. Calculate I_T from V_R/R; X_C from V_C/I, and Z from V_T/I_T.	I_T = ________ mA X_C = ________ ohms Z = ________ ohms	Does Z equal the arithmetic sum of R and X_C? ________. We may conclude that since V_A is not equal to $V_R + V_C$ and Z is not equal to $R + X_C$, these values are the resultant of two out-of-phase vectors. We can solve for the resultant vectors by means of the Pythagorean theorem or trigonometry. Is the voltage across the resistor in phase with the current through it? ________. Since a capacitor opposes a change in voltage, we might assume that the I_C (*leads, lags*) ________ V_C. In a perfect capacitor I_C (*leads, lags*) ________ V_C by 90 degrees. We should also conclude that since this is a series circuit, the larger the amount of X_C (the smaller the C), the more like a purely (*resistive, capacitive*) ________ circuit the circuit will act. This means the larger the X_C compared to the R, the (*greater, smaller*) ________ will be the resultant phase angle between V_A and I_T.

PROJECT
56
CONTINUED

RC CIRCUITS IN AC
V, I, R, Z, and θ Relationships in a Series *RC* Circuit *(Continued)*

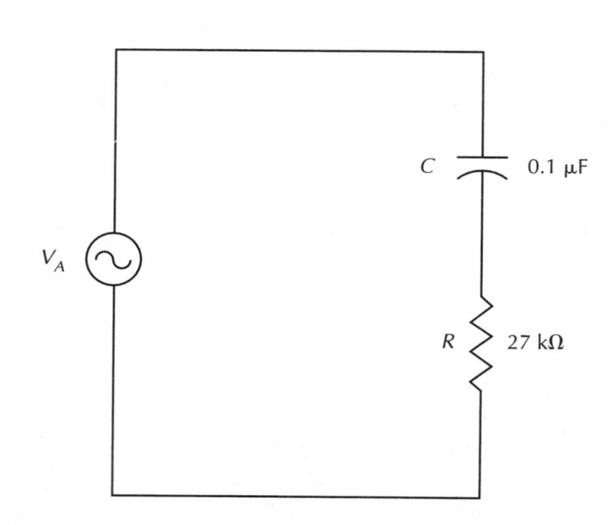

FIGURE 75

PROJECT PURPOSE

To demonstrate the key electrical parameter relationships in a series *RC* circuit. To observe that due to out-of-phase elements, simple dc analysis techniques cannot be used to determine circuit parameters in ac circuits containing reactive components. Furthermore, to provide practice in using simple ac analysis techniques and in drawing ac circuit vector diagrams.

PARTS NEEDED

- ☐ DMM/VOM (2)
- ☐ Function generator or audio oscillator
- ☐ Dual-trace oscilloscope
- ☐ CIS
- ☐ Capacitor(s):
  0.1 μF
- ☐ Resistor(s):
  27 kΩ

ACTIVITY	OBSERVATION	CONCLUSION
4. Calculate Z using the Pythagorean approach. 5. Draw an impedance diagram for the circuit in the Observation column.	Z calculated = ________ ohms	—
6. Use the Z diagram and trigonometry and determine the phase angle.	θ = ________ degrees	—
7. Repeat the previous steps 2–6. This time set f = 200 Hz and keep V_A at 3 volts.	V_A = ________ volts V_R = ________ volts V_C = ________ volts I_T = ________ mA X_C = ________ ohms Z = ________ ohms θ = ________ degrees	Increasing frequency while keeping all other factors the same caused Z to ________; X_C to ________; θ to ________.

EXTRA CREDIT STEP(S)

8. Use the measured and calculated data in the previous steps 2 and 3, and draw a *V-I* vector diagram in the Observation column.		
9. Use trigonometry and determine the phase angle.	θ = ________ degrees	Does the phase angle from the *V-I* vector diagram agree reasonably with the angle you determined from the Z diagram in step 5? ________.

PROJECT
56
CONTINUED

RC CIRCUITS IN AC
V, I, R, Z, and θ Relationships in a Series *RC* Circuit *(Continued)*

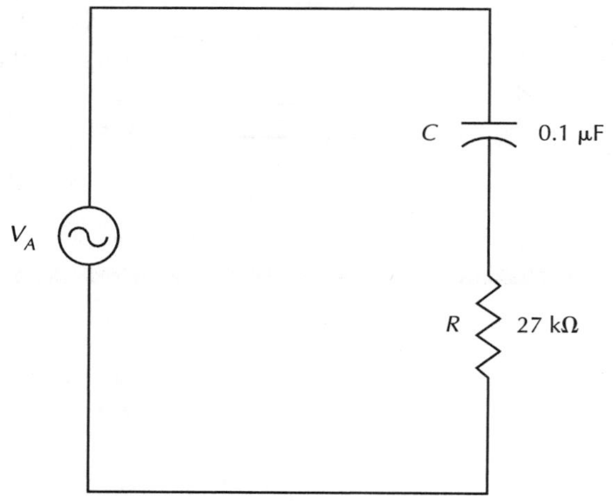

V_A C 0.1 μF R 27 kΩ

FIGURE 75

PROJECT PURPOSE

To demonstrate the key electrical parameter relationships in a series *RC* circuit. To observe that due to out-of-phase elements, simple dc analysis techniques cannot be used to determine circuit parameters in ac circuits containing reactive components. Furthermore, to provide practice in using simple ac analysis techniques and in drawing ac circuit vector diagrams.

PARTS NEEDED

- ☐ DMM/VOM (2)
- ☐ Function generator or audio oscillator
- ☐ Dual-trace oscilloscope
- ☐ CIS
- ☐ Capacitor(s):
  0.1 μF
- ☐ Resistor(s):
  27 kΩ

EXTRA CREDIT STEP(S)

10. Again, set the source for 100 Hz and a V_A of 3 volts. Use a dual-trace oscilloscope and perform a phase comparison of V_A and circuit current, (represented by the voltage across the resistor). **CAUTION:** Be sure the signal source ground and the scope ground(s) are connected to the same end (the bottom end in the diagram) of the resistor when making the measurements to prevent the grounds from shorting out a portion of the circuit! Determine the phase difference between the two signals.

θ determined by the scope phase comparison:
θ = ____________________ degrees

Do the scope phase measurements and the phase angle calculations agree reasonably with your earlier findings? (Considering tolerances in components, source and scope frequency calibration tolerances, etc.) __________.

PROJECT 57

RC CIRCUITS IN AC
V, I, R, Z, and *θ* Relationships in a Parallel *RC* Circuit

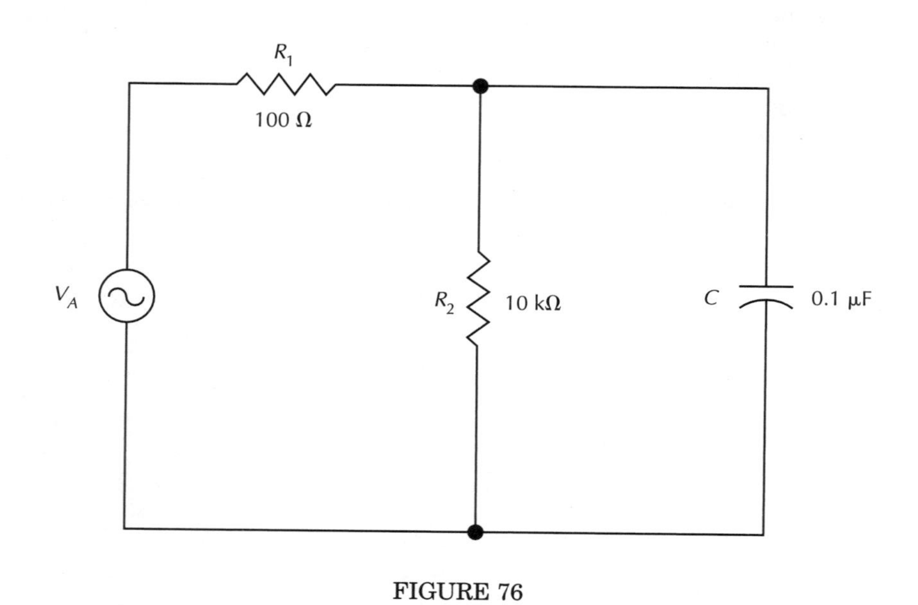

FIGURE 76

PROJECT PURPOSE

To demonstrate the key electrical parameter relationships in a parallel *RC* circuit. To observe that due to out-of-phase elements, simple dc analysis techniques cannot be used to determine circuit parameters in ac circuits containing reactive components. To compare results of computing circuit parameters using Ohm's law and the Pythagorean theorem.

PARTS NEEDED

- ☐ DMM/VOM (2)
- ☐ Function generator or audio oscillator
- ☐ CIS
- ☐ Capacitor(s): 0.1 μF
- ☐ Resistor(s): 100 Ω, 10 kΩ

By way of review, recall that in a purely resistive parallel ac circuit, the following conditions exist:

> $Z = R_T$; $\theta = 0$ degrees; I_T = arithmetic sum of branch currents; and branch voltage is in phase with branch current for a purely resistive circuit.

In this project, you will examine parameters in a parallel *RC* circuit. It should be noted again that for this project we will be using the 100 Ω resistor in series with the main line as a circuit current indicator. (Current through R_1 equals ten times its voltage drop.)

ACTIVITY	OBSERVATION	CONCLUSION
1. Connect the initial circuit as shown in Figure 76.	—	—
2. Set the audio oscillator to a frequency of 500 Hz and V_A to 3 volts. Measure the circuit voltages.	V_A = ________ volts V_1 = ________ volts V_2 = ________ volts V_C = ________ volts	Does the addition of the voltages around any closed loop equal V_A? ________. We conclude from this that V_R and V_C are (*in phase, out of phase*) ________.
3. Calculate I_T from V_1. Calculate I_2 by Ohm's law. Calculate X_C by using the X_C formula, then calculate I_C using the formula: $I_C = V_C/X_C$. Also, calculate θ using the arctan of $\frac{IC}{IR}$	I_T = ________ mA I_2 = ________ mA X_C = ________ ohms I_C = ________ mA θ = ________ degrees	Does the total current equal the arithmetic sum of the branch currents? ________. This is because the branch currents are ________ ________ ________. Is the current through R_2 in phase with V_2? ________. The current through the capacitor branch (*leads, lags*) ________ the voltage across the capacitor by close to ________ degrees. Since I_T is the vector resultant of I_R and I_C, we would logically conclude that I_T (*leads, lags*) ________ V_A by an angle that is between ________ and ________ degrees.
4. Calculate the circuit Z_T from V_T/I_T.	Z_T = ________ ohms	It is interesting to note that the impedance of a circuit consisting of two parallel 10-kΩ branches, whose currents are 90 degrees out of phase, would ≅ 7.07 kΩ. Does our demonstration circuit result compare closely to this? ________ Why not? ________

PROJECT 57 CONTINUED

RC CIRCUITS IN AC *V*, *I*, *R*, *Z*, and θ Relationships in a Parallel *RC* Circuit *(Continued)*

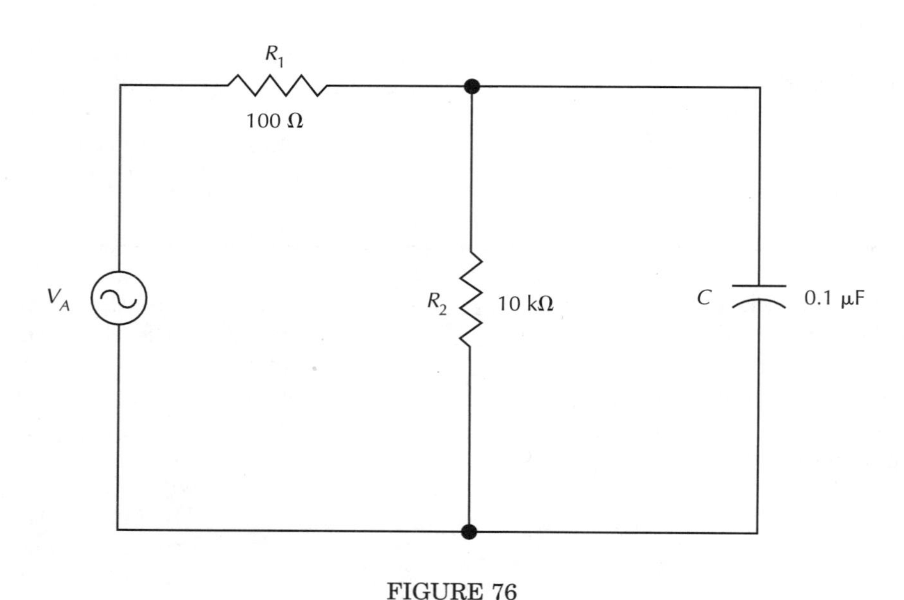

FIGURE 76

PROJECT PURPOSE

To demonstrate the key electrical parameter relationships in a parallel *RC* circuit. To observe that due to out-of-phase elements, simple dc analysis techniques cannot be used to determine circuit parameters in ac circuits containing reactive components. To compare results of computing circuit parameters using Ohm's law and the Pythagorean theorem.

PARTS NEEDED

- ☐ DMM/VOM (2)
- ☐ Function generator or audio oscillator
- ☐ CIS
- ☐ Capacitor(s):
  0.1 μF
- ☐ Resistor(s):
  100 Ω
  10 kΩ

ACTIVITY	OBSERVATION	CONCLUSION
5. Use I_2 and I_C values and determine I_T by means of Pythagorean theorem.	I_T calculated = ________ mA	—
6. Repeat the previous steps 2–5. This time set f = 250 Hz and keep V_A at 3 volts.	V_A = ________ volts V_1 = ________ volts V_2 = ________ volts V_C = ________ volts I_T = ________ mA I_2 = ________ mA X_C = ________ ohms I_C = ________ mA Z_T = ________ ohms θ = ________ degrees	Decreasing frequency caused Z to ________; X_C to ________; θ to ________.

RC CIRCUITS IN AC

Complete the following review questions, indicating the appropriate response by placing a check in the box next to the correct answer.

1. The higher the frequency of V_A applied to any *RC* circuit, either series or parallel, the ____________________________ the circuit impedance will be.

   ☐ higher
   ☐ lower

2. In a purely capacitive series circuit, the arithmetic sum of the individual voltage drops equals V_A.

   ☐ True
   ☐ False

3. In a series *RC* circuit, the arithmetic sum of the individual voltage drops equals V_A.

   ☐ True
   ☐ False

4. In a parallel *RC* circuit, the arithmetic sum of the branch currents equals I_T.

   ☐ True
   ☐ False

5. In a series *RC* circuit, if frequency, resistance, or capacitance increases, the circuit phase angle will:

   ☐ increase
   ☐ decrease
   ☐ remain the same

6. In a parallel *RC* circuit, if frequency, resistance, or capacitance increases, the circuit phase angle will:

   ☐ increase
   ☐ decrease
   ☐ remain the same

7. As the resistance in a parallel *RC* circuit is increased, the circuit will become more:

   ☐ resistive
   ☐ capacitive

8. As the capacitance in a series *RC* circuit is increased, the circuit will become more:

   ☐ resistive
   ☐ capacitive

9. A circuit that is capacitive is one where the circuit voltage is lagging the circuit current, and I_c and V_c are 90 degrees out of phase.

   ☐ True
   ☐ False

10. The impedance of a parallel *RC* circuit will increase as frequency is decreased.

   ☐ True
   ☐ False

SERIES RESONANCE

Objectives

You will connect several ac *RLC* circuits that illustrate the important characteristics of series resonant circuits.

In completing these projects, you will connect circuits, make measurements, perform calculations, draw conclusions, and be able to answer questions about the following items related to series resonant circuits.

- Definition of resonance
- Relationships of V, I, R, Z, θ, Q, bandwidth, and frequency in series *RLC* circuits
- Finding a circuit's resonant frequency
- Outstanding characteristics of a circuit at series resonance

PROJECT/TOPIC CORRELATION INFORMATION

PROJECT		TEXT CHAPTER	SECTION	RELATED TEXT TOPIC(S)
58	X_L and X_C Relationships to Frequency	22	22-1	X_L, X_C, and Frequency
59	V, I, R, Z, and θ Relationships when $X_L = X_C$	22	22-2	Series Resonance Characteristics
60	Q and Voltage in a Series Resonant Circuit	22	22-5	Q and Resonant Rise of Voltage
61	Bandwidth Related to Q	22	22-10	Selectivity, Bandwidth, and Bandpass

PROJECT 58

SERIES RESONANCE X_L and X_C Relationships to Frequency

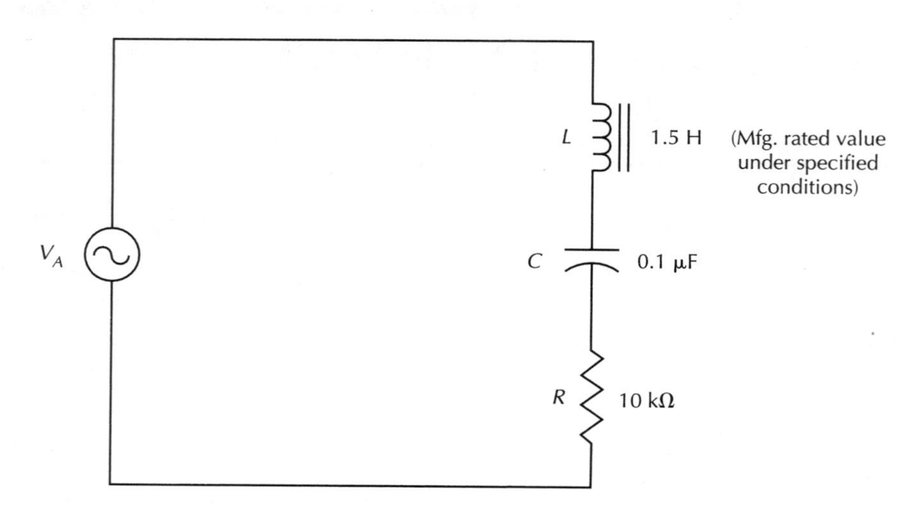

FIGURE 77

PROJECT PURPOSE

To verify the opposite effects on X_L and X_C when the frequency of signal applied to a circuit containing both is changed.

PARTS NEEDED

- ☐ DMM/VOM (2)
- ☐ Function generator or audio oscillator
- ☐ CIS
- ☐ Capacitor(s):
  0.1 μF
- ☐ Inductor(s):
  1.5 H, 95 Ω (or approximate)
- ☐ Resistor(s):
  10 kΩ

ACTIVITY	OBSERVATION	CONCLUSION
1. Connect the initial circuit as shown in Figure 77.	—	—
2. Set the function generator to a frequency of 150 Hz and V_A to 3 volts. Measure V_R and calculate I_T. Measure V_L and V_C and calculate X_L and X_C by Ohm's law.	V_A = ______ volts V_R = ______ volts I_T = ______ mA V_L = ______ volts V_C = ______ volts X_L = ______ ohms X_C = ______ ohms	If the X_C formula were used, the X_C of a 0.1 μF capacitor at 150 Hz would be calculated as ______ ohms. Can the difference between the Ohm's law results and the X_C formula be attributed to meter loading effects, capacitor tolerance, and audio oscillator dial calibration tolerance? ______.
3. Change the function generator frequency to 300 Hz and keep V_A at 3 volts. Measure V_R and calculate I_T. Measure V_L and V_C and calculate X_L and X_C by Ohm's law.	V_A = ______ volts V_R = ______ volts I_T = ______ mA V_L = ______ volts V_C = ______ volts X_L = ______ ohms X_C = ______ ohms	When frequency was increased from 150 Hz to 300 Hz (or doubled), the X_L *(increased, decreased)* ______ by approximately ______ times, and the X_C *(increased, decreased)* ______ by about ______ times, compared to their values at 150 Hz. We conclude from these results that X_L is *(directly, inversely)* ______ proportional to frequency, and X_C is *(directly, inversely)* ______ proportional to frequency. In a circuit that has both inductance and capacitance, there must be some frequency where X_L and X_C would be ______. In our circuit, would the frequency where this would occur have to be higher or lower than 300 Hz? ______.

PROJECT 59

SERIES RESONANCE
V, I, R, Z, and θ Relationships when $X_L = X_C$

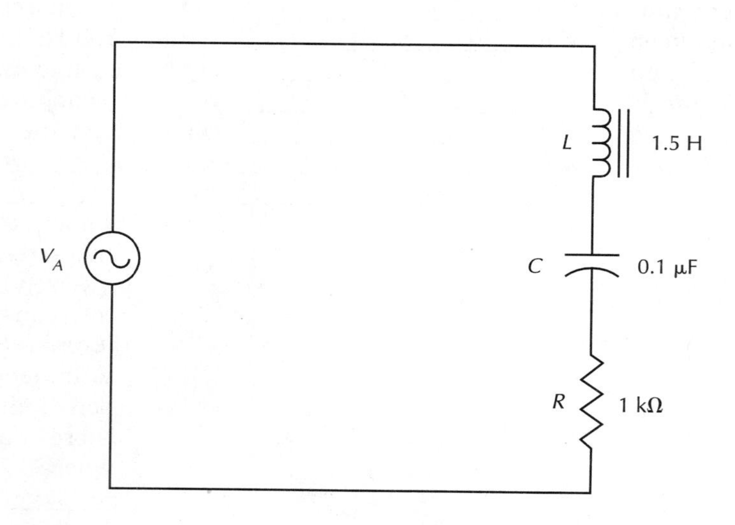

FIGURE 78

PROJECT PURPOSE

To demonstrate the resonance effects at a frequency where $X_L = X_C$. To show that the series *RLC* circuit basically acts resistively at resonance, inductively above resonance, and capacitively below resonance.

PARTS NEEDED

- ☐ DMM/VOM (2)
- ☐ Function generator or audio oscillator
- ☐ CIS
- ☐ Capacitor(s):
  0.1 μF
- ☐ Inductor(s):
  1.5 H, 95 Ω (or approximate)
- ☐ Resistor(s):
  1 kΩ

ACTIVITY	OBSERVATION	CONCLUSION
1. Connect the initial circuit as shown in Figure 78.	—	—
2. Set V_A at 3 volts and adjust the function generator frequency while monitoring V_R until V_R is maximum. Measure V_R and calculate I_T. Measure V_L and V_C. Calculate X_L and X_C by Ohm's law.	V_A = ______ volts Frequency = ______ Hz (approximately) V_R = ______ volts I_T = ______ mA V_L = ______ volts V_C = ______ volts X_L = ______ ohms X_C = ______ ohms	Are X_L and X_C close to being equal at the frequency of maximum V_R? ______. Would they be equal if circuit and measurement conditions were perfect? ______. The voltage across the coil (*leads, lags*) ______ the current by about ______ degrees; whereas, the voltage across the capacitor (*leads, lags*) ______ the current by close to ______ degrees. This means that V_L and V_C are close to ______ degrees out of phase with each other. Since I is the same throughout the series circuit and the $I \times X_L$ drop is essentially ______ degrees out of phase with the $I \times X_C$ drop, we may conclude that X_L and X_C reactances are opposite (vectorially) and at resonance are equal, therefore, cancel each other's effects on Z. Notice that V_L and V_C both are greater than V_A. This is due to the cancelling effect of the reactances. If X_L and X_C perfectly cancelled (as would be true with perfect equal and opposite reactances) the circuit Z would equal ______. The resultant circuit current would be (*in phase, out of phase*) ______ with V_A and the circuit would be acting purely ______.
3. Change the frequency to a new frequency that is well above resonance. Measure V_L and V_C and determine whether the circuit is now acting inductively or capacitively.	V_L is now ______ than V_C.	The circuit is now acting like an (*RL, RC*) ______ circuit, since X ______ is more than cancelling out X ______. Thus, the circuit is acting like R in series with a resultant X ______. The value of this resultant X is equal to a value of X ______ total minus X ______. The result is that I_T (*leads, lags*) ______ V_A by an angle between ______ and ______ degrees; thus the circuit is (*inductive, capacitive*) ______.

PROJECT

59

CONTINUED

SERIES RESONANCE

V, I, R, Z, and θ Relationships when $X_L = X_C$ *(Continued)*

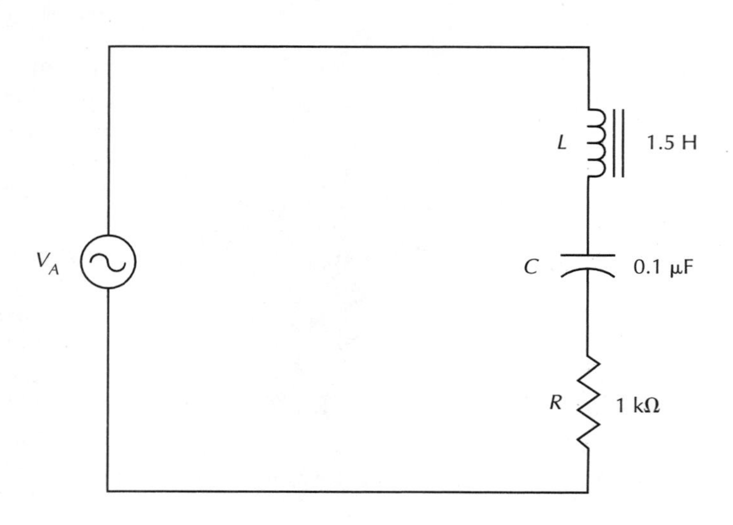

FIGURE 78

PROJECT PURPOSE

To demonstrate the resonance effects at a frequency where $X_L = X_C$. To show that the series *RLC* circuit basically acts resistively at resonance, inductively above resonance, and capacitively below resonance.

PARTS NEEDED

- ☐ DMM/VOM (2)
- ☐ Function generator or audio oscillator
- ☐ CIS
- ☐ Capacitor(s):
  0.1 μF
- ☐ Inductor(s):
  1.5 H, 95 Ω (or approximate)
- ☐ Resistor(s):
  1 kΩ

ACTIVITY	OBSERVATION	CONCLUSION
4. Keep the same circuit as shown in Figure 78.	—	—
5. Change the input frequency to a new frequency that is well below the resonant frequency determined in the earlier steps. Measure V_L and V_C and determine whether the circuit is now acting inductively or capacitively.	New frequency is lower than ____ Hz. V_L is now ________________ than V_C.	Since V _________ is now greater than V _________ and the current is the same through both reactances, it indicates that X _______ is greater than X _________. This means that X __________ is more than cancelling X __________ and the circuit is equivalent to an (*RL, RC*) _________ circuit. The circuit impedance therefore would equal the vector resultant of R and a series X _____ whose value equals the difference between _________ and _________. We may conclude from the preceding that at a frequency where $X_L = X_C$ (resonance), the circuit acts essentially like a pure ________________ circuit. At frequencies below resonance, a series *RLC* circuit will act equivalent to a simple R _______ circuit. At frequencies above resonance, a *series RLC* circuit will act equivalent to a simple R ______ circuit. We may also conclude that the larger the value of C or L, the (*lower, higher*) _____________ will be the resonant frequency and vice versa. Summarizing our observations for a series resonant circuit, we conclude that at resonance Z is (*minimum, maximum*) ______________________ since X_L and X_C cancel.

PROJECT 60

SERIES RESONANCE
Q and Voltage in a Series Resonant Circuit

FIGURE 79

PROJECT PURPOSE

To show, via circuit measurements, that the voltage across reactive components in a series resonant circuit is greater than V applied. To demonstrate that the amount of this voltage across each reactive component is related to the circuit Q factor.

PARTS NEEDED

- ☐ DMM/VOM (2)
- ☐ Function generator or audio oscillator
- ☐ CIS
- ☐ Capacitor(s):
  0.1 μF
- ☐ Inductor(s):
  1.5 H, 95 Ω (or approximate)
  2.5 mH (or approximate)

The figure of merit for a resonant circuit is called "Q." In general terms, the higher the ratio of reactance at resonance to series resistance, the higher the Q. If we had a perfect inductor and capacitor tuned to resonance and there were zero resistance in the circuit, the circuit impedance would be zero. Thus, current would be infinite. However, this is impossible in practical circuits. Therefore, Q is considered to be the relationship X_L/R, or X_C/R for series resonant circuits. Also, because of the cancelling effect of X_L and X_C allowing high circuit current at resonance, there is a voltage magnification across the reactive components at resonance. During this project we will illustrate some of these facts.

ACTIVITY	OBSERVATION	CONCLUSION
1. Connect the initial circuit as shown in Figure 79.	—	—
2. Set V_A at 3 volts and adjust the frequency, while monitoring V_L, for maximum V_L. Measure V_A and V_L and calculate the circuit Q from the ratio of V_L to V_A.	V_A = ________ volts V_L = ________ volts (approximately) Q = ________ (approximately) f = ________	The ratio of X_L/R for this circuit must be approximately ________ ________.
3. With the same V_A and frequency as the previous step, measure V_C and calculate the circuit Q from V_C/V_A.	V_A = ________ volts V_C = ________ volts (approximately) Q = ________ (approximately) f = ________	The ratio of X_C/R for this circuit must be approximately ________ ________. What is the X_C of the capacitor according to the X_C formula? Approximately ________ ohms. Knowing the ratio of X_C/R, what must be the approximate effective R in this circuit? ________. From the preceding, we may conclude that V_L or V_C equals ________ times V_A; that Q equals the ratio of ________ or ________ to R; and the higher the Q, the (*higher*, *lower*) ________ the circuit current will be at resonance.
4. If time permits, replace the 1.5-H choke with a 2.5-mH inductor and perform the same measurements and calculations called for in step 2 above.	V_A = ________ volts V_L = ________ volts Q = ________ f = ________	The ratio of X_L/R for this circuit must be approximately ________.

PROJECT
60
CONTINUED

SERIES RESONANCE
Q and Voltage in a Series Resonant Circuit *(Continued)*

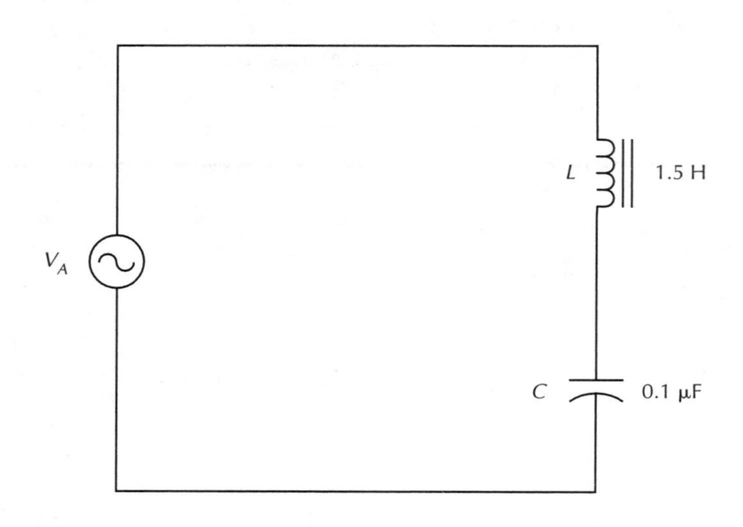

FIGURE 79

PROJECT PURPOSE

To show, via circuit measurements, that the voltage across reactive components in a series resonant circuit is greater than V applied. To demonstrate that the amount of this voltage across each reactive component is related to the circuit Q factor.

PARTS NEEDED

- ☐ DMM/VOM (2)
- ☐ Function generator or audio oscillator
- ☐ CIS
- ☐ Capacitor(s):
  0.1 μF
- ☐ Inductor(s):
  1.5 H, 95 Ω (or approximate)
  2.5 mH (or approximate)

EXTRA CREDIT STEP(S)

5. Assuming that the 0.1-μF capacitor is very close to its rated value, determine what the resonant frequency of the circuit would be if the initial circuit inductor were acting at its rated value of 1.5-H.

If the inductor really was acting as a 1.5-H inductor, the resonant frequency would be approximately ________ Hz.

—

6. Based on the measured resonant frequency in step 2, how much apparent inductance does this inductor have under the operating conditions used?

"Apparent" inductance is approximately ________ H.

It may be concluded that for a given capacitance value, the higher the inductance value used in conjunction with the capacitor, the (*higher, lower*) ________ the resonant frequency will be.

7. Perform appropriate calculations and compare the Q calculated from data in step 2 of Project 59 to the Q determined in step 2 of this Project (Project 60).

Q calculated from step 2 of Project 59 = ________.
Q determined in step 2 of this project = ________.

Since the same values of L, C, and V_A are used in both projects, what accounts for the difference in Q values? ____.

PROJECT 61

SERIES RESONANCE
Bandwidth Related to Q

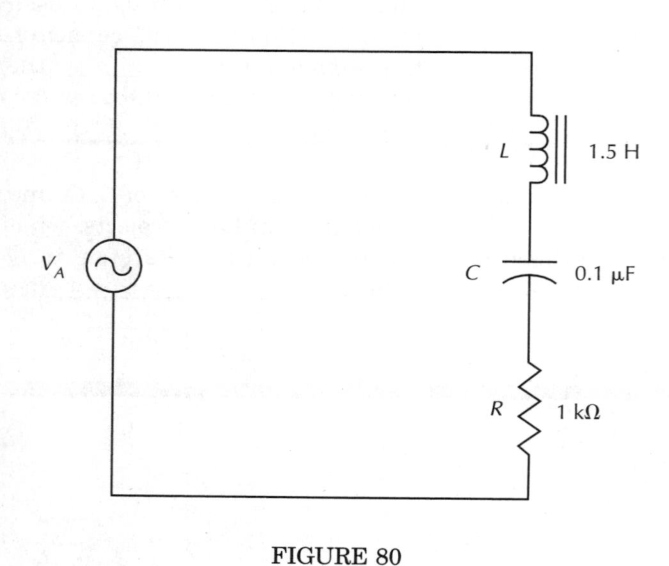

FIGURE 80

(Plot with 1 kΩ R)

V_R

f_r

f

(Plot with 10 kΩ R)

V_R

f_r

f

FIGURE 81

NOTE: Label f_r and upper and lower bandpass frequencies on each graph.

PROJECT PURPOSE

To provide practical experience and practice in making measurements to find "half-power points" and from these measurements to determine the series *RLC* circuit bandwidth.

PARTS NEEDED

- ☐ DMM/VOM (2)
- ☐ Function generator or audio oscillator
- ☐ CIS
- ☐ Capacitor(s):
  0.1 µF
- ☐ Inductor(s):
  1.5 H, 95 Ω (or approximate)
  2.5 mH (or approximate)
- ☐ Resistor(s):
  1 kΩ
  10 kΩ

Bandwidth for a series resonant circuit may be defined as the difference in frequency between the two frequencies (one below resonance and one above) at which the circuit current is 0.707 (70.7%) of the maximum current (which occurs at resonance). It is interesting to note that the higher the Q of a circuit, the higher the maximum current will be at resonance for any given V_A. If a graph is made of current versus frequency, it will be shown that the higher the Q, (and thus the I), the steeper will be the slope of the resonance curve and the smaller the bandwidth. Since Q and bandwidth are related, one formula for bandwidth is: Bandwidth = f_r (resonant frequency)/Q and therefore $Q = f_r$/bandwidth.

ACTIVITY	OBSERVATION	CONCLUSION
1. Connect the initial circuit as shown in Figure 80.	—	—
2. Set V_A to 3 volts. While monitoring V_R, set the frequency of the audio oscillator for resonance (maximum V_R). Measure V_R, V_L, and V_C. Calculate I at resonance.	V_A = ________ volts V_R = ________ volts (approximately) V_L = ________ volts (approximately) V_C = ________ volts (approximately) I = ________ mA (approximately)	The approximate Q of this circuit (V_L/V_A) = ________. According to the formula Bandwidth = f_r/Q, the bandwidth of this circuit should be approximately ________ Hz. If circuit current decreased to 70.7% of I_{max}, the current would be approximately ________ mA, and then V_R would equal ________ volts.
3. Keeping V_A at 3 volts, adjust frequency to a frequency below resonance where V_R equals 0.707 of $V_{R_{max}}$. Note this frequency, then change the frequency above resonance until V_R equals 0.707 of $V_{R_{max}}$ and note this frequency. From these two frequencies determine the measured bandwidth.	Bandwidth = ________ Hz (approximately) f below resonance where V_R at 70% ________ Hz f above resonance where V_R at 70% ________ Hz	Does the measured bandwidth approximate the value calculated from the bandwidth formula in step 2? ________.
4. If time permits, change R to a 10-kΩ resistor and repeat the steps above. Note whether the Q and bandwidth increased or decreased with the higher R value.	Q ________ Bandwidth = ________ Hz	The higher the Q, the ________ the bandwidth. The higher the circuit R, the ________ the bandwidth.

PROJECT **61** CONTINUED

SERIES RESONANCE
Bandwidth Related to Q *(Continued)*

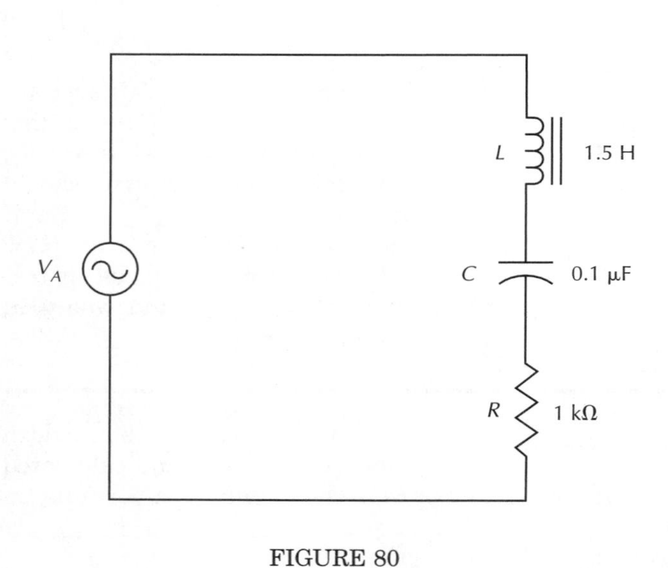

FIGURE 80

(Plot with 1 kΩ R)

V_R

f_r

f

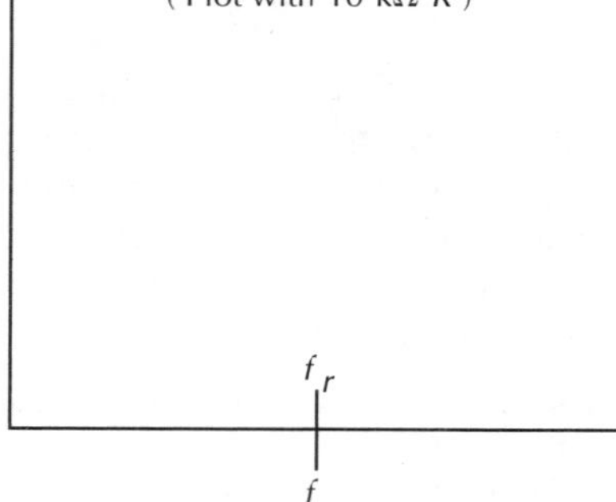

FIGURE 81

NOTE: Label f_r and upper and lower bandpass frequencies on each graph.

PROJECT PURPOSE

To provide practical experience and practice in making measurements to find "half-power points" and from these measurements to determine the series *RLC* circuit bandwidth.

PARTS NEEDED

- ☐ DMM/VOM (2)
- ☐ Function generator or audio oscillator
- ☐ CIS
- ☐ Capacitor(s):
  0.1 µF
- ☐ Inductor(s):
  1.5 H, 95 Ω (or approximate)
  2.5 mH (or approximate)
- ☐ Resistor(s):
  1 kΩ
  10 kΩ

ACTIVITY	OBSERVATION	CONCLUSION
5. Make graphic plots of the data collected in steps 2 through 4 above, on the coordinates shown in Figure 81.	—	—
6. Replace the 1.5-H choke with the 2.5-mH inductor and repeat steps 2 and 3.	V_A = ______ volts V_R = ______ volts V_L = ______ volts V_C = ______ volts I = ______ mA B.W. = ______ Hz f_{low} = ______ Hz f_{high} = ______ Hz	The approximate Q of this circuit (V_L/V_A) = ______. According to the formula Bandwidth = f_r/Q, the bandwidth of this circuit should be approximately ______ Hz. If circuit current decreased to 70.7% of I_{max}, the current would be approximately ______ mA, and then V_R would equal ______ volts. Does the measured bandwidth approximate the value calculated from the bandwidth formula in step 2?

SERIES RESONANCE

Complete the following review questions, indicating the appropriate response by placing a check in the box next to the correct answer.

1. Resonance is sometimes defined as the circuit condition when:

   - ☐ $R = X_L$
   - ☐ $R = X_C$
   - ☐ $X_L = Z$
   - ☐ $X_C = Z$
   - ☐ $X_C = X_L$

2. Three circuit factors that determine whether an *RLC* circuit will be equivalent to an *RC*, an *RL*, or a resistive circuit are:

   - ☐ V, I, and Z
   - ☐ f, R, and V_A
   - ☐ f, L, and C
   - ☐ none of these

3. If a series *RLC* circuit is at resonance and then f is increased, the circuit will begin to act:

   - ☐ capacitively
   - ☐ inductively
   - ☐ resistively

4. What is the phase relationship between V_C and V_L in a series *RLC* circuit?

   - ☐ 0 degrees
   - ☐ 45 degrees
   - ☐ 90 degrees
   - ☐ 180 degrees
   - ☐ none of these

5. What is the phase relationship between I_C and I_L in a series *RLC* circuit?

   - ☐ 0 degrees
   - ☐ 45 degrees
   - ☐ 90 degrees
   - ☐ 180 degrees
   - ☐ none of these

6. As the resonant frequency is approached for any *RLC* circuit, the phase angle will:

   - ☐ increase
   - ☐ decrease
   - ☐ remain the same

7. Is it possible for an *RLC* circuit to change from capacitive to inductive simply by changing the frequency of the applied voltage?

   ☐ Yes
   ☐ No

8. Impedance in a series *RLC* circuit at resonance is:

   ☐ maximum
   ☐ minimum
   ☐ neither of these

9. The voltage across either reactive component in a series resonant *RLC* circuit is equal to:

   ☐ 0 volts
   ☐ $R \times C$
   ☐ $Q \times V_T$
   ☐ $V_T - V_R$
   ☐ none of these

10. If the series resistance in a series resonant *RLC* circuit is decreased, the Q of the circuit will:

    ☐ increase
    ☐ decrease
    ☐ remain the same

PARALLEL RESONANCE

Objectives

You will connect several ac *RLC* circuits that illustrate the important characteristics of parallel resonant circuits.

In completing these projects, you will connect circuits, make measurements, perform calculations, draw conclusions, and be able to answer questions about the following items related to parallel resonant circuits.

- Definition of resonance
- Relationships of V, I, R, Z, θ, Q, bandwidth, and frequency in parallel *RLC* circuits
- Finding a circuit's resonant frequency
- Outstanding characteristics of a circuit at parallel resonance

PROJECT/TOPIC CORRELATION INFORMATION

PROJECT		TEXT CHAPTER	SECTION	RELATED TEXT TOPIC(S)
62	V, I, R, Z, and θ Relationships when $X_L = X_C$	22	22-6	Parallel Resonance Characteristics
63	Q and Impedance in a Parallel Resonant Circuit	22	22-8	Effect of a Coupled Load on the Tuned Circuit Q and the Resonant Rise of Impedance
64	Bandwidth Related to Q	22	22-10	Selectivity, Bandwidth, and Bandpass

PROJECT 62

PARALLEL RESONANCE
V, I, R, Z, and θ Relationships when $X_L = X_C$

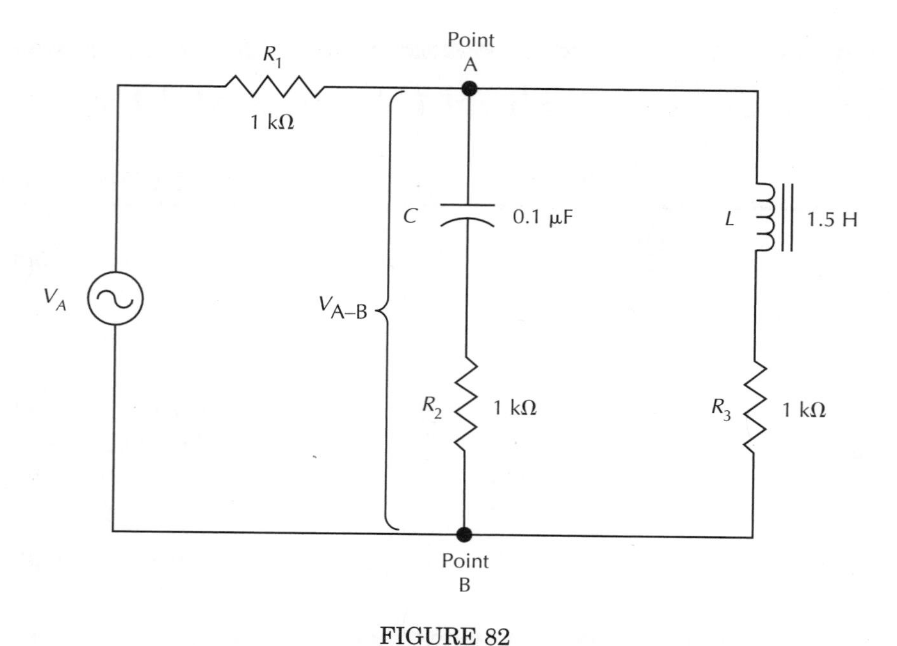

FIGURE 82

PROJECT PURPOSE

To provide experience in performing analysis of a parallel resonant circuit relative to the key electrical parameters. Additionally, to verify that a parallel resonant circuit acts inductively, below the resonant frequency, and capacitively, above the resonant frequency.

PARTS NEEDED

- ☐ DMM/VOM (2)
- ☐ Function generator or audio oscillator
- ☐ CIS
- ☐ Capacitor(s):
  0.1 μF
- ☐ Inductor(s):
  1.5 H, 95 Ω (or approximate)
- ☐ Resistor(s):
  1 kΩ (3)

Quickly reviewing, recall that X_L is directly proportional to frequency, and X_C is inversely proportional to frequency. For any given circuit containing L and C, there is a frequency where $X_L = X_C$, often referred to as the *resonant frequency*. Some interesting phenomena occur at resonance, and we will observe some of these for the parallel LC circuit. For convenience we will use 1-kΩ resistors as current indicators. They do have an effect on the operation of the circuit; however, not to the extent that we cannot observe some of the resonance phenomena.

ACTIVITY	OBSERVATION	CONCLUSION
1. Connect the initial circuit as shown in Figure 82.	—	—
2. Adjust V_A for 3 volts V_{A-B} *between points A and B*. Monitor V_1 with a voltmeter and adjust the frequency for *minimum* V_1. Check that V_{A-B} is still 3 volts. Use the measured V_1 and calculate I_T. Measure V_2 and calculate I_C. Measure V_3 and calculate I_L.	V_{A-B} = ______ volts V_1 = ______ volts I_T = ______ mA V_2 = ______ volts I_C = ______ mA V_3 = ______ volts I_L = ______ mA $f \cong$ ______ Hz	Are I_C and I_L close to being equal at the frequency of minimum V_1? ______. Would they be precisely equal if we had a pure L and C and ideal measurement conditions? ______. Does I_T equal the sum of the branch currents? ______. This is because I_C and I_L are ______ ______. Since I_C leads V_C by 90 degrees, and I_L lags V_L by about 90 degrees, then I_C and I_L must be about ______ degrees out of phase with each other. Theoretically, if I_C and I_L were exactly equal and 180 degrees out of phase, the resultant I_T would be ______. This means that Z would be ______. In practical circuits, however, Z is maximum at resonance, and its value depends on the Q of the circuit. $Z = Q \times X_L$ (or X_C).
3. Calculate the Z of the parallel resonant circuit using the resultant I_T of the branches as determined from V_1 and the 3 volts V_{A-B} ($Z = V/I$).	Z = ______ ohms (approximately)	The X_C of the capacitor at the resonant frequency, according to the X_C formula, is about ______ ohms. If $X_L = X_C$, then X_L is about ______ ohms. Notice that Z is greater than either branch's opposition to current. This is in contrast to a purely resistive parallel circuit, where Z is less than the least R branch.

PROJECT 62 CONTINUED

PARALLEL RESONANCE V, I, R, Z, and θ Relationships when $X_L = X_C$ (Continued)

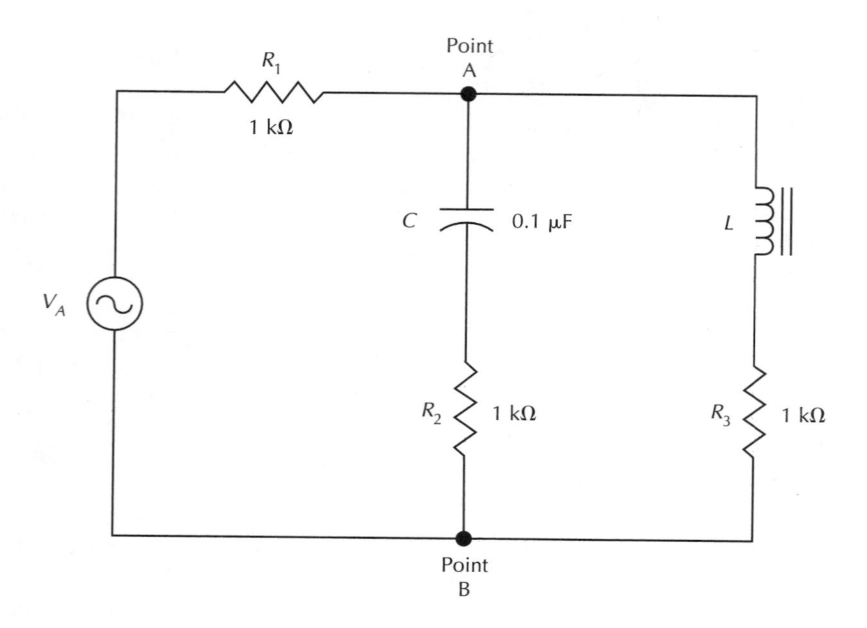

FIGURE 83

PROJECT PURPOSE

To provide experience in performing analysis of a parallel resonant circuit relative to the key electrical parameters. Additionally, to verify that a parallel resonant circuit acts inductively, below the resonant frequency, and capacitively, above the resonant frequency.

PARTS NEEDED

- ☐ DMM/VOM (2)
- ☐ Function generator or audio oscillator
- ☐ CIS
- ☐ Capacitor(s): 0.1 μF
- ☐ Inductor(s): 1.5 H, 95 Ω (or approximate)
- ☐ Resistor(s): 1 kΩ (3)

ACTIVITY	OBSERVATION	CONCLUSION
4. Keep the same circuit as shown in Figure 83.	—	—
5. Change the frequency to a new frequency that is well above f_r. Measure V_2 and V_3. Calculate I_C and I_L and determine whether the circuit is now acting inductively or capacitively.	I_C is now ____________ than I_L.	Since (I_C, I_L) ____________ is now greater than (I_C, I_L) ____________, it will cancel it out as far as I_T is concerned. Therefore, the resultant I_T is (*leading, lagging*) ____________. The circuit at this frequency is acting equivalent to a simple R ______ circuit. This means a parallel LC circuit at a frequency that is above resonance will act (*inductively, capacitively*) ____________. This is in contrast to a series LC circuit, which above resonance acts ____________.
6. Change the input frequency to a new frequency that is well below the resonant frequency of the circuit. Determine I_C and I_L and note whether the circuit is now acting inductively or capacitively.	I_C is now ____________ than I_L.	The circuit is now acting ________, since the predominant current is ____________. Thus, we conclude that a parallel LC circuit acts ____________ below resonance. This is in contrast to a series LC circuit, which acts ____________ below resonance. Summarizing our observations for a parallel resonant circuit (at resonance), we conclude that Z is (*minimum, maximum*) ____________, since the branch currents cancel, causing I_T to be (*minimum, maximum*) ________ at resonance. Since the reactive currents cancel, then the resultant circuit current is (*resistive, capacitive, inductive*) ____________ and (*in phase, out of phase*) ________ with V_A. This means the circuit phase angle is ___ degrees. In order for the branch currents to completely cancel, they must be ______ degrees out of phase with each other. At resonance, the parallel LC circuit acts (*resistively, capacitively, inductively*) ________; above resonance acts (*resistively, capacitively, inductively*) ____________; and below resonance acts (*resistively, capacitively, inductively*) ____________.

PROJECT
63

PARALLEL RESONANCE
Q and Impedance in a Parallel Resonant Circuit

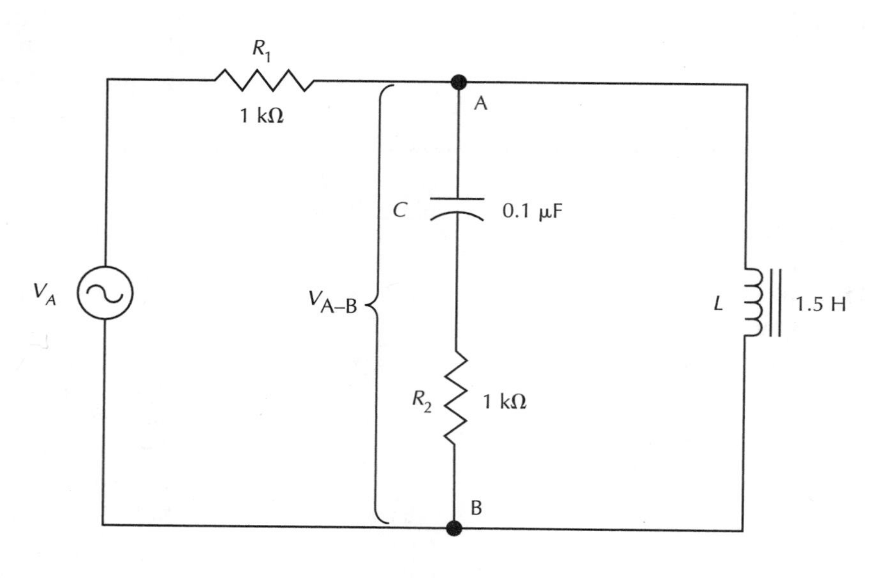

FIGURE 84

PROJECT PURPOSE

To verify that Z is maximum at resonance due to the approximate 180-degree phase-relationship of opposite reactance parallel branch currents. To show that the Z of the parallel resonant network is related to the circuit Q.

PARTS NEEDED

- ☐ DMM/VOM (2)
- ☐ Function generator or audio oscillator
- ☐ CIS
- ☐ Capacitor(s):
  0.1 μF
- ☐ Inductor(s):
  1.5 H, 95 Ω (or approximate)
- ☐ Resistor(s):
  1 kΩ (2)
  10 kΩ

It should be noted that the Q of a parallel resonant circuit is often considered as the ratio of Z/X_L (or Z/X_C), and that the ratio of branch current to I_T is approximately this same number. For convenience, we will use the ratio of I_C to I_T for determining Q in this project. We are once again using 1-kΩ resistors for easy current monitoring.

ACTIVITY	OBSERVATION	CONCLUSION
1. Connect the initial circuit as shown in Figure 84.	—	—
2. Adjust V_A for 3 volts *between points A and B*. Monitor V_1 and adjust the frequency for minimum V_1 (resonance). Check that V_{A-B} is still 3 volts. Determine I_T and I_C from V_1 and V_2, respectively.	V_{A-B} = ________ volts V_1 = ________ volts I_T = ________ mA V_2 = ________ volts I_C = ________ mA	I_C is approximately ________ times greater than I_T. This indicates a Q of about ________. The X_C of the capacitor at the resonant frequency, if calculated by the X_C formula, is about ________ ohms. If the circuit Z is calculated from the formula $Z = Q \times X_C$, the value of Z is ________ ohms. If we calculate Z by Ohm's law using the measured V_A and calculated I_T, the result is Z = ________ ohms. Are the two results reasonably comparable considering the measurement errors, and so on? ____.
3. If time permits, connect a 10-kΩ resistor in parallel with points A and B and repeat the procedures of step 2 to determine the new Q and Z. Note if Q and Z increased or decreased with the shunt 10-kΩ present in the circuit.	Q ________ Z ________ ohms	We may conclude that the lower the R in shunt with a parallel LC circuit, the (*higher, lower*) ________ the Q and Z will be.

PROJECT 64

PARALLEL RESONANCE
Bandwidth Related to Q

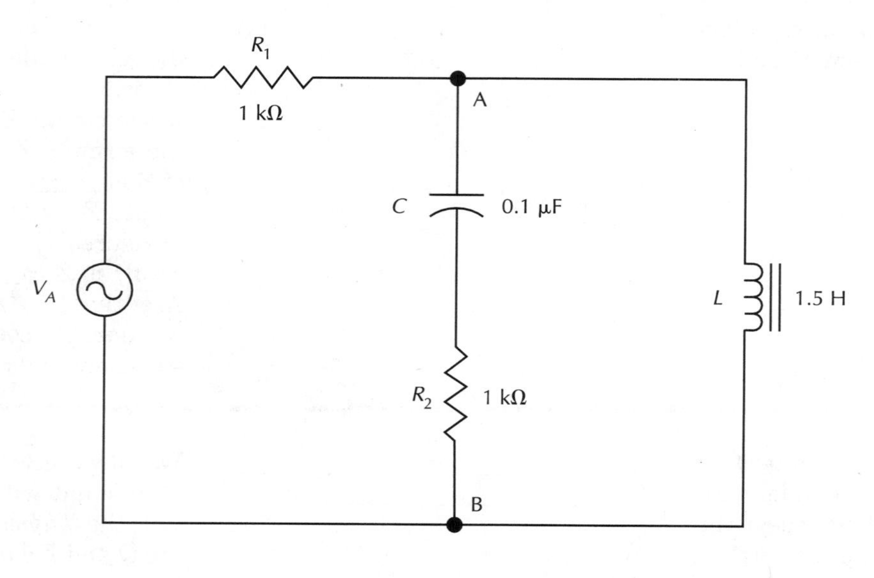

FIGURE 85

PROJECT PURPOSE

To provide experience in making measurements and performing analysis of a parallel resonant circuit relative to bandwidth and Q. To note how Q and Z are changed when a shunt resistance is introduced to the circuit.

PARTS NEEDED

- ☐ DMM/VOM (2)
- ☐ Function generator or audio oscillator
- ☐ CIS
- ☐ Capacitor(s):
  0.1 μF
- ☐ Inductor(s):
  1.5 H, 95 Ω (or approximate)
- ☐ Resistor(s):
  1 kΩ (2)
  27 kΩ

Bandwidth for a parallel resonant circuit might be defined as the difference in frequency between two frequencies (one below resonance and one above), at which the circuit impedance is 70.7% of the maximum impedance, which occurs at resonance. When the circuit Z is 0.707 of maximum, the circuit current will be 1.414 times I minimum. For convenience, we will again use the 1-kΩ current-indicating resistors in the circuit and will utilize the 1.4 concept discussed in the previous sentence to determine where Z is 70.7% of maximum.

ACTIVITY	OBSERVATION	CONCLUSION
1. Connect the initial circuit as shown in Figure 85.	—	—
2. Adjust V_A for 3 volts *between points A and B*. Monitor V_1 and adjust the frequency for minimum V_1 (resonance). Check that $V_{A\text{–}B}$ is still 3 volts. Determine I_T and I_C from V_1 and V_2, respectively. Also determine Q from the formula: $Q = I_C/I_T$.	$V_{A\text{–}B}$ = ______ volts V_1 = ______ volts I_T = ______ mA V_2 = ______ volts I_C = ______ mA Q = ______ (approximately)	Since I_T is the resultant of the two reactive branch currents, we may determine the Z at resonance of the parallel resonant circuit by dividing the 3.0 volts $V_{A\text{–}B}$ by I_T. Z for the parallel resonant circuit thus equals approximately ______ ohms.
3. Calculate what the current will be at the two frequencies when $Z = 0.707 \times Z_{max}$ by multiplying the I_T at resonance by 1.4.	I at 70.7% Z points = ______ mA.	If the impedance is 70.7% of Z at resonance, then the current will be 1.4 times I_{min} because 1 divided by 0.707 = ______.
4. Keeping $V_{A\text{–}B}$ at 3 volts, find the two frequencies at which Z is 0.707 of Z_{max} by monitoring I_T (V_1). One frequency should be above the resonant frequency, one below. Determine the bandwidth from f_{hi}–f_{lo}.	Bandwidth = ______ Hz (approximately)	If bandwidth is determined from the bandwidth = f_r/Q formula, using the value of Q from step 2, the value would be approximately ______ Hz. Does the bandwidth in this step reasonably correlate to the f_r/Q formula results, all things considered? ______.
5. If time permits, connect a 27-kΩ resistor in shunt with the parallel LC circuit (between points A and B) and repeat the procedures of step 4 to determine bandwidth.	The bandwidth with a shunt 27-kΩ is (*more, less*) ______ than without it.	If the bandwidth is greater, the Q of the circuit must have (*increased, decreased*) ______ according to the Bandwidth = f_r/Q formula.

PARALLEL RESONANCE

Complete the following review questions, indicating the appropriate response by placing a check in the box next to the correct answer.

1. The impedance of a parallel LC circuit at resonance is:

   - ☐ maximum
   - ☐ minimum
   - ☐ neither of these

2. The resultant total current of a parallel LC circuit at resonance is:

   - ☐ maximum
   - ☐ minimum
   - ☐ neither of these

3. The current "through" either reactive branch of a parallel LC circuit at resonance is equal to:

   - ☐ V_A/Z_T
   - ☐ $Q \times X_L$
   - ☐ $Q \times I_T$
   - ☐ f_r/Q
   - ☐ none of these

4. The impedance of a parallel LC circuit at resonance is equal to:

   - ☐ V_A/I_C
   - ☐ $Q \times X_L$
   - ☐ X_C
   - ☐ X_L
   - ☐ $X_L + X_C$
   - ☐ none of these

5. At a frequency higher than the resonant frequency of a parallel LC circuit, the circuit acts somewhat:

   - ☐ inductive
   - ☐ capacitive

6. At resonance, the phase angle between V_A and I_T for a parallel LC circuit is:

   - ☐ 0 degrees
   - ☐ 45 degrees
   - ☐ 90 degrees
   - ☐ 180 degrees
   - ☐ none of these

7. If the Q of a parallel resonant circuit is increased, the bandwidth will:

   ☐ increase
   ☐ decrease
   ☐ remain the same

8. If the Q of a parallel resonant circuit is decreased, the Z will:

   ☐ increase
   ☐ decrease
   ☐ remain the same

9. At a frequency lower than the resonant frequency, a parallel LC circuit acts somewhat:

   ☐ inductive
   ☐ capacitive

10. If the value of shunt resistance in parallel with a parallel LC circuit is changed, will the resonant frequency change?

   ☐ Yes
   ☐ No

THE SEMICONDUCTOR DIODE

Objectives

You will connect circuits that illustrate the conduction characteristics of semiconductor diodes and the operation of diode clipper circuits. You will also have a chance to test semiconductor diodes with a VOM ohmmeter and a DMM diode test.

In completing these projects, you will connect circuits, make measurements, perform calculations, draw conclusions, and answer questions about the following items related to semiconductor diodes:

- Forward and reverse bias
- Diode conduction characteristics
- The semiconductor diode symbol
- Typical forward-biased voltage drop
- VOM ohmmeter and DMM tests for semiconductor diodes
- Operation of series and parallel diode clippers

PROJECT/TOPIC CORRELATION INFORMATION

PROJECT	TEXT CHAPTER	SECTION	RELATED TEXT TOPIC(S)
65 Forward and Reverse Bias, and I vs. V	24	24-1	P-N Junction Diodes
66 VOM Ohmmeter Diode Tests and DMM Diode Checker Tests	24	24-2	Rectifier Diodes
67 Diode Clipper Circuits	24	24-3	Diode Clipper Circuits

PROJECT 65

THE SEMICONDUCTOR DIODE
Forward and Reverse Bias, and I vs. V

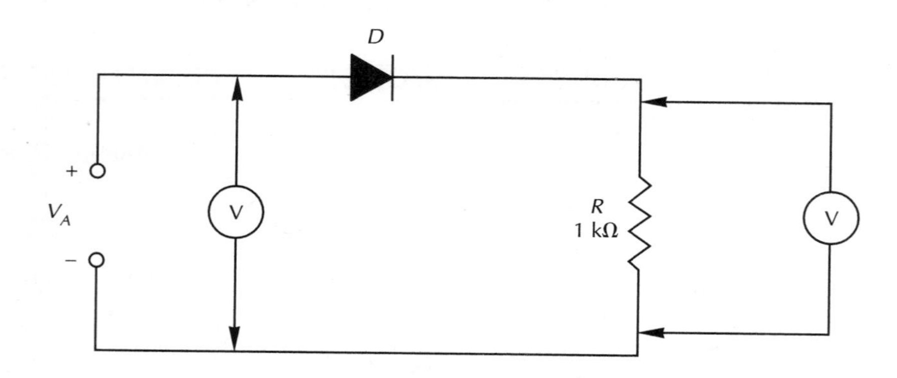

FIGURE 86

PROJECT PURPOSE

To demonstrate the current-controlling characteristics of a diode when forward-biased and reverse-biased.

PARTS NEEDED

- ☐ DMM/VOM (2)
- ☐ VVPS (dc)
- ☐ CIS
- ☐ Diode: Silicon, 1-amp (such as 1N4002)
- ☐ Resistor:
  1 kΩ

For this project you will use the voltage drop across a 1-kΩ resistor in series with the diode as a means for monitoring the current. Since this is a 1-kΩ resistor, the amount of current in milliamperes is the same as the values you read for the voltage drop across the resistor. Example: If you read 22 volts across the 1-kΩ resistor, the current is equal to 22 mA.

Ideally, you will use two meters for this experiment: one for monitoring the value of V_A and the second for measuring the voltages across the resistor and diode.

ACTIVITY	OBSERVATION	CONCLUSION
1. Connect the initial circuit as shown in Figure 86.	—	—
2. Apply 5 volts to the circuit with the anode connected to the positive side of the source. Determine the circuit current from V_R for the forward-biased diode.	I = ______________ mA (approximately)	Is the diode conducting with the anode positive with respect to the cathode? __________. The voltage drop across the forward-biased diode is approximately ______ volts.
3. Reverse the connections to the diode so the positive side of the source will be connected to the cathode of the diode. Determine the circuit current for V_R for the reverse-biased diode.	I = ______________ mA	Is the diode conducting with the cathode positive with respect to the anode? ______________. For the reverse-biased condition, the diode is acting like (*an open*, *a short*) __________ circuit. The resistance of the reverse-biased diode is virtually __________. Since V_R = 0 volts, the voltage across the diode must equal __________.
4. Reverse the diode connections once more in order to forward bias it, then increase V_A in small steps, as indicated in the Observation column, and note the current at each step.	For V_A = 0.5 V: I = __________ mA For V_A = 1 V: I = __________ mA V_A = 2 V: I = __________ mA V_A = 3 V: I = __________ mA V_A = 4 V: I = __________ mA V_A = 5 V: I = __________ mA V_A = 6 V: I = __________ mA V_A = 7 V: I = __________ mA V_A = 8 V: I = __________ mA V_A = 9 V: I = __________ mA V_A = 10 V: I = __________ mA	As V_A was increased, the current (*increased*, *decreased*) __________. We can deduce, therefore, that diode conduction is (*directly*, *inversely*) ____________ proportional to V_A when forward biased. What is the approximate dc resistance of the diode when V_A is 7 volts? __________ ohms. We conclude from the preceding that a diode will conduct when the anode is (*positive*, *negative*) __________ with respect to the cathode.

PROJECT

65

CONTINUED

THE SEMICONDUCTOR DIODE
Forward and Reverse Bias, and *I* vs. *V* *(Continued)*

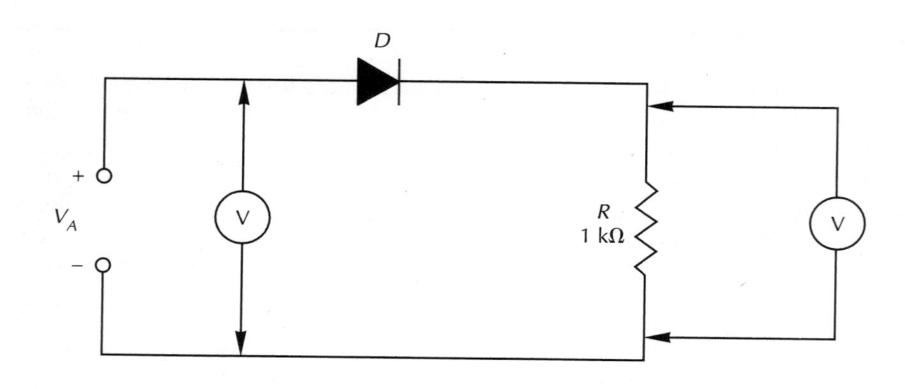

FIGURE 86

PROJECT PURPOSE

To demonstrate the current-controlling characteristics of a diode when forward-biased and reverse-biased.

PARTS NEEDED

- ☐ DMM/VOM (2)
- ☐ VVPS (dc)
- ☐ CIS
- ☐ Diode: Silicon, 1-amp (such as 1N4002)
- ☐ Resistor:
  1 kΩ

ACTIVITY	OBSERVATION	CONCLUSION
5. With the diode forward biased, measure the voltage drop across the diode for each increase in V_A as shown in the Observation column.	$V_A = 0.5$ V: V_D = _______________ V $V_A = 1$ V: V_D = _______________ V $V_A = 2$ V: V_D = _______________ V $V_A = 4$ V: V_D = _______________ V $V_A = 10$ V: V_D = _______________ V	As V_A reached a certain level of forward biasing, the voltage drop across the diode *(decreased, increased, stayed about the same)* _______________. This implies that the resistance of the diode changed *(linearly, nonlinearly)* __________ as V_A was increased.

PROJECT 66

THE SEMICONDUCTOR DIODE
VOM Ohmmeter Diode Tests and DMM Diode Checker Tests

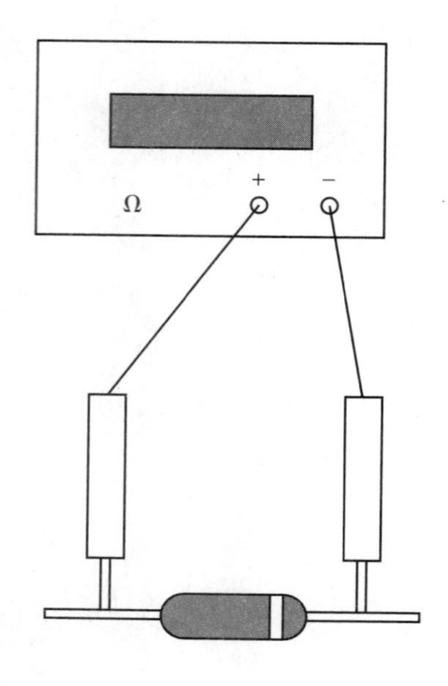

FIGURE 87

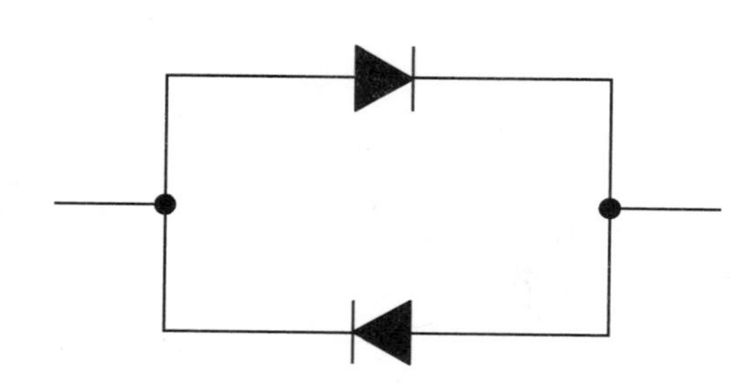

FIGURE 88

PROJECT PURPOSE

For this project you will use a VOM ohmmeter and a DMM diode checker to test the operating condition of silicon diodes. A good diode is one that has a reverse resistance that is many times greater than its forward resistance. For the VOM tests, you will see that it is the ratio of the reverse-to-forward resistance that is important, and not the actual resistance values. For the DMM diode checker, you will use the criteria defined in the operator's manual for your DMM.

PARTS NEEDED

- ☐ DMM/VOM
- ☐ Diodes: Silicon, 1-amp rating (4)

ACTIVITY	OBSERVATION	CONCLUSION
1. Set the VOM ohmmeter range selector switch for a mid-range setting: approximately the $R \times 100$ range. Devise a way to identify your four silicon diodes as Diode #1, Diode #2, Diode #3, and Diode #4.	—	—
2. Connect the VOM ohmmeter leads to Diode #1, positive lead to the anode and negative lead to the cathode (Figure 87). Determine this resistance reading, $R_{forward}$ for Diode #1. Reverse the ohmmeter connections to the diode so that the positive lead is now to the cathode and the negative lead to the anode. Note this resistance as $R_{reverse}$ for Diode #1. Finally, calculate the ratio of $R_{reverse}$-to-$R_{forward}$ as R_{ratio}: $R_{ratio} = R_{reverse} \div R_{forward}$ Repeat these measurements for each of the four diodes.	Diode #1: $R_{forward}$ = ________ Ω $R_{reverse}$ = ________ Ω R_{ratio} = ________ Diode #2: $R_{forward}$ = ________ Ω $R_{reverse}$ = ________ Ω R_{ratio} = ________ Diode #3: $R_{forward}$ = ________ Ω $R_{reverse}$ = ________ Ω R_{ratio} = ________ Diode #4: $R_{forward}$ = ________ Ω $R_{reverse}$ = ________ Ω R_{ratio} = ________	A good diode should have a (*high, low*) ________ reverse-to-forward resistance ratio. All good diodes of the same type should have exactly the same set of resistance readings and resistance ratios (*True* or *False*) ________.
3. Set the VOM ohmmeter for a high range, such as the $R \times 1\ M\Omega$ range.	—	—
4. Determine the $R_{forward}$ and $R_{reverse}$ values for Diode #1. Use these values to calculate R_{ratio}.	Diode #1 (high-resistance ohmmeter range): $R_{forward}$ = ________ Ω $R_{reverse}$ = ________ Ω R_{ratio} = ________	Diode forward and reverse resistance readings for a good diode will be (*substantially the same, different*) ________ when measured on a mid-range and on a high-range ohmmeter scale.

PROJECT **66** CONTINUED

THE SEMICONDUCTOR DIODE
VOM Ohmmeter Diode Tests and DMM Diode Checker Tests
(Continued)

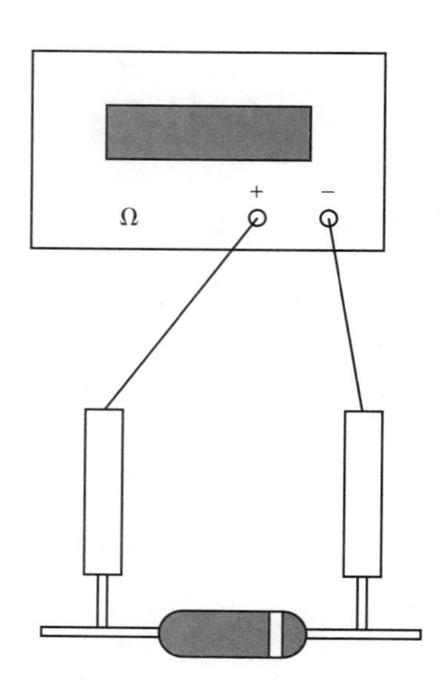

FIGURE 87

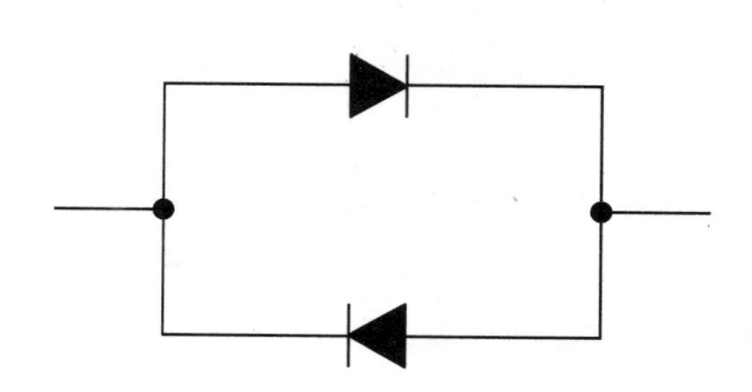

FIGURE 88

PROJECT PURPOSE

For this project you will use a VOM ohmmeter and a DMM diode checker to test the operating condition of silicon diodes. A good diode is one that has a reverse resistance that is many times greater than its forward resistance. For the VOM tests, you will see that it is the ratio of the reverse-to-forward resistance that is important, and not the actual resistance values. For the DMM diode checker, you will use the criteria defined in the operator's manual for your DMM.

PARTS NEEDED

- ☐ DMM/VOM
- ☐ Diodes: Silicon, 1-amp rating (4)
- ☐ Shorted diode
- ☐ 1 MΩ resistor

ACTIVITY	OBSERVATION	CONCLUSION
5. Set the VOM ohmmeter range selector switch for a lower-range setting, such as the $R \times 10$ range.	—	—
6. Determine the $R_{forward}$ and $R_{reverse}$ values for Diode #1. Use these values to calculate R_{ratio}.	Diode #1 (low-resistance ohmmeter range): $R_{forward}$ = ____________ Ω $R_{reverse}$ = ____________ Ω R_{ratio} = ____________	Diode forward and reverse resistance readings for a good diode will be (*substantially the same*, *different*) ____________ when measured on a mid-range and on a low-range ohmmeter scale. Given the choice of testing a diode on a low-range, mid-range, or high-range scale on an ohmmeter, explain which would provide the most reliable results. Why? ____________ ____________
7. To simulate a shorted diode, connect two diodes in inverse parallel (back-to-back) as shown in Figure 88.	—	The ratio of reverse-to-forward resistance of a shorted diode is a (*negative*, *low*, *high*) ____________ value.
8. Using the VOM ohmmeter setting that provided the highest R_{ratio} results in steps 1 through 4, determine the values of $R_{forward}$ and $R_{reverse}$, and calculate R_{ratio}.	Simulated shorted diode: $R_{forward}$ = ____________ Ω $R_{reverse}$ = ____________ Ω R_{ratio} = ____________	—

Perform the following steps only if you have access to a DMM that has a diode testing function.

ACTIVITY	OBSERVATION	CONCLUSION
9. Set the DMM to the diode test position. Select a good diode that you previously tested. Connect the diode to the DMM, with the negative lead to the cathode and the positive lead to the anode. Record the DMM reading.	Forward voltage for a good diode: $V_{forward}$ = ____________ V	For a good diode, the forward DMM reading is relatively (*high*, *low*) ____________.
10. Switch the leads to the diode, so that the negative lead of the DMM is connected to the anode and the positive lead to the cathode. Record the DMM reading.	Reverse voltage for a good diode: $V_{reverse}$ = ____________ V	For a good diode, the reverse DMM reading is relatively (*high*, *low*) ____________.

PROJECT 66 CONTINUED

THE SEMICONDUCTOR DIODE
VOM Ohmmeter Diode Tests and DMM Diode Checker Tests
(Continued)

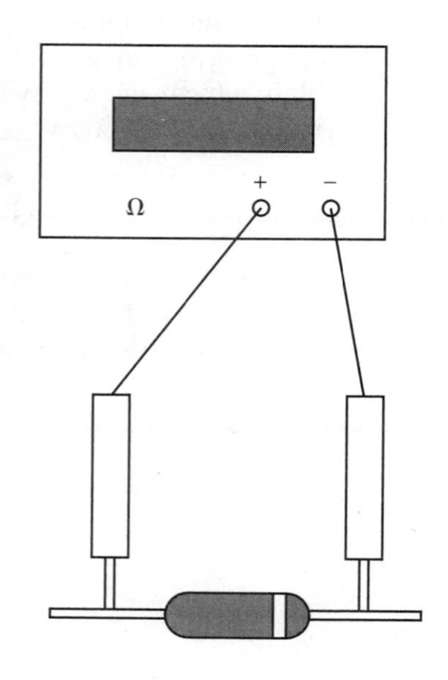

FIGURE 87

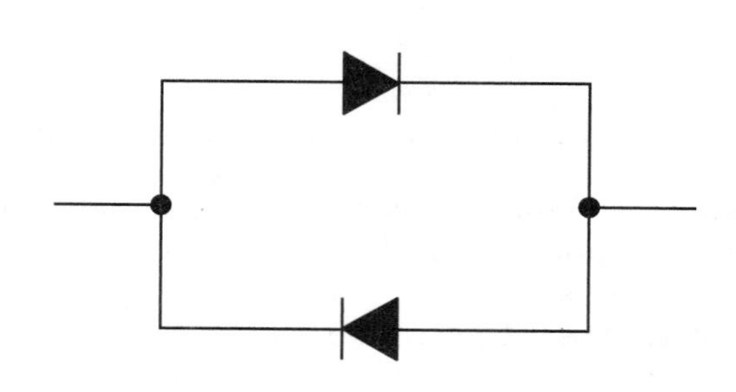

FIGURE 88

PROJECT PURPOSE

For this project you will use a VOM ohmmeter and a DMM diode checker to test the operating condition of silicon diodes. A good diode is one that has a reverse resistance that is many times greater than its forward resistance. For the VOM tests, you will see that it is the ratio of the reverse-to-forward resistance that is important, and not the actual resistance values. For the DMM diode checker, you will use the criteria defined in the operator's manual for your DMM.

PARTS NEEDED

- ☐ DMM/VOM
- ☐ Diodes: Silicon, 1-amp rating (4)
- ☐ Shorted diode
- ☐ 1 MΩ resistor

ACTIVITY	OBSERVATION	CONCLUSION
11. Replace the diode with one that is known to be shorted. (Do not use the "simulated diode" previously described in step 7.) Connect the diode to check the forward voltage and record the results. Then connect the diode to check the reverse voltage and record the results.	Voltages for a shorted diode: $V_{forward}$ = ____________ V $V_{reverse}$ = ____________ V	For a shorted diode: The forward DMM reading is relatively (*high, low*) ____________. The reverse DMM reading is relatively (*high, low*) ____________.
12. Use a 1 MΩ resistor to simulate an open diode. Connect this "open diode" to the DMM and check the forward voltage drop. Record the result. Reverse the leads to the "open diode," read the reverse voltage, and record the result.	Voltages for an open diode: $V_{forward}$ = ____________ V $V_{reverse}$ = ____________ V	For an open diode: The forward DMM reading is relatively (*high, low*) ____________. The reverse DMM reading is relatively (*high, low*) ____________.

PROJECT 67

THE SEMICONDUCTOR DIODE
Diode Clipper Circuits

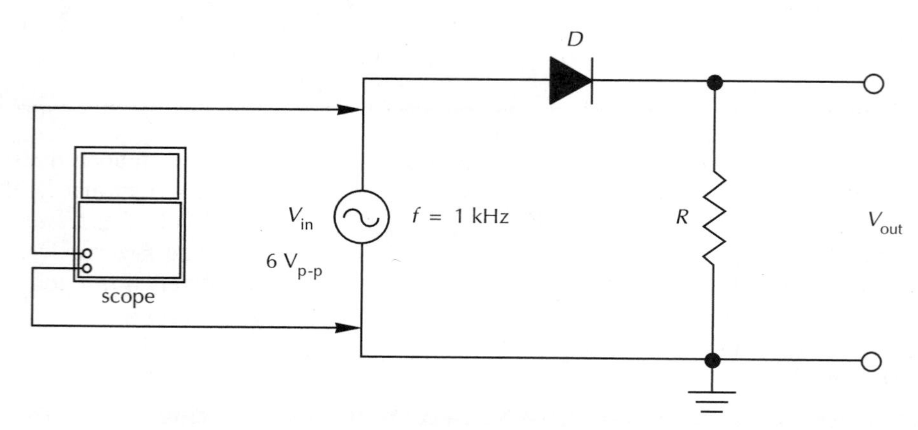

FIGURE 89

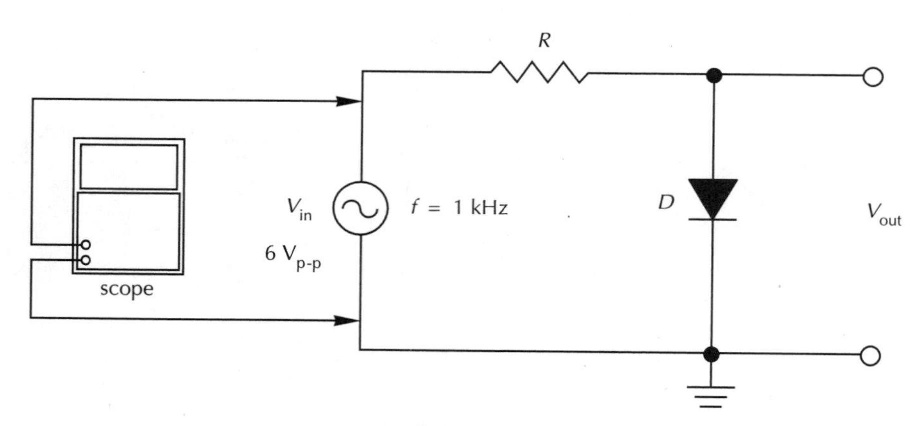

FIGURE 90

PROJECT PURPOSE

In this project you will observe the actual operation of all four possible types of diode clipper circuits.

PARTS NEEDED

- ☐ Dual-trace oscilloscope
- ☐ Function generator or audio oscillator
- ☐ CIS
- ☐ Diodes: Silicon (2)
- ☐ Resistor:
  27 kΩ

ACTIVITY	OBSERVATION	CONCLUSION
1. Connect the circuit exactly as shown in Figure 89.	—	—
2. Set the function generator (sine-wave mode) or audio oscillator to a frequency of 1 kHz and V_{in} of 6 V (peak-to-peak). Draw the waveform for V_{in}, clearly indicating the positive and negative peak voltage values.	Waveform: + 0 – V_{in} positive peak voltage = _____ V V_{in} negative peak voltage = _____ V	—
3. Draw the waveform observed at V_{out}. Again, clearly indicate the positive and negative peak voltage values.	Waveform: + 0 – V_{out} positive peak voltage = _____ V V_{out} negative peak voltage = _____ V	This circuit is a (*series*, *parallel*) ________________ diode clipper. The output waveform indicates this circuit is a (*positive*, *negative*) ____________ clipper. During the positive half-cycle of V_{in}, the diode is (*forward*, *reverse*) ___________ biased and current (*is*, *is not*) ____________ flowing through the resistor.
4. Reverse the direction of the diode from that shown in Figure 89. Draw the waveform observed at V_{out}, and show the positive and negative peak voltage values.	Waveform: + 0 – V_{out} positive peak voltage = _____ V V_{out} negative peak voltage = _____ V	This circuit is a (*series*, *parallel*) ________________ diode clipper. The output waveform indicates this circuit is a (*positive*, *negative*) ____________ clipper. During the positive half-cycle of V_{in}, the diode is (*forward*, *reverse*) ___________ biased and current (*is*, *is not*) ____________ flowing through the resistor.

PROJECT
67
CONTINUED

THE SEMICONDUCTOR DIODE
Diode Clipper Circuits *(Continued)*

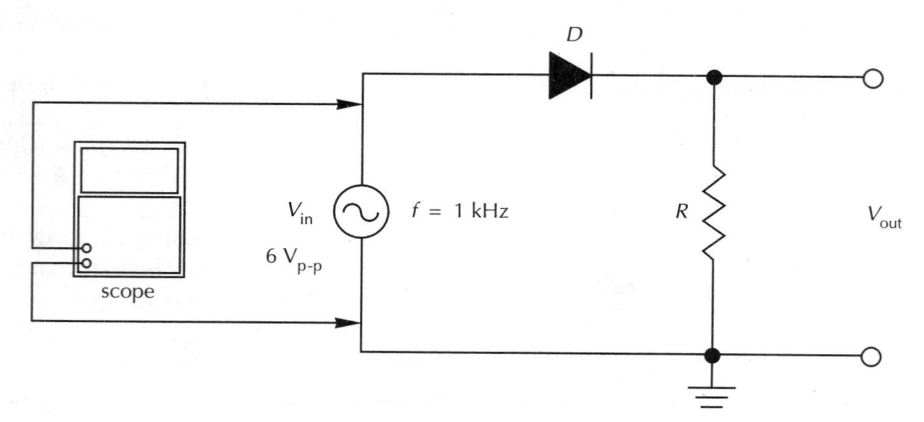

FIGURE 89

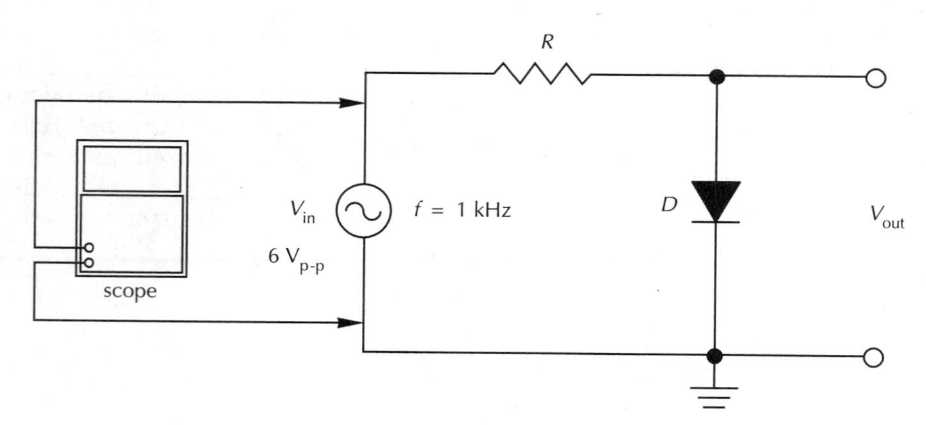

FIGURE 90

PROJECT PURPOSE

In this project you will observe the actual operation of all four possible types of diode clipper circuits.

PARTS NEEDED

- ☐ Dual-trace oscilloscope
- ☐ Function generator or audio oscillator
- ☐ CIS
- ☐ Diodes: Silicon (2)
- ☐ Resistor:
  27 kΩ

ACTIVITY	OBSERVATION	CONCLUSION
5. Connect the circuit exactly as shown in Figure 90. Make sure the input waveform is 1 kHz at 6 V peak-to-peak.	—	—
6. Draw the waveform observed at V_{out}, clearly indicating the positive and negative peak voltage values.	Waveform: + 0 – V_{out} positive peak voltage = _____ V V_{out} negative peak voltage = ____ V	This circuit is a (*series*, *parallel*) ________________ diode clipper. The output waveform indicates this circuit is a (*positive*, *negative*) ___________ clipper. During the positive half-cycle of V_{in}, the diode is (*forward*, *reverse*) ___________ biased and current (*is*, *is not*) ___________ flowing through the resistor.
7. Reverse the direction of the diode from that shown in Figure 90. Draw the waveform observed at V_{out}, and show the positive and negative peak voltage values.	Waveform: + 0 – V_{out} positive peak voltage = _____ V V_{out} negative peak voltage = ____ V	This circuit is a (*series*, *parallel*) ________________ diode clipper. The output waveform indicates this circuit is a (*positive*, *negative*) ___________ clipper. During the positive half-cycle of V_{in}, the diode is (*forward*, *reverse*) ___________ biased and current (*is*, *is not*) ___________ flowing through the resistor.

THE SEMICONDUCTOR DIODE

Complete the following review questions, indicating the appropriate response by placing a check in the box next to the correct answer.

1. In analyzing the symbol for the semiconductor diode, the

   - ☐ arrow is the cathode and the "flat bar" is the anode
   - ☐ arrow is the anode and the "flat bar" is the cathode
   - ☐ neither of these

2. When a diode is forward biased, the

   - ☐ anode is negative with respect to the cathode
   - ☐ cathode is positive with respect to the anode
   - ☐ anode is positive with respect to the cathode
   - ☐ none of these

3. A diode in which the cathode is more negative than the anode is

   - ☐ forward biased
   - ☐ reverse biased
   - ☐ neither of these

4. As the forward bias on a diode is increased, the current will

   - ☐ increase
   - ☐ decrease
   - ☐ remain the same

5. A reverse-biased rectifier diode acts like

   - ☐ a short
   - ☐ an open
   - ☐ a 10-kΩ resistor

6. The voltage drop across a forward-conducting silicon diode is approximately:

   - ☐ 0.25 V
   - ☐ 0.5 V
   - ☐ 0.7 V
   - ☐ 1.0 V
   - ☐ 2.0 V

7. When testing a good rectifier diode with an ohmmeter

   - ☐ the ratio of forward resistance to reverse resistance is high
   - ☐ the ratio of reverse resistance to forward resistance is high
   - ☐ the ratio of reverse resistance to forward resistance is low
   - ☐ none of these

8. When testing a shorted rectifier diode with an ohmmeter

   - ☐ the forward resistance and reverse resistance are both fairly low
   - ☐ the forward resistance and reverse resistance are both fairly high
   - ☐ the forward resistance and reverse resistance are significantly different
   - ☐ none of these

9. In a series diode clipper circuit, the voltage output is taken across the

   - ☐ power source
   - ☐ resistor
   - ☐ diode

10. Reversing the polarity of the diode in a parallel diode clipper circuit

   - ☐ reverses the clipping polarity
   - ☐ changes it to a series clipper circuit
   - ☐ shorts out the power source
   - ☐ has no effect on the circuit

SPECIAL-PURPOSE DIODES

Objectives

You will connect circuits that illustrate the special voltage-regulating and voltage-clipping characteristics of zener diodes, and the light-emitting qualities of light-emitting diodes.

In completing these projects, you will connect circuits, make measurements, perform calculations, draw conclusions, and answer questions about the following items related to zener diodes and light-emitting diodes (LEDs).

- Reverse-bias characteristics of zener diodes
- Zener-diode action when the applied voltage is below the rated zener breakdown voltage level
- Zener diode action when the applied voltage is above the rated zener breakdown voltage level
- The zener diode used in a clipping circuit
- Forward-bias characteristics of LEDs
- Values of series-limiting resistors

PROJECT/TOPIC CORRELATION INFORMATION

PROJECT	TEXT CHAPTER	SECTION	RELATED TEXT TOPIC(S)
68 Zener Diodes	24	24-4	Zener Diodes
69 Light-Emitting Diodes	24	24-5	Light-Emitting Diodes

PROJECT 68

SPECIAL-PURPOSE DIODES
Zener Diodes

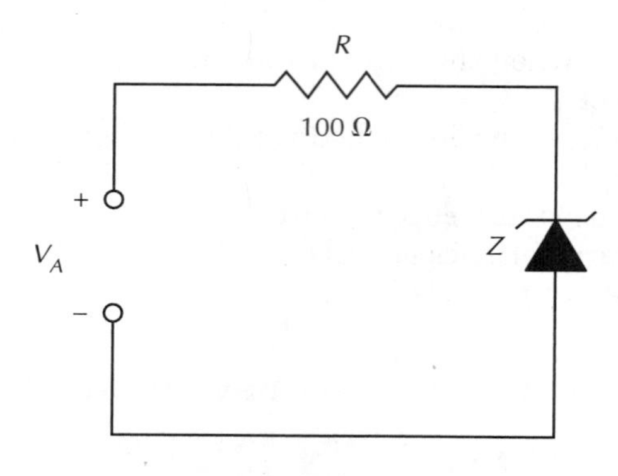

FIGURE 91

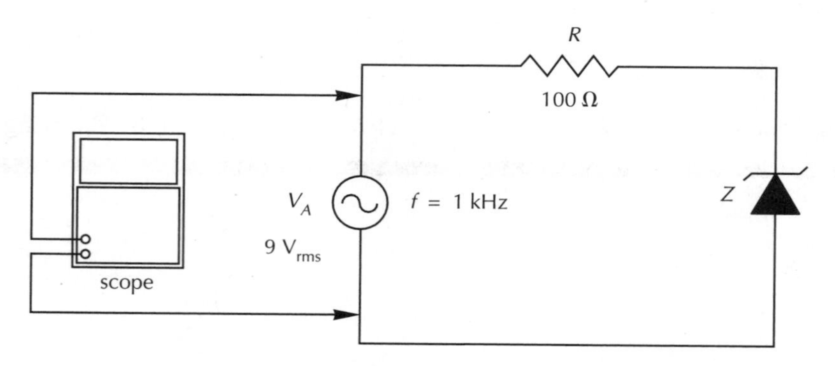

FIGURE 92

PROJECT PURPOSE

In this project you will observe the regulating and clipping action of a zener diode.

PARTS NEEDED

- ☐ DMM/VOM (2)
- ☐ Dual-trace oscilloscope
- ☐ VVPS (dc)
- ☐ CIS
- ☐ Function generator or audio oscillator
- ☐ Zener diode: 5.1 V, 1 W (1N4733 or equivalent)
- ☐ Resistor:
  100 Ω

You can closely approximate the amount of current flowing through the circuit (in mA) by measuring the voltage across the 100-Ω resistor and multiplying the value by 10. Example: If you read 2.2 volts across the resistor, the current through the circuit will be close to 22 mA. Ideally, you will use two meters for this experiment: one for monitoring V_A, and the second for taking measurements across the resistor and across the diode.

ACTIVITY	OBSERVATION	CONCLUSION
1. Connect the circuit shown in Figure 91.	—	The zener diode in this circuit is properly (*forward, reverse*) ______ biased. What is the rated zener voltage (V_Z) for this zener diode? ______ V.
2. Set V_A to 3 volts. Measure the voltage across the zener diode (V_D) and the voltage across the resistor (V_R), and determine the current through the circuit (I_R). Increase V_A in small steps, as indicated in the Observation column, and determine the values of V_D, V_R, and I_R for each step.	For V_A = 3 V: V_D = ______ V, V_R = ______ V, I_R = ______ mA For V_A = 4 V: V_D = ______ V, V_R = ______ V, I_R = ______ mA For V_A = 5 V: V_D = ______ V, V_R = ______ V, I_R = ______ mA For V_A = 6 V: V_D = ______ V, V_R = ______ V, I_R = ______ mA For V_A = 7 V: V_D = ______ V, V_R = ______ V, I_R = ______ mA	As long as V_A remains below V_Z, the voltage across the zener diode (*equals V_A, remains close to the diode's V_Z value, remains very close to zero volts*) ______. While V_A is above V_Z, the voltage across the zener diode (*equals V_A, remains close to the diode's V_Z value, remains very close to zero volts*) ______. While V_A remains below V_Z, the current through the zener diode (*changes with V_A, remains at a regulated level, remains very close to zero*) ______. Once V_A rises above V_Z, the current through the zener diode (*changes with V_A, remains at a regulated level, remains very close to zero*) ______.
3. Return V_A to 0 volts. Determine the exact zener breakdown voltage (V_Z) by gradually increasing the value of V_A while observing the voltage across the zener diode. Note the values where further increasing V_A no longer causes a significant increase in the zener voltage.	V_Z = ______ V	The voltage across a zener diode remains at the fixed V_Z level as long as V_A is (*below, above*) ______ V_Z.

PROJECT 68 CONTINUED

SPECIAL-PURPOSE DIODES
Zener Diodes *(Continued)*

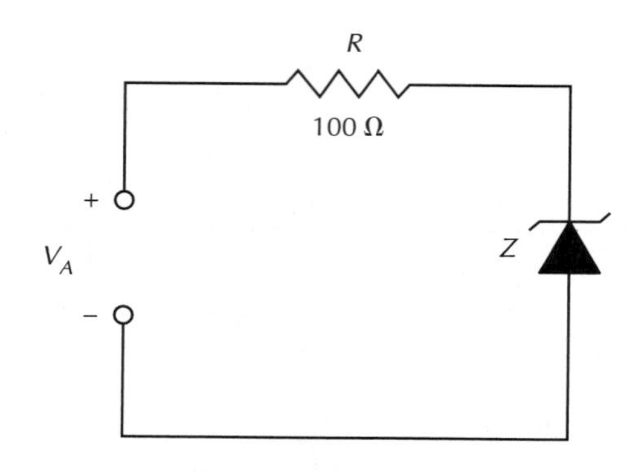

FIGURE 91

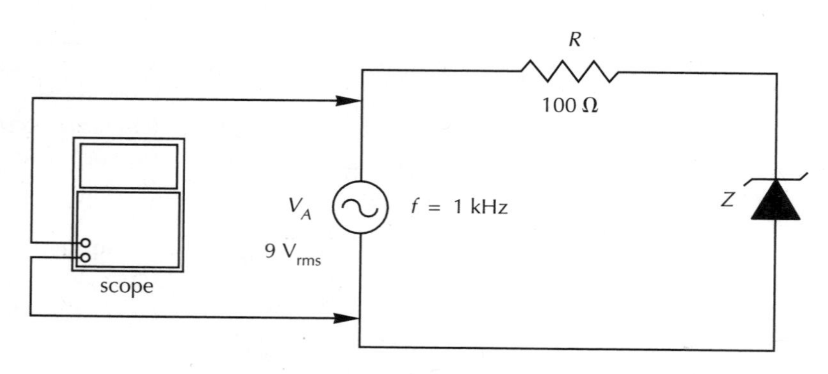

FIGURE 92

PROJECT PURPOSE

In this project you will observe the regulating and clipping action of a zener diode.

PARTS NEEDED

- ☐ DMM/VOM (2)
- ☐ Dual-trace oscilloscope
- ☐ VVPS (dc)
- ☐ CIS
- ☐ Function generator or audio oscillator
- ☐ Zener diode: 5.1 V, 1 W (1N4733 or equivalent)
- ☐ Resistor: 100 Ω

ACTIVITY	OBSERVATION	CONCLUSION
4. Replace the VVPS for V_A with the function generator or audio generator (Figure 92). Set the function generator (sine-wave mode) to 1 kHz at 9 V rms.	—	—
5. Sketch the waveform you find at V_A, indicating the positive and negative peak values.	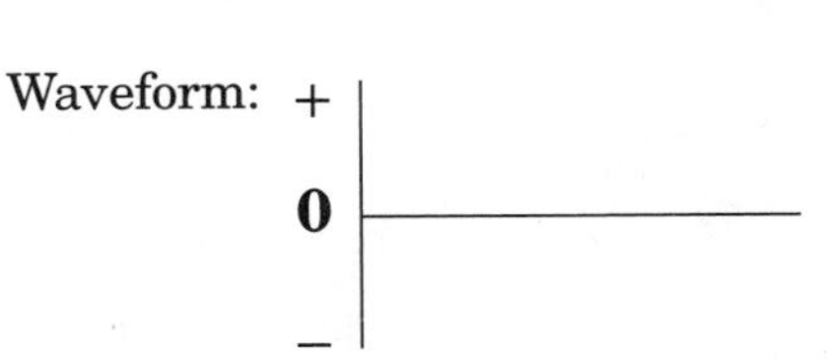 Positive peak V_A = ______ V Negative peak V_A = ______ V	—
6. Sketch the waveform you find across the zener diode. Indicate the peak positive and peak negative voltages.	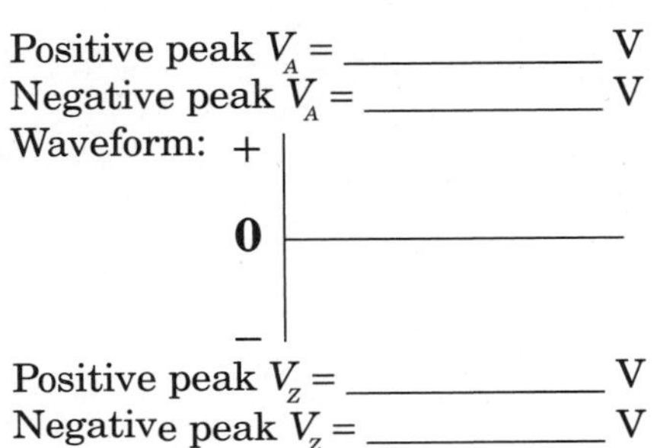 Positive peak V_Z = ______ V Negative peak V_Z = ______ V	Explain the shape of the waveform that occurs across the zener diode during the positive half-cycle of the input waveform. Account for the shape of the waveform that occurs across the zener diode during the negative half-cycle of the input waveform. Describe how this circuit could be considered a clipping circuit. (**NOTE**: Use space below for explanations.)

PROJECT 69

SPECIAL-PURPOSE DIODES
Light-Emitting Diodes

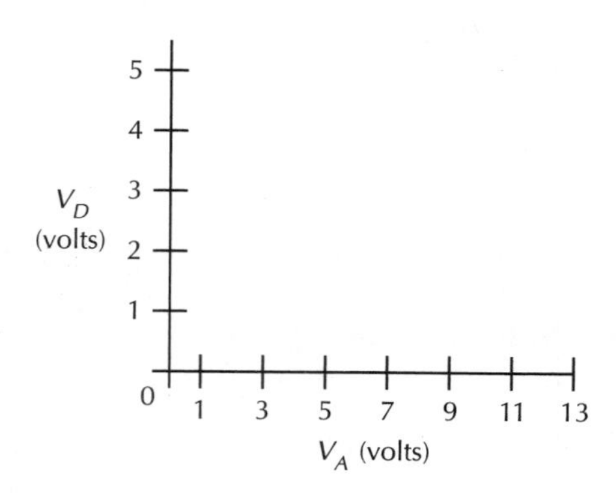

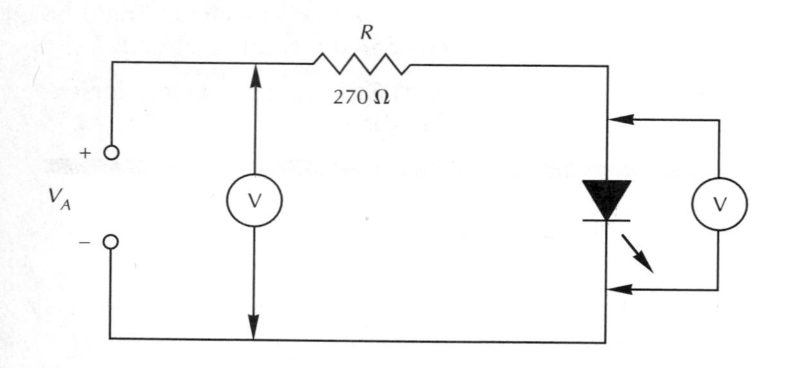

FIGURE 93

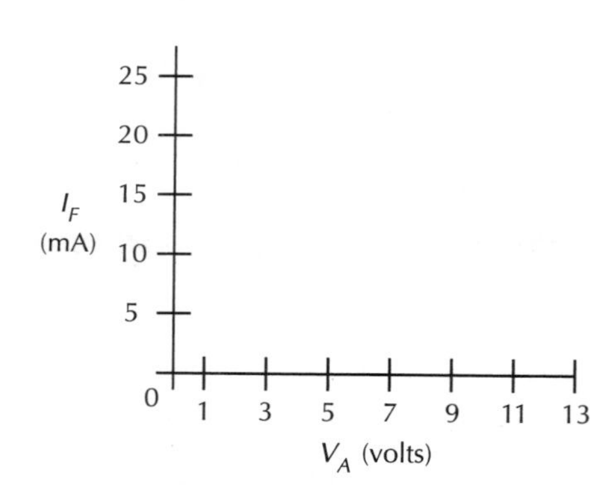

FIGURE 94

PROJECT PURPOSE

In this experiment you will observe the operation of a typical LED. Ideally, you will use two meters for this experiment: one to monitor the applied dc voltage (V_A) and a second to monitor the forward voltage drop across the LED (V_D).

PARTS NEEDED

- ☐ DMM/VOM (2)
- ☐ VVPS (dc)
- ☐ CIS
- ☐ LED: gallium-arsenide red, 20 mA
- ☐ Resistors:
  - 47 Ω,
  - 220 Ω
  - 270 Ω
  - 330 Ω

You will need the following formula to calculate the amount of forward current through the LED.

$$I_F = (V_A - V_D)/R$$

where:

I_F is the forward current through the diode
V_A is the voltage applied to the circuit
V_D is the forward voltage drop across the LED
R is the value of the resistor connected in series with the LED

ACTIVITY	OBSERVATION	CONCLUSION
1. Connect the circuit as shown in Figure 93.	—	—
2. Set V_A to 1 Vdc. Measure the voltage across the LED, calculate the amount of current flowing through the circuit, and note the intensity of the light (none, dim, moderately bright, bright, very bright). Increase V_A in steps indicated in the Observation column. Determine V_D and I_F, and note the brightness level in each case.	V_A = 1 V; V_D = ______ V, I_F = ____ mA, brightness = ____ V_A = 3 V; V_D = ______ V, I_F = ____ mA, brightness = ____ V_A = 5 V; V_D = ______ V, I_F = ____ mA, brightness = ____ V_A = 7 V; V_D = ______ V, I_F = ____ mA, brightness = ____ V_A = 9 V; V_D = ______ V, I_F = ____ mA, brightness = ____ V_A = 11 V; V_D = ______ V, I_F = ____ mA, brightness = ____ V_A = 13 V; V_D = ______ V, I_F = ____ mA, brightness = ____	Complete the graphs in Figure 94 which show how V_D increases with V_A, and how I_F increases with V_A. The brightness of the LED seems to correlate better with the amount (V_D, I_F) ______ than with the amount of (V_D, I_F) ______. Based on the data you observed, the forward junction potential for this LED is about ______ V.
3. Set V_A to 3 Vdc, replace R with a 47-Ω resistor, and complete the blanks in the Observation column. Repeat the steps for the values of V_A and R indicated.	V_A = 3 V; R = 47 Ω, I_F = ____ mA, brightness = ______ V_A = 6 V; R = 220 Ω, I_F = ____ mA, brightness = ______ V_A = 9 V; R = 330 Ω, I_F = ____ mA, brightness = ______	It is possible to use a given LED with just about any amount of applied dc as long as you select the correct value of series resistor for approximately 20 mA of forward current. (*True*, *False*) ______.

SPECIAL-PURPOSE DIODES

Complete the following review questions, indicating the appropriate response by placing a check in the box next to the correct answer.

1. A zener diode is normally connected into a dc circuit in a direction that causes it to

   - ☐ reverse biased
   - ☐ forward biased

2. When a zener diode is properly connected into a variable-voltage dc circuit, the zener conducts whenever the dc source

   - ☐ is greater than zero volts
   - ☐ exceeds the zener's forward junction potential
   - ☐ exceeds the zener's reverse breakdown voltage

3. A forward-biased zener diode behaves much the same as a forward-biased silicon rectifier diode.

   - ☐ True
   - ☐ False

4. When a zener diode rated at 12 V is operating from an 18-V source and the series resistor has a value of 100 Ω, the current through the zener diode is approximately

   - ☐ 12 mA
   - ☐ 18 mA
   - ☐ 60 mA
   - ☐ 120 mA

5. Used as a voltage clipper, a zener diode clips one polarity at the forward junction potential and the opposite polarity at the reverse breakdown potential.

   - ☐ True
   - ☐ False

6. The schematic symbols for an LED and a photodiode

   - ☐ are the same except the arrows point in different directions
   - ☐ are exactly the same
   - ☐ bear no resemblance to one another

7. An LED is connected into a dc circuit in a direction that causes it to

   - ☐ be reverse biased
   - ☐ gather light energy
   - ☐ be forward biased
   - ☐ break down and conduct backwards

8. The forward junction potential for an LED tends to be ________________ that of a rectifier diode.

   - ☐ less than
   - ☐ about the same as
   - ☐ greater than

9. The reverse breakdown voltage for an LED tends to be ________________ that of a rectifier diode.

   - ☐ less than
   - ☐ about the same as
   - ☐ greater than

10. Suppose you want to operate an LED from a 12-Vdc source, and the known specifications for the LED are $V_D = 1.7$ V when $I_F = 15$ mA. Which of the following practical resistors is closest to the correct value for the series resistor you should use?

    - ☐ 270 Ω
    - ☐ 330 Ω
    - ☐ 470 Ω
    - ☐ 680 Ω
    - ☐ 1 kΩ

POWER SUPPLIES

Objectives

You will connect half-wave and bridge rectifier circuits and observe their operation with and without capacitor filtering. You will also demonstrate the operation of a voltage doubler power supply.

In completing these projects, you will connect circuits, make measurements, perform calculations, draw conclusions, and be able to answer questions about the following items related to power supplies.

- Relationship of output ripple frequency to ac input frequency
- Relationship of output voltage to ac input voltage
- Comparison of filtered and unfiltered outputs
- Distinction between half- and full-wave rectification
- Relationship of the amount of output ripple to the amount of capacitor filtering

PROJECT/TOPIC CORRELATION INFORMATION

PROJECT	TEXT CHAPTER	SECTION	RELATED TEXT TOPIC(S)
70 Half-Wave Rectifier	25	25-2 25-3	Half-Wave Rectifier Circuits Capacitance Filters
71 Bridge Rectifier	25	25-2 25-3	The Bridge Rectifier Capacitance Filters
72 Voltage Multiplier	25	25-5	Basic Voltage Multiplier Circuits

PROJECT
70

POWER SUPPLIES
Half-Wave Rectifier

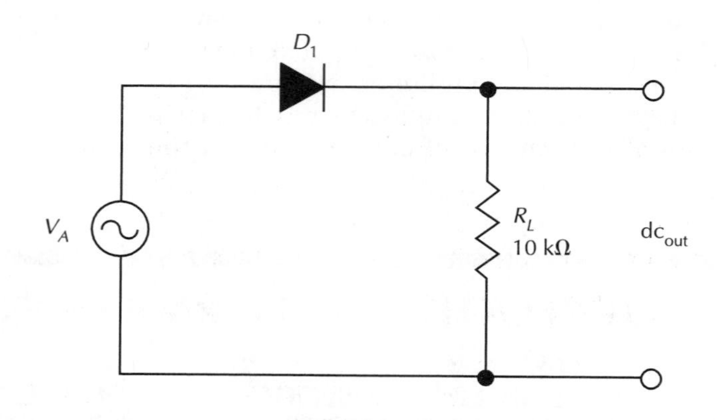

FIGURE 95

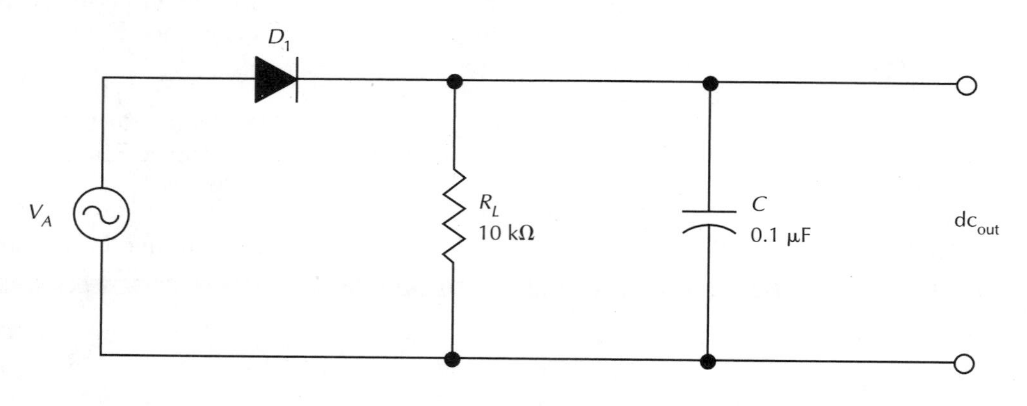

FIGURE 96

PROJECT PURPOSE

The purposes of this project are to demonstrate the action of a simple half-wave rectifier and the effects of capacitor filtering.

PARTS NEEDED

- ☐ DMM/VOM
- ☐ Oscilloscope
- ☐ CIS
- ☐ Function generator or audio oscillator
- ☐ Diode: Silicon, 1-amp rating
- ☐ Resistor:
  10 kΩ
- ☐ Capacitors:
  0.1 μF
  1.0 μF

The following formulas will be helpful for drawing proper conclusions about the results of your work:
Effective ac (rms) = $0.707 \times V_{max}$ (V_p)
Average value = $0.637 \times V_{max}$ (V_p)
Peak value = $1.414 \times V_{eff}$ (rms)
Peak value = V_{eff} (rms) ÷ 0.707

ACTIVITY	OBSERVATION	CONCLUSION
1. Connect the initial circuit as shown in Figure 95.	—	—
2. Set the function generator (sine-wave mode) or audio oscillator to a frequency of 100 Hz and V_A to 3 V_{rms}. Measure the dc voltage output across R_L (dc_{out}).	ac_{in} = ______ V dc_{out} = ______ V	The dc output voltage should be approximately what percentage of the rms input voltage? Approximately ______ %. Since the average voltage value over one ac alternation is $0.637 \times V_p$ and the effective value is $0.707 \times V_p$, then V_{avg} must be about ______ tenths of V_{eff}, because 0.637 is about ______ tenths of 0.707. Since the diode can conduct only half the time with ac applied, the average dc_{out} for the half-cycle the diode *does not* conduct is ______ V.
3. Connect the oscilloscope across R_L. Sketch the waveform for two complete cycles in the Observation column. Indicate the maximum and minimum voltage levels.	Waveform: + 0 –	For the half-cycle the diode does conduct, the peak V_{out} should be about ______ times V_{eff} (neglecting the small diode voltage drop). The end result is that the average dc_{out} over the *entire cycle* of ac input is about ______ times V_{eff} (again, neglecting the diode voltage drop).

PROJECT 70 CONTINUED

POWER SUPPLIES
Half-Wave Rectifier *(Continued)*

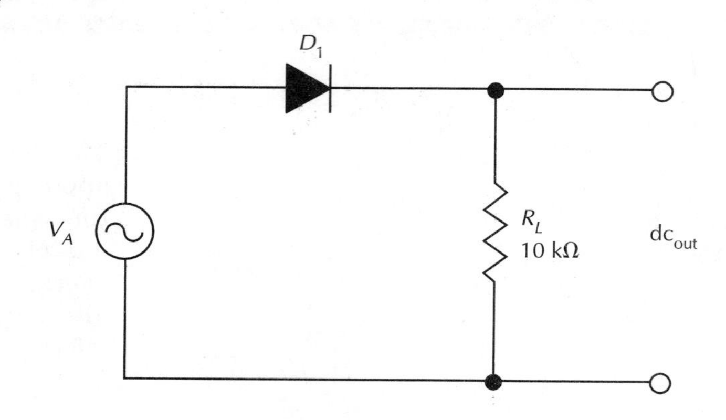

FIGURE 95

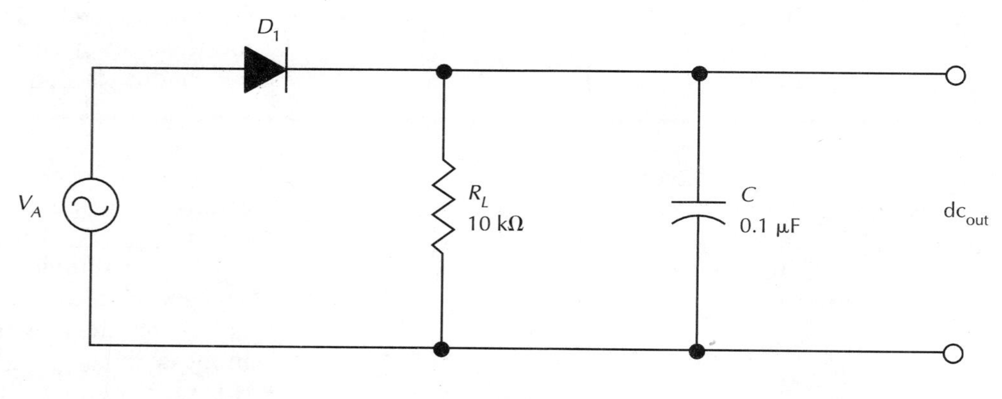

FIGURE 96

PROJECT PURPOSE

The purposes of this project are to demonstrate the action of a simple half-wave rectifier and the effects of capacitor filtering.

PARTS NEEDED

- ☐ DMM/VOM
- ☐ Oscilloscope
- ☐ CIS
- ☐ Function generator or audio oscillator
- ☐ Diode: Silicon, 1-amp rating
- ☐ Resistor:
  10 kΩ
- ☐ Capacitors:
  0.1 μF
  1.0 μF

ACTIVITY	OBSERVATION	CONCLUSION
4. Connect the 0.1-μF filter capacitor in parallel R_L as shown in Figure 96.	—	—
5. Measure and record the ac voltage at the input (ac_{in}) and the dc voltage output across R_L (dc_{out}). Sketch two complete ac cycles of the oscilloscope waveform found across R_L.	ac_{in} = ______ V dc_{out} = ______ V Waveform: + 0 –	Is the dc output voltage higher or lower than the effective ac input voltage? ______. This can be explained by the fact that the filter capacitor charges to the ______ value of the input voltage, rather than the average or effective values. Since the charge path for the capacitor is through the low resistance of the forward-conducting diode, the capacitor has time to charge to the ______ value of the input voltage. But when the input voltage begins to decrease from its ______ value, the capacitor begins to ______ slowly through the load resistor, R_L. Before the capacitor can discharge completely, the next cycle of ac will reach a point where it is higher than the voltage remaining on the capacitor. The diode thus begins to conduct again and charge the capacitor to the ______ voltage value again.
6. Replace the 0.1-μF filter capacitor with a 1.0-μF capacitor.	—	—
7. Record the dc_{out} and sketch two complete ac cycles the oscilloscope waveform found across R_L.	dc_{out} = ______ V Waveform: + 0 –	Is dc_{out} higher or lower with the 1.0-μF filter capacitor? This is because the *RC* discharge time with the 1.0-μF filter capacitor is much ______ than with the 0.1-μF capacitor. The larger the value of filter capacitor, the (*higher*, *lower*) ______ the amount of discharge during each nonconducting half-cycle for the diode. This means the average dc output voltage (*increases*, *decreases*) ______ with increasing values of filter capacitance.

PROJECT 71

POWER SUPPLIES
Bridge Rectifier

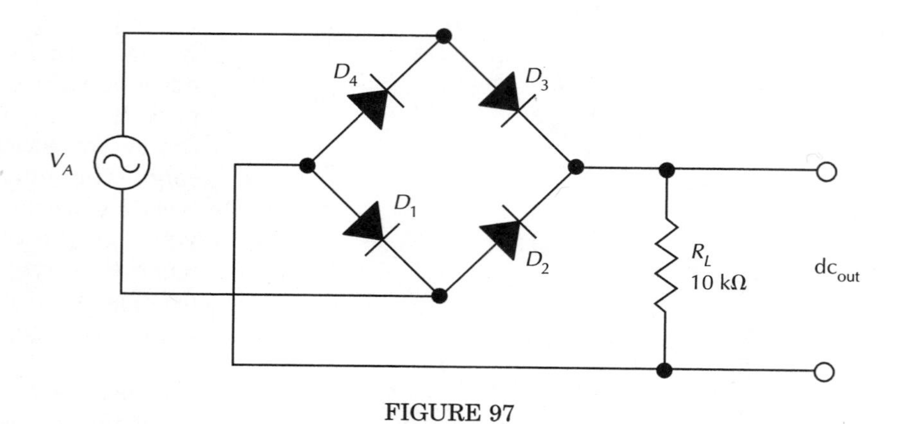

FIGURE 97

FIGURE 98

PROJECT PURPOSE

In this project you will study the action of a bridge rectifier and demonstrate the effects of capacitor filtering.

PARTS NEEDED

- ☐ DMM/VOM
- ☐ Oscilloscope
- ☐ CIS
- ☐ Function generator or audio oscillator
- ☐ Diodes: Silicon, 1-amp rating (4)
- ☐ Resistor: 10 kΩ
- ☐ Capacitor: 1.0 µF

ACTIVITY	OBSERVATION	CONCLUSION
1. Connect the initial circuit as shown in Figure 97.	—	—
2. Set the function generator (sine-wave mode) or audio oscillator to a frequency of 100 Hz and V_A to 3 V_{rms}. Measure the dc voltage output across R_L (dc_{out}).	ac_{in} = ______ V dc_{out} = ______ V	Current (*flows, does not flow*) ______ through R_L on both alternations of each ac input cycle. This means the bridge rectifier is a (*half-wave, full-wave*) ______ rectifier. According to theory, if there were no diode voltage drops, the average dc output of the bridge circuit without filtering should be ______ × V_{eff} of the applied ac. Does your measured value of dc_{out} for this step agree reasonably with theory? ______. How do you account for most of the difference, if any? ______
3. Connect the oscilloscope across R_L. Sketch the waveform for two complete cycles in the Observation column. Indicate the maximum and minimum voltage levels.	Waveform: + / 0 / –	The frequency of the waveform across the output resistor is (*one-half, equal to, double*) ______ the ac input frequency.

PROJECT
71
CONTINUED

POWER SUPPLIES
Bridge Rectifier *(Continued)*

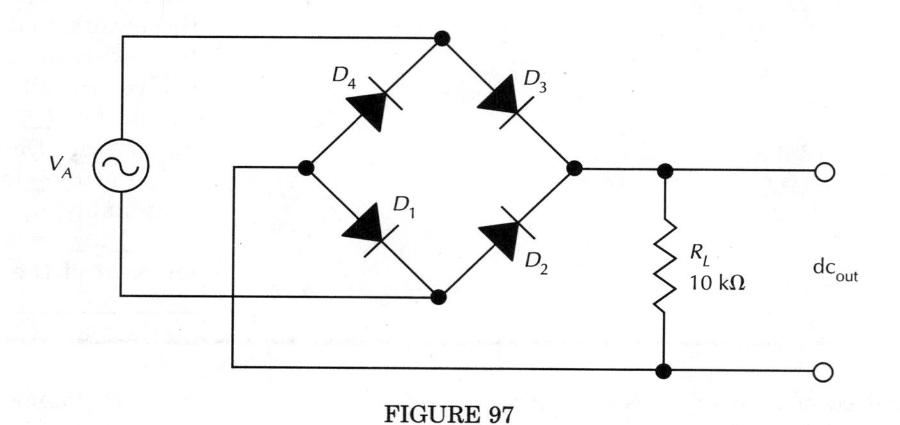

FIGURE 97

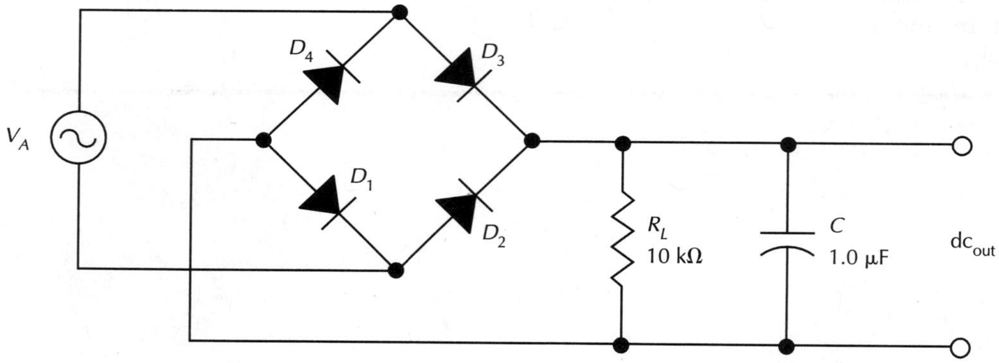

FIGURE 98

PROJECT PURPOSE

In this project you will study the action of a bridge rectifier and demonstrate the effects of capacitor filtering.

PARTS NEEDED

- ☐ DMM/VOM
- ☐ Oscilloscope
- ☐ CIS
- ☐ Function generator or audio oscillator
- ☐ Diodes: Silicon, 1-amp rating (4)
- ☐ Resistor: 10 kΩ
- ☐ Capacitor: 1.0 μF

ACTIVITY	OBSERVATION	CONCLUSION
4. Connect the 1.0-μF filter capacitor across R_L as shown in Figure 98.	—	—
5. Measure and record the ac voltage at the input (ac_{in}) and the dc voltage output across R_L (dc_{out}). Sketch two complete ac cycles of the oscilloscope waveform found across R_L.	ac_{in} = ____________ V dc_{out} = ____________ V Waveform: + 0 –	What is the highest voltage output we could get from this circuit with 7.0 volts applied and no load current? About __________ volts. What is the value of load current according to our measured value of V_{out}? __________ mA. The filter capacitor causes the dc level of V_{out} to be (*higher*, *lower*) __________ than if it were not in the circuit. This filter capacitor has (*more*, *less*) __________ time to discharge between charging pulses in a bridge circuit than in a half-wave rectifier circuit. The reason for this answer is that the bridge rectifier has __________ charging pulses for each complete cycle of the ac input waveform. The capacitor in a bridge rectifier will not discharge to as low a value of voltage between charging pulses as in a half-wave rectifier. When compared to a half-wave rectifier, the average voltage output of a bridge rectifier will be (*higher*, *lower*) __________ and remain closer to (*peak*, *effective*, *average*) __________ of V_A than for a half-wave rectifier having the same R_L and filter capacitor.

PROJECT

72

POWER SUPPLIES
Voltage Multiplier

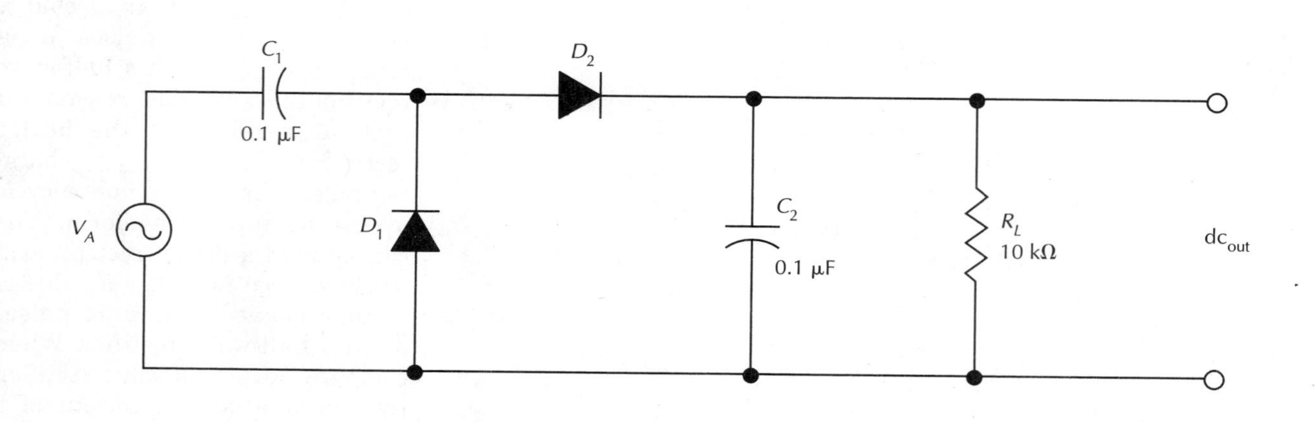

FIGURE 99

PROJECT PURPOSE

In this project you will observe the action of a half-wave cascade voltage doubler.

PARTS NEEDED

- ☐ DMM/VOM
- ☐ Oscilloscope
- ☐ CIS
- ☐ Function generator or audio oscillator
- ☐ Diodes: Silicon, 1-amp rating (2)
- ☐ Resistor: 10 kΩ
- ☐ Capacitors:
  - 0.1 μF (2)
  - 1.0 μF

ACTIVITY	OBSERVATION	CONCLUSION
1. Connect the circuit shown in Figure 99.	—	—
2. Set the function generator (sine-wave mode) or audio oscillator to 100 Hz and V_A to 6 V_{rms}. Connect the oscilloscope across V_A and measure the peak voltage value.	V_A (peak) = ________ V	If you have used a meter to set V_A to 6 V_{rms}, the peak value as measured with the oscilloscope should be ________ times the V_{rms} value. Does your measured value for V_A (peak) match up reasonably well with that ratio? ________.
3. Connect the oscilloscope across R_L. Sketch the waveform for two complete cycles in the Observation column. Indicate the maximum and minimum voltage values, and note the peak-to-peak ripple of the output voltage: Peak-to-peak ripple of V_{out} = V_{out} (max) – V_{out} (min)	Waveform: + 0 – V_{out} (max) = ________ V V_{out} (min) = ________ V Peak-to-peak ripple of V_{out} = ____ V	The peak value of the output voltage should be about ______ times the peak value of the ac input voltage. Do your measurements confirm this? ________. The frequency of the ripple voltage at the output of this circuit is (*one-half, equal to, twice*) ________ the frequency of the input waveform.
4. Replace capacitor C_2 with the 1.0-μF capacitor. If this is an electrolytic capacitor, make sure you connect the positive terminal of the capacitor to the positive output terminal (at the cathode of diode D_2).	—	—
5. Connect the oscilloscope across R_L. Measure and record the maximum and minimum voltage values. Calculate the peak-to-peak ripple of the output voltage.	V_{out} (max) = ________ V V_{out} (min) = ________ V Peak-to-peak ripple of V_{out} = ____ V	Increasing the value of the output capacitor (*increased, decreased*) ________ the output ripple voltage. Increasing the value of the output capacitor increased the average dc output voltage. How can you account for this? ________

POWER SUPPLIES

Complete the following review questions, indicating the appropriate response by placing a check in the box next to the correct answer.

1. The dc output voltage of a half-wave unfiltered rectifier is ____________ if you neglect the forward voltage drop across the diode.

   - ☐ $0.707 \times V_{eff}$
   - ☐ $0.637 \times V_{eff}$
   - ☐ $0.45 \times V_{eff}$
   - ☐ none of these

2. The average dc output voltage of a half-wave rectifier with capacitor filtering is higher than the dc output of a half-wave rectifier without a filter.

   - ☐ True
   - ☐ False

3. A typical voltage drop across a conducting silicon diode is approximately

   - ☐ 0.0 V
   - ☐ 0.25 V
   - ☐ 0.5 V
   - ☐ 0.7 V
   - ☐ 0.9 V

4. The ripple frequency at the output of a half-wave rectifier is

   - ☐ half the ac input frequency
   - ☐ the same as the ac input frequency
   - ☐ twice the ac input frequency
   - ☐ none of these

5. The bridge rectifier is one form of

   - ☐ half-wave rectification
   - ☐ full-wave rectification
   - ☐ voltage multiplication
   - ☐ none of these

6. For one cycle of ac input, a bridge rectifier yields

   - ☐ one-half pulse of output
   - ☐ one pulse of output
   - ☐ two pulses of output
   - ☐ four pulses of output

7. Assuming a half-wave and a full-wave rectifier have identical values of output load resistance and filter capacitance, which will yield the lesser amount of output ripple voltage?

   ☐ The half-wave rectifier
   ☐ The full-wave rectifier

8. In a bridge rectifier circuit, how many diodes are conducting during a given alternation of the input ac waveform?

   ☐ One
   ☐ Two
   ☐ Three
   ☐ All four

9. Neglecting the forward voltage drop across the diodes, the peak output voltage of a half-wave cascade voltage doubler is ________________ the peak voltage of the ac input waveform.

   ☐ close to one-half
   ☐ very nearly equal to
   ☐ about twice
   ☐ somewhat more than twice

10. In a half-wave cascade voltage doubler, the frequency of the output ripple is ______________ times the frequency of the ac input waveform.

    ☐ one-half
    ☐ one
    ☐ two
    ☐ more than two

BJT CHARACTERISTICS

Objectives

You will connect a BJT with proper biasing, measure the voltage levels, and calculate other voltages and currents. You will also determine the dc beta and alpha ratios for a BJT, and use an ohmmeter to test for shorted junctions in a BJT.

In completing these projects, you will connect circuits, make measurements, perform calculations, draw conclusions, and be able to answer questions about the following items related to BJT biasing.

- The polarity of connections for the emitter-base junction of a properly operating NPN transistor circuit
- The polarity of connections for the base-collector junction of a properly operating NPN transistor circuit
- Determining the currents in a BJT circuit, given certain voltage measurements and values of resistors
- Relative values of dc beta and alpha of a good BJT
- Relative resistance readings for good and shorted junctions in a BJT

PROJECT/TOPIC CORRELATION INFORMATION

PROJECT	TEXT CHAPTER	SECTION	RELATED TEXT TOPIC(S)
73 BJT Biasing	26	26-2	BJT Bias Voltages and Currents
74 Transistor Alpha and Beta	26	26-2	BJT Bias Voltages and Currents
75 Ohmmeter Tests for BJTs	26	26-5	Information for the Technician

PROJECT
73

BJT CHARACTERISTICS
BJT Biasing

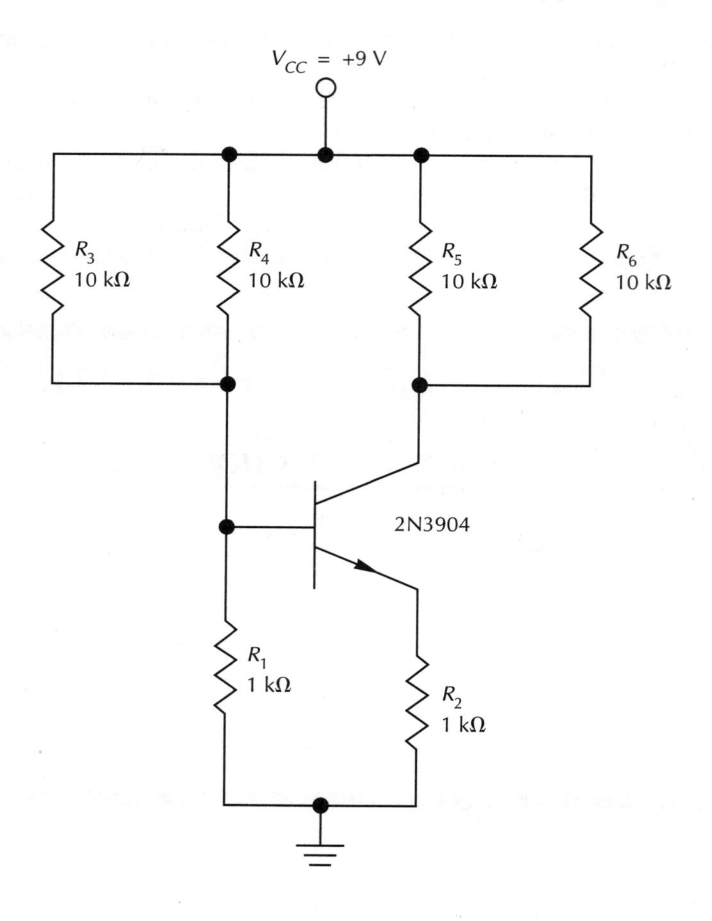

FIGURE 100

PROJECT PURPOSE

In this project you will observe how changes in bias current of a BJT affect its operation.

PARTS NEEDED

- ☐ DMM/VOM
- ☐ VVPS (dc)
- ☐ CIS
- ☐ NPN silicon transistor: 2N3904 (or equivalent)
- ☐ Resistors:
  1 kΩ (2)
  10 kΩ (4)

A normally biased BJT has the emitter-base junction forward biased, and the collector-base junction reverse biased. Forward biasing, in this case, means the N-type material is connected to the negative side of a dc source and the P-type material is connected to the positive side of the same dc source. So with an NPN transistor, this means the base should be positive with respect to the emitter (emitter-base bias), and the collector should be positive with respect to the base (collector-base bias).

The typical emitter-base junction voltage of a normally operating silicon transistor is about 0.7 volts. The collector-base junction voltage for the same transistor depends on the source voltage and values of the external components.

The following formulas can be helpful for completing the work in this project.

Formula 1 $V_{EB} = V_B - V_E$
where:
V_{EB} is the emitter-base voltage
V_B is the voltage measured from base to common
V_E is the voltage measured from emitter to common

Formula 2 $I_E = V_E/R_E$
where:
I_E is the emitter current
V_E is the voltage from emitter to common
R_E is the value of the emitter resistor

Formula 3 $V_{RC} = V_{CC} - V_C$
where:
V_{RC} is the voltage across the collector resistor
V_{CC} is the amount of supply voltage
V_C is the voltage measured from the collector to common

Formula 4 $I_C = V_{RC}/R_C$
where:
I_C is the collector current
V_{RC} is the voltage across the collector resistor
R_C is the value of the collector resistor

ACTIVITY	OBSERVATION	CONCLUSION
1. Connect the circuit shown in Figure 100.	—	The total resistance between the base and $+V_{CC}$ is _____ kΩ. The total collector resistance (combination of R_5 and R_6) is _______ kΩ.

PROJECT 73 CONTINUED

BJT CHARACTERISTICS
BJT Biasing *(Continued)*

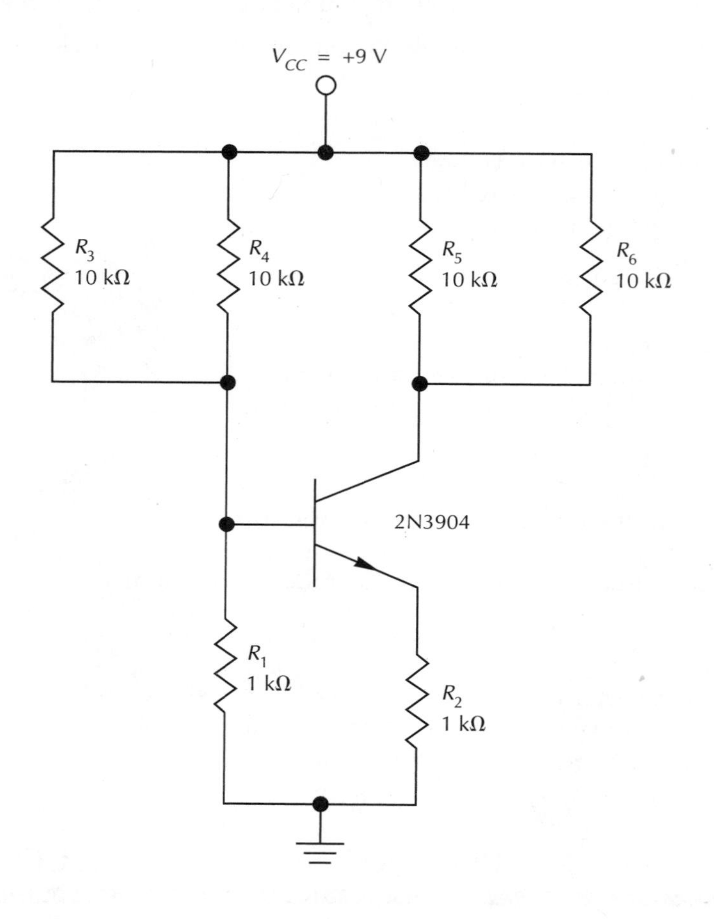

FIGURE 100

PROJECT PURPOSE

In this project you will observe how changes in bias current of a BJT affect its operation.

PARTS NEEDED

- ☐ DMM/VOM
- ☐ VVPS (dc)
- ☐ CIS
- ☐ NPN silicon transistor: 2N3904 (or equivalent)
- ☐ Resistors:
  1 kΩ (2)
  10 kΩ (4)

ACTIVITY	OBSERVATION	CONCLUSION
2. Set the dc source voltage to 9 V. Consider the negative side of the source as common, and measure the circuit voltages cited in the Observation column.	V_{CC} = ________ volts Emitter to common = ______ volts Base to common = ________ volts Collector to common = _____ volts	The base is (*positive*, *negative*) ____________ with respect to the emitter. What is the difference in potential between the emitter and base? ___________ V. What is the potential difference between the collector and emitter? ________ V. How much current must be passing through the emitter resistor? __________ mA. What is the voltage drop across the collector load resistance? _______ V. This means the collector current must be approximately ___________ mA. The collector is more (*positive*, *negative*) ____________ than the base. This means the collector-base junction is (*forward*, *reverse*) ______________________ biased.
3. Remove resistor R_4 from the base circuit.	—	The total resistance between the base and $+V_{CC}$ is ___________ kΩ.
4. Make sure the dc source voltage is set at 9 V, and measure the circuit voltages cited in the Observation column.	Emitter to common = ______ volts Base to common = ________ volts Collector to common = _____ volts	The total resistance between the base and $+V_{CC}$ is _______ kΩ. What is the difference in potential between the emitter and base? ________ V. What is the potential difference between the collector and emitter? _____ V. How much current must be passing through the emitter resistor? _______ mA. What is the voltage drop across the collector load resistance? _______ V. This means the collector current must be approximately ___________ mA. Has changing the value of the base-to-V_{CC} resistance changed: a. The emitter-to-collector voltage? _______ If so, how? __________ b. The current? ______ If so, how? ______________ c. The emitter-to-base voltage? ______. If so, how much? ___ V. The emitter current? ________. If so, how? ________________.

PROJECT 73 CONTINUED

BJT CHARACTERISTICS
BJT Biasing *(Continued)*

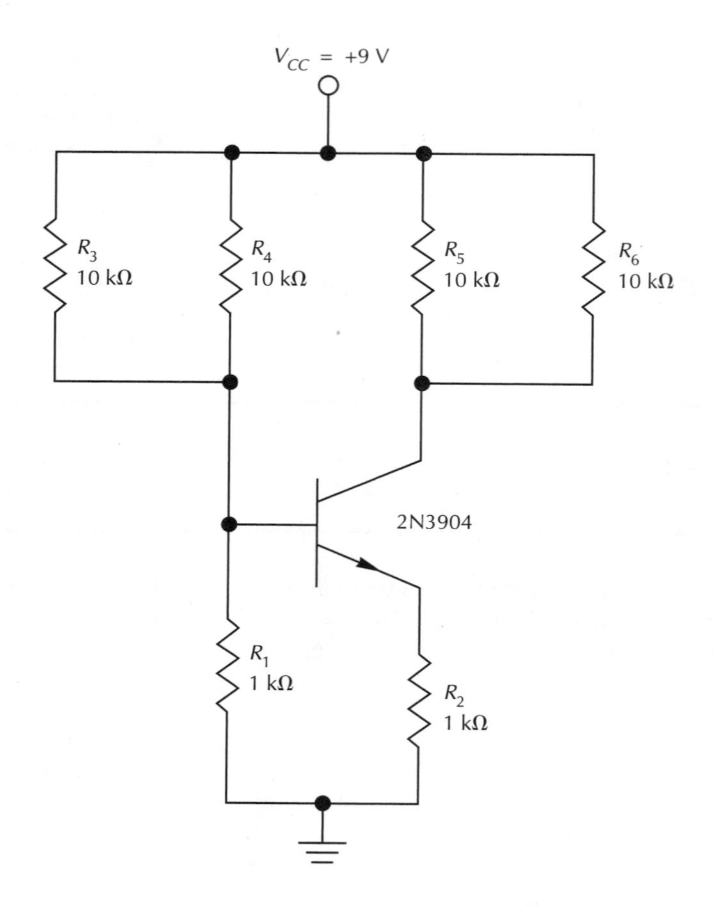

FIGURE 100

PROJECT PURPOSE

In this project you will observe how changes in bias current of a BJT affect its operation.

PARTS NEEDED

- ☐ DMM/VOM
- ☐ VVPS (dc)
- ☐ CIS
- ☐ NPN silicon transistor: 2N3904 (or equivalent)
- ☐ Resistors:
  1 kΩ (2)
  10 kΩ (4)

ACTIVITY	OBSERVATION	CONCLUSION
5. Replace resistor R_4 as shown in Figure 100, and remove resistor R_6 from the collector circuit.	—	The collector resistance is now _________________ kΩ.
6. After making sure the dc source voltage is set at 9 V, measure the circuit voltages cited in the Observation column.	Emitter to common = _____ volts Base to common = ________ volts Collector to common = _____ volts	The total resistance between the base and $+V_{CC}$ is ______ kΩ. What is the difference in potential between the emitter and base? _______ V. What is the potential difference between the collector and emitter? _____ V. How much current must be passing through the emitter resistor? _______ mA. What is the voltage drop across the collector load resistance? _____ V. This means the collector current must be approximately ____________ mA. Has changing the value of the collector resistance (compared with your results in Step 2) changed: a. The emitter-to-collector voltage? ______. If so, how? __________. b. The current? _____. If so, how? ________________________. c. The emitter-to-base voltage? ___________. If so, how much? ______________________ V. d. The emitter current? ________. If so, how? ________________.

PROJECT

74

BJT CHARACTERISTICS
Transistor Alpha and Beta

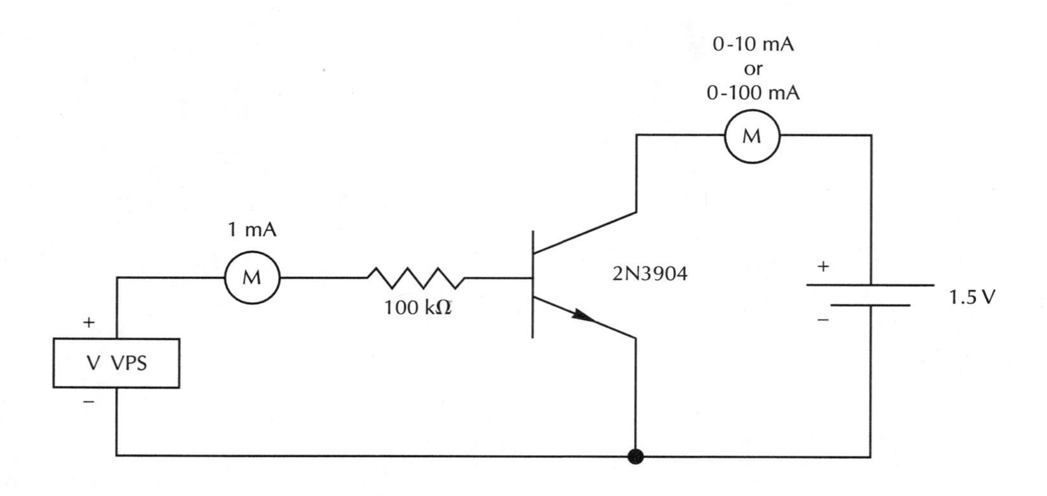

FIGURE 101

PROJECT PURPOSE

In this project, you will make the readings that are necessary for determining the actual α and β of a BJT.

PARTS NEEDED

- ☐ DMM/VOM (2)
- ☐ VVPS (dc)
- ☐ 1.5-V cell
- ☐ NPN silicon transistor: 2N3904 (or equivalent)
- ☐ CIS
- ☐ Resistor: 100 kΩ

The alpha (α) and beta (β) of a BJT refer to specific ratios of currents. These ratios have been found useful in analyzing BJTs and BJT circuits.

Alpha can be defined as the ratio of collector current to emitter current:

$$\alpha = I_C/I_E$$

Typical values of a range from about 0.96 to 0.99. This means that approximately 96–99 percent of the emitter current becomes collector current. The remaining few percent flow as emitter-base current.

Beta (specifically dc beta) can be defined as the ratio of collector current to base current:

$$\beta_{dc} = I_C/I_B$$

Typical values of β_{dc} range from 50–100 for small signal transistors. As you can see, the transistor exhibits a current gain from base to collector. This gain characteristic allows the transistor to be used as the active element in amplifiers, oscillators, and many other practical electrical and electronic circuits.

Ideally, you will use two meters for this experiment: one for monitoring I_B, and a second for monitoring I_C.

ACTIVITY / OBSERVATION / CONCLUSION

1. Connect the circuit as shown in Figure 101.

Observation: —

Conclusion: —

2. Using the 1.5-V cell for the collector supply voltage, carefully adjust the VVPS voltage until 0.1 mA of base current is flowing as indicated by the 1-mA meter in the base circuit. Note the amount of collector current that is flowing while the base current is at 0.1 mA.

V_{CC} = ______ V
I_B = ______ mA
I_C = ______ mA

The collector current is about ______ times greater than the base current. Since beta is defined as the ratio of collector current to base current, we conclude that the beta of this transistor under these operating conditions is ______.

3. Carefully readjust the VVPS voltage setting in order to obtain 0.2 mA of base current. Note the corresponding amount of collector current.

V_{CC} = ______ V
I_B = ______ mA
I_C = ______ mA

The beta of the transistor with 0.2 mA of base current is ______. Since the emitter current equals the sum of the collector current and base current, the emitter current in this step must be about ______ mA. The α for this step is about ______ which means that ______ percent of the emitter current flows as collector current.

PROJECT
74
CONTINUED

BJT CHARACTERISTICS
Transistor Alpha and Beta *(Continued)*

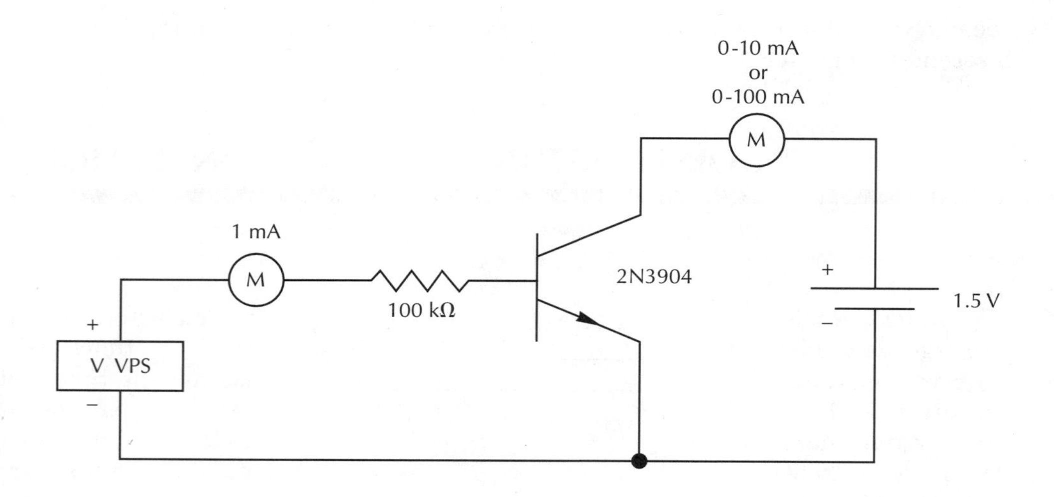

FIGURE 101

PROJECT PURPOSE

In this project, you will make the readings that are necessary for determining the actual α and β of a BJT.

PARTS NEEDED

- ☐ DMM/VOM (2)
- ☐ VVPS (dc)
- ☐ 1.5-V cell
- ☐ NPN silicon transistor: 2N3904 (or equivalent)
- ☐ CIS
- ☐ Resistor: 100 kΩ

ACTIVITY	OBSERVATION	CONCLUSION
4. Carefully readjust the VVPS voltage setting in order to obtain 0.3 mA of base current. Note the corresponding amount of collector current.	V_{CC} = ______ V I_B = ______ mA I_C = ______ mA	The dc beta for this step is about ______. We conclude that the dc beta is somewhat dependent on the operating parameters and (*is, is not*) ______ constant.

PROJECT 75

BJT CHARACTERISTICS
Ohmmeter Tests for BJTs

FIGURE 102
Basing Diagram (To be drawn by student)

PROJECT PURPOSE

For this project you will use an ohmmeter to test for a shorted junction in an NPN transistor.

PARTS NEEDED

- ☐ DMM/VOM
- ☐ NPN silicon transistor: 2N3904 (or equivalent)

A nonshorted transistor junction has a reverse resistance that is many times greater than its forward resistance. As with silicon diodes, the ratio of the reverse-to-forward resistance is more important than actual resistance values.

Recall from your textbook assignment that current can flow between the emitter and collector terminal of a BJT only while the emitter-base junction is forward biased. When testing the emitter-collector portion of a BJT with an ohmmeter, you will find that a good transistor will show a high resistance, regardless of the polarity of the test leads. A shorted transistor will show a relatively low resistance between the emitter and collector terminals.

Throughout this project, keep the ohmmeter set on a mid-range scale: approximately the $R \times 100$ scale for a VOM (analog) ohmmeter or the 20-kΩ range for a DMM (digital) ohmmeter.

ACTIVITY	OBSERVATION	CONCLUSION
1. Determine the basing diagram for your NPN transistor and make a sketch of the basing diagram in the space provided for Figure 102. Be sure you clearly identify the emitter, base, and collector terminals.	—	—
2. Connect the ohmmeter to the emitter and base terminals on the NPN transistor. Make sure you have the positive test lead connected to the base terminal and the negative test lead connected to the emitter terminal. Record the measured resistance. Reverse the polarity of the ohmmeter's test leads and repeat this base-emitter measurement. Calculate the high-to-low resistance ratio.	Resistance when base is positive and emitter is negative = _______ Resistance when base is negative and emitter is positive = _______ Ratio of the high resistance reading to low resistance reading = _______	When the positive terminal of an ohmmeter is connected to the base terminal of an NPN transistor and the negative terminal is connected to the emitter, the base-emitter junction is said to be (*forward*, *reverse*) __________ biased, and the resistance should be relatively (*high*, *low*) __________. When the ohmmeter terminals are reversed, as in the second part of this Activity step, the emitter-base junction is (*forward*, *reverse*) __________ biased, and the resistance should be relatively (*high*, *low*) __________. You can determine that the emitter-base junction is **not** shorted when you find that the ratio of the higher reading to the lower reading is (*rather high*, *close to 1*) __________.

PROJECT
75
CONTINUED

BJT CHARACTERISTICS
Ohmmeter Tests for BJTs *(Continued)*

PROJECT PURPOSE

For this project you will use an ohmmeter to test for a shorted junction in an NPN transistor.

PARTS NEEDED

- ☐ DMM/VOM
- ☐ NPN silicon transistor: 2N3904 (or equivalent)

ACTIVITY	OBSERVATION	CONCLUSION
3. Connect the ohmmeter to the base and collector terminals on the NPN transistor. Begin with the positive ohmmeter lead connected to the base terminal and the negative test lead to the collector terminal. Record the measured resistance. Reverse the polarity of the ohmmeter's test leads and repeat the resistance measurement between the base and collector. Calculate the high-to-low resistance ratio.	Resistance when base is positive and collector is negative = _______ Resistance when base is negative and collector is positive = _______ Ratio of the high resistance reading to low resistance reading = _______	When the positive terminal of an ohmmeter is connected to the base terminal of an NPN transistor and the negative terminal is connected to the collector, the base-collector junction is said to be (*forward, reverse*) ____________ biased, and the resistance should be relatively (*high, low*) __________. When the ohmmeter terminals are reversed, the base-collector junction is (*forward, reverse*) __________ biased, and the resistance should be relatively (*high, low*) ________. You can determine that the base-collector junction is **not** shorted when you find that the ratio of the higher reading to the lower reading is (*rather high, close to 1*) __________.
4. Connect the ohmmeter to the emitter and collector terminals on the NPN transistor. For the first measurement, fix the positive lead of the ohmmeter to the emitter terminal and the negative lead to the collector terminal. Record the measured resistance. Reverse the polarity of the ohmmeter's test leads and repeat the resistance measurement between the emitter and collector. Calculate the high-to-low resistance ratio.	Resistance when emitter is positive and collector is negative = _______ Resistance when emitter is negative and collector is positive = _______ Ratio of the high resistance reading to low resistance reading = _______	When the positive terminal of an ohmmeter is connected to the emitter terminal of a good NPN transistor and the negative terminal is connected to the collector, the resistance should be fairly (*high, low*) _______. When the ohmmeter terminals are reversed, the emitter-collector resistance should be relatively (*high, low*) ____________________.

BJT CHARACTERISTICS

Complete the following review questions, indicating the appropriate response by placing a check in the box next to the correct answer.

1. The emitter-base junction of an NPN transistor is said to be forward biased when

   - ☐ the base is more positive than the emitter
   - ☐ the emitter is more positive than the base
   - ☐ the emitter and base have the same voltage

2. The base-collector junction of an NPN transistor is properly biased for normal transistor operation when

   - ☐ the base is more positive than the collector
   - ☐ the collector is more positive than the base
   - ☐ the base and collector have the same applied voltage

3. For the proper operation of a BJT, the emitter-base and base-collector junctions must be forward biased at the same time.

   - ☐ True
   - ☐ False

4. When a properly operating BJT circuit has a resistor of known value connected between the emitter and common, you can determine the amount of emitter current by

   - ☐ measuring the emitter-to-common voltage and dividing by the amount of emitter resistance.
   - ☐ measuring the base-to-common voltage, measuring the emitter-to-common voltage, and dividing the difference by the amount of emitter resistance.
   - ☐ measuring the positive supply voltage, measuring the emitter-to-common voltage, and dividing the difference by the value of the emitter resistance.

5. V_{CE} for a properly operating BJT circuit can be determined by taking the difference between the voltage measured between the collector and common and the voltage measured between the emitter and common.

   - ☐ True
   - ☐ False

6. The dc beta of a good BJT is

   - ☐ always much greater than 1
   - ☐ always a bit less than 1

7. The alpha of a good BJT is

   - ☐ always much greater than 1
   - ☐ always a bit less than 1

8. When the emitter-base junction of an NPN transistor is shorted, an ohmmeter test will show

   - ☐ low resistance in one direction and high resistance when the leads are reversed
   - ☐ low resistance in both directions
   - ☐ high resistance in both directions

9. When the base-collector junction of an NPN transistor is in good working order, an ohmmeter test will show

   - ☐ low resistance in one direction and high resistance when the leads are reversed
   - ☐ low resistance in both directions
   - ☐ high resistance in both directions

10. The resistance between the emitter and collector of a good BJT will be very high in one direction and relatively low when the ohmmeter leads are reversed.

    - ☐ True
    - ☐ False

BJT AMPLIFIER CONFIGURATIONS

Objectives

You will connect common-emitter (CE) and common-collector (CC) BJT amplifiers and note their ac signal-amplifying characteristics.

In completing these projects, you will connect circuits, make measurements, compare input and output ac waveforms on a dual-trace oscilloscope, perform calculations, draw conclusions, and answer questions about the following items related to CE and CC amplifiers.

- The amount of phase difference between the output ac waveform and the input waveform
- The amount of gain that is characteristic of CE and CC amplifiers
- The effects that changing resistance values and supply voltage have upon the gain of the circuits
- Distinguishing a schematic diagram of a CE amplifier from that of a CC amplifier

PROJECT/TOPIC CORRELATION INFORMATION

PROJECT	TEXT CHAPTER	SECTION	RELATED TEXT TOPIC(S)
76 Common-Emitter Amplifier	27	27-2	The Common-Emitter (CE) Amplifier
77 Common-Collector Amplifier	27	27-2	Common-Collector (CC) Amplifiers

PROJECT 76 BJT AMPLIFIER CONFIGURATIONS
Common-Emitter Amplifier

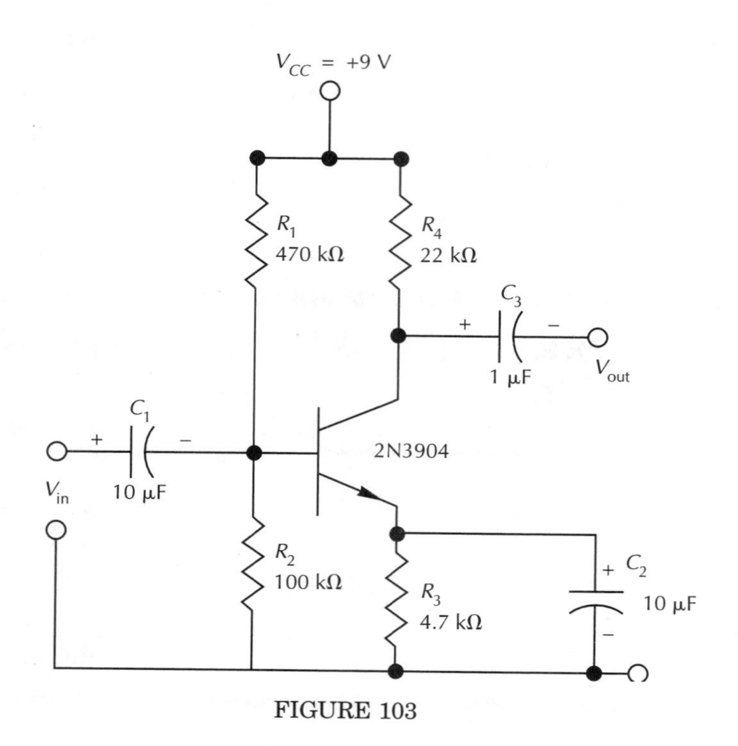

FIGURE 103

PROJECT PURPOSE

For this project you will construct a common-emitter amplifier circuit, and you will use the DMM/VOM and dual-trace oscilloscope to determine the circuit's operating features.

PARTS NEEDED

- ☐ DMM/VOM
- ☐ VVPS (dc)
- ☐ CIS
- ☐ Dual-trace oscilloscope
- ☐ Function generator or audio oscillator
- ☐ NPN silicon transistor: 2N3904 (or equivalent)
- ☐ Resistors:
  - 4.7 kΩ
  - 22 kΩ
  - 47 kΩ
  - 100 kΩ
  - 470 kΩ
- ☐ Capacitors:
  - 1 μF
  - 10 μF (2)

You will need the following formula.

$$A_V = V_{out}/V_{in}$$

where:

A_V is the voltage gain of an amplifier
V_{out} is the signal voltage level at the output of the amplifier (usually peak-to-peak)
V_{in} is the signal voltage level at the input of the amplifier (usually peak-to-peak)

ACTIVITY	OBSERVATION	CONCLUSION
1. Connect the circuit as shown in Figure 103. Connect the signal source (function generator in the sine-wave mode or the audio oscillator) to V_{in}, but make sure its output is at zero volts.	—	—
2. Set the dc source for a V_{CC} of +9 V. Consider the negative side of the source as common, and measure the dc voltages required in the Observation column.	V_E (voltage from emitter to common) = ________ Vdc V_C (voltage from collector to common) = ________ Vdc V_B (voltage from base to common) = ________ Vdc	—
3. Connect one channel of the oscilloscope to V_{in}, and connect the second channel of the oscilloscope to V_{out}. Adjust the signal source for an input signal of 1 kHz at 0.5 V peak-to-peak. If the signal you find at V_{out} is clipped ("flattened") on either or both alternations, **reduce the input signal level until the clipping action is no longer observed**. Sketch the waveforms and determine the readings specified in the Observation column.	V_{in} waveform: 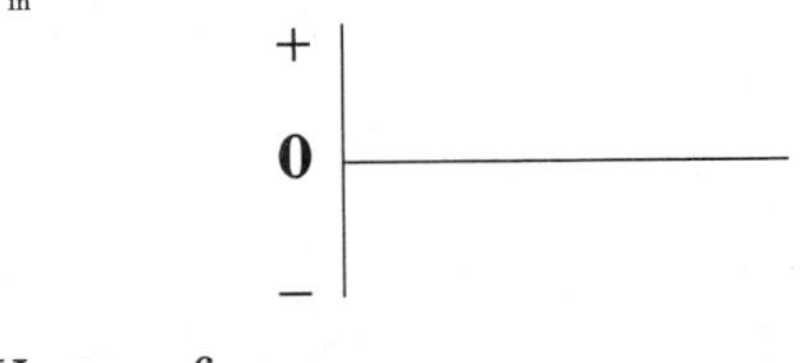 V_{out} waveform: 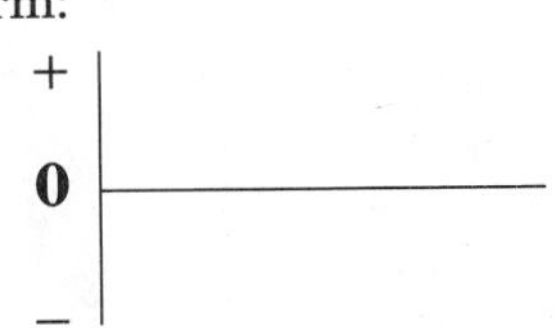 V_{in} = ________ V peak-to-peak V_{out} = ________ V peak-to-peak	The oscilloscope traces indicate that the output voltage from a common-emitter amplifier is (*in phase, 180° out of phase*) ________ with its input waveform. Use the readings you found for V_{in} and V_{out} to determine the voltage gain (A_V) of this circuit. A_V = ________. You have seen that the voltage gain of a common-emitter amplifier can be greater than 1. (*True, False*) ________

PROJECT **76** CONTINUED

BJT AMPLIFIER CONFIGURATIONS
Common-Emitter Amplifier *(Continued)*

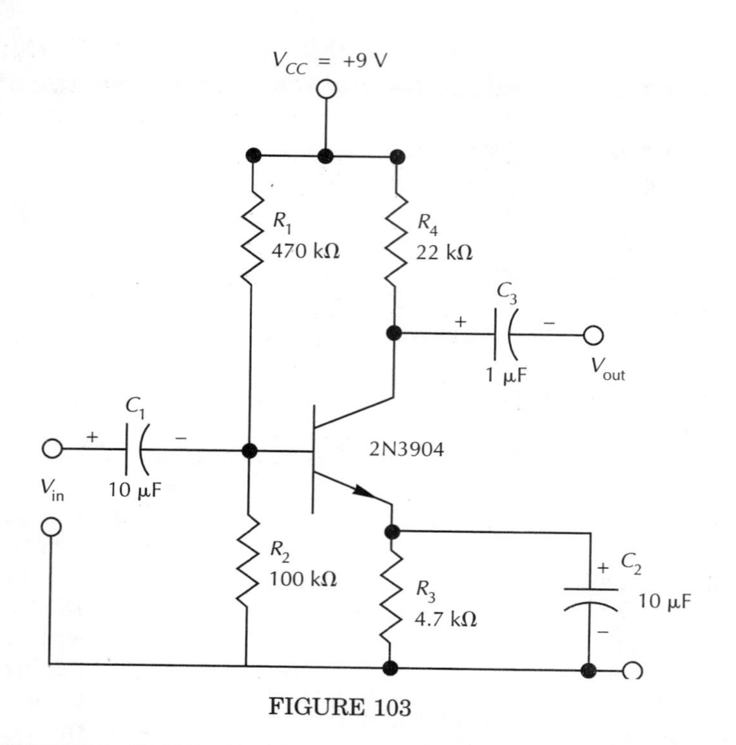

FIGURE 103

PROJECT PURPOSE

For this project you will construct a common-emitter amplifier circuit, and you will use the DMM/VOM and dual-trace oscilloscope to determine the circuit's operating features.

PARTS NEEDED

- ☐ DMM/VOM
- ☐ VVPS (dc)
- ☐ CIS
- ☐ Dual-trace oscilloscope
- ☐ Function generator or audio oscillator
- ☐ NPN silicon transistor: 2N3904 (or equivalent)
- ☐ Resistors:
  - 4.7 kΩ
  - 22 kΩ
  - 47 kΩ
  - 100 kΩ
  - 470 kΩ
- ☐ Capacitors:
  - 1 μF
  - 10 μF (2)

ACTIVITY	OBSERVATION	CONCLUSION
4. Remove the 22-kΩ collector resistor (R_4) and replace it with a 47-kΩ resistor.	—	—
5. If either peak of the waveform at V_{out} is clipped, reduce the input signal level until the clipping effect is no longer apparent. Gather the information requested in the Observation column.	V_E (voltage from emitter to common) = ______ Vdc V_C (voltage from collector to common) = ______ Vdc V_B (voltage from base to common) = ______ Vdc V_{in} waveform: + 0 – V_{out} waveform: + 0 – V_{in} = ______ V peak-to-peak V_{out} = ______ V peak-to-peak	Use the readings you found for V_{in} and V_{out} to determine the voltage gain (A_V) of this circuit. A_V = ______. When you increased the value of the load resistor (R_4), did you notice a significant change in the amount of voltage gain? ______ If so, describe the amount and direction of change. ______ ______ ______ ______.

PROJECT 77

BJT AMPLIFIER CONFIGURATIONS

Common-Collector Amplifier

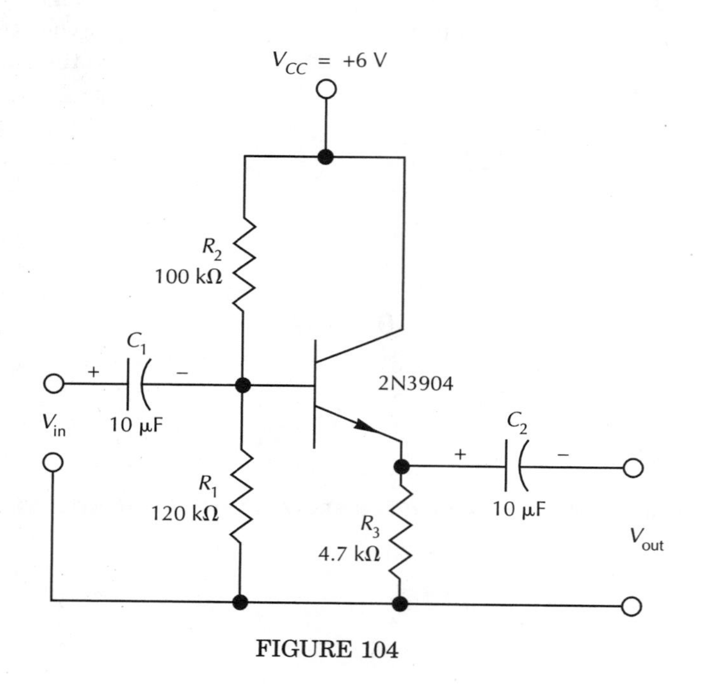

FIGURE 104

PROJECT PURPOSE

For this project you will construct a common-collector amplifier circuit, and you will use the DMM/VOM and dual-trace oscilloscope to determine the circuit's main operating features.

PARTS NEEDED

- ☐ DMM/VOM
- ☐ VVPS (dc)
- ☐ CIS
- ☐ Dual-trace oscilloscope
- ☐ Function generator or audio oscillator
- ☐ NPN silicon transistor: 2N3904 (or equivalent)
- ☐ Resistors:
  4.7 kΩ
  100 kΩ
  120 kΩ
- ☐ Capacitors:
  10 µF (2)

ACTIVITY	OBSERVATION	CONCLUSION
1. Connect the circuit as shown in Figure 104. Connect the signal source (function generator in the sine-wave mode or the audio oscillator) to V_{in}, but make sure its output is set for zero volts.	—	—
2. Set the dc source for a V_{CC} of +6 V. Consider the negative side of the source as common, and measure the dc voltages required in the Observation column.	V_E (voltage from emitter to common) = ______ Vdc V_B (voltage from base to common) = ______ Vdc	—
3. Connect one channel of the oscilloscope to V_{in}, and connect the second channel of the oscilloscope to V_{out}. Adjust the signal source for an input signal of 1 kHz at 0.5 V peak-to-peak. If the signal you find at V_{out} is clipped ("flattened") on either or both alternations, **reduce the input signal level until the clipping action is no longer observed**. Sketch the waveforms and determine the readings specified in the Observation column.	V_{in} waveform: + 0 – V_{out} waveform: + 0 – V_{in} = ______ V peak-to-peak V_{out} = ______ V peak-to-peak	The oscilloscope traces indicate that the output voltage from a common-collector amplifier is (*in phase, 180° out of phase*) ______ with its input waveform. Use the readings you found for V_{in} and V_{out} to determine the voltage gain (A_V) of this circuit. A_V = ______. (Recall that $A_V = V_{out}/V_{in}$.) You have seen in this circuit that the voltage gain of a common-collector amplifier can be greater than 1. (*True, False*) ______
4. Increase the setting of the dc source for a V_{CC} of +9 V.	—	—
5. If either peak of the waveform at V_{out} is clipped, reduce the input signal level until the clipping effect is no longer apparent. Gather the information requested in the Observation column.	V_E (voltage from emitter to common) = ______ Vdc V_B (voltage from base to common) = ______ Vdc V_{in} waveform: + 0 – V_{out} waveform: + 0 – V_{in} = ______ V peak-to-peak V_{out} = ______ V peak-to-peak	Use the readings you found for V_{in} and V_{out} to determine the voltage gain (A_V) of this circuit. A_V = ______. When you increased the value of V_{CC}, did you notice a significant change in the amount of voltage gain? ______. If so, describe the amount and direction of change; if not, describe which readings did change. ______

BJT AMPLIFIER CONFIGURATIONS

Complete the following review questions, indicating the appropriate response by placing a check in the box next to the correct answer.

1. The output ac waveform from a common-emitter amplifier is shifted ________ degrees relative to the input ac waveform.

   - ☐ 0
   - ☐ 90
   - ☐ 180
   - ☐ 360

2. The voltage gain of a common-emitter amplifier can be greater than 1.

   - ☐ True
   - ☐ False

3. Increasing the value of the collector resistor in a common-emitter amplifier ____________ the voltage gain of the circuit.

   - ☐ has no effect on
   - ☐ reduces
   - ☐ increases

4. For the common-emitter circuit you used in Project 76, the dc voltage readings show that the emitter-base junction is

   - ☐ using forward bias
   - ☐ using reverse bias
   - ☐ using no bias
   - ☐ negatively biased

5. As you decrease the amount of ac signal applied at V_{in} for a common-emitter amplifier, a corresponding decrease in V_{out} indicates that gain of the circuit is also decreasing.

   - ☐ True
   - ☐ False

6. You can identify a common-emitter amplifier by noting that

   - ☐ the emitter is connected to the negative terminal of the power source
   - ☐ the input and output terminals are both connected to the emitter
   - ☐ neither the signal input nor the signal output is connected to the emitter
   - ☐ none of these

7. The output ac waveform from a common-collector amplifier is shifted ________ degrees relative to the input ac waveform.

   - ☐ 0
   - ☐ 90
   - ☐ 180
   - ☐ 360

8. A common-collector amplifier is also known as a voltage follower because the output voltage "follows" the input voltage.

   - ☐ True
   - ☐ False

9. Increasing the value of V_{CC} for a common-collector amplifier

   - ☐ increases the voltage gain
   - ☐ has little, if any, effect on the voltage gain

10. You can identify an NPN common-collector amplifier by noting that

    - ☐ the collector is connected directly to the positive terminal of the power source
    - ☐ the collector is grounded, or connected to circuit common
    - ☐ the output is taken from the collector
    - ☐ none of these

BJT AMPLIFIER CLASSES OF OPERATION

Objectives

You will connect each of the three main classes of BJT amplifiers and investigate their dc biasing and ac signal-amplifying characteristics.

In completing these projects, you will connect circuits, make measurements, compare input and output ac waveforms on a dual-trace oscilloscope, perform calculations, draw conclusions, and answer questions about the following items related to BJT amplifier classes of operation.

- The nature of base-emitter dc biasing for the three main classes of BJT amplifier circuits
- The way the three main classes of BJT amplifiers affect the input ac waveform
- Comparisons of amplifier efficiency and distortion

PROJECT/TOPIC CORRELATION INFORMATION

PROJECT	TEXT CHAPTER	SECTION	RELATED TEXT TOPIC(S)
78 BJT Class A Amplifier	27	27-4	Classification by Class of Operation Class A Amplifier Operation
79 BJT Class B Amplifier	27	27-4	Class B Amplifier Operation
80 BJT Class C Amplifier	27	27-4	Class C Amplifier Operation

PROJECT 78

BJT AMPLIFIER CLASSES OF OPERATION

BJT Class A Amplifier

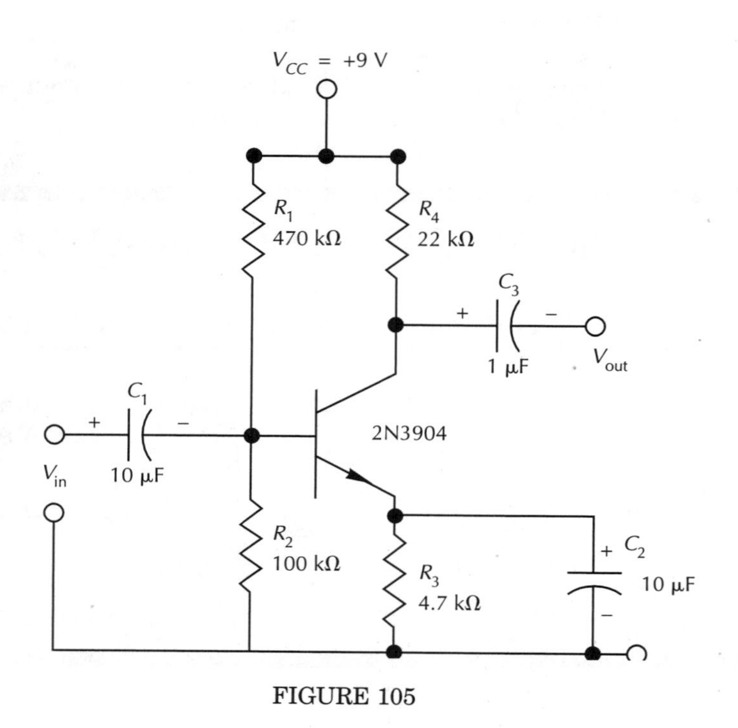

FIGURE 105

PROJECT PURPOSE

The purpose of this project is to demonstrate the dc biasing and ac amplification of a Class A, common-emitter BJT amplifier circuit.

PARTS NEEDED

- ☐ DMM/VOM
- ☐ VVPS (dc)
- ☐ CIS
- ☐ Dual-trace oscilloscope
- ☐ Function generator or audio oscillator
- ☐ NPN silicon transistor: 2N3904 (or equivalent)
- ☐ Resistors:
  - 4.7 kΩ
  - 22 kΩ
  - 100 kΩ
  - 470 kΩ
- ☐ Capacitors:
  - 1 μF
  - 10 μF (2)

The following formula will be helpful:

$$A_V = V_{out}/V_{in}$$

where:

A_V is the voltage gain of an amplifier
V_{out} is the signal voltage level at the output of the amplifier (usually peak-to-peak)
V_{in} is the signal voltage level at the input of the amplifier (usually peak-to-peak)

ACTIVITY	OBSERVATION	CONCLUSION
1. Connect the circuit as shown in Figure 105.	—	—
2. Set the dc source for a V_{CC} of +9 V. Make sure the signal source (function generator in the sine-wave mode or an audio oscillator) is disconnected from the circuit or set for 0 V output. Consider the negative side of the source as common, and measure the circuit's dc voltages cited in the Observation column.	V_E (voltage from emitter to common) = ______ Vdc V_C (voltage from collector to common) = ______ Vdc V_B (voltage from base to common) = ______ Vdc	Use the data from Observation step 2 to determine: V_{CE} (voltage between collector and emitter) = ______ Vdc. V_{EB} (voltage between the emitter and base) = ______ Vdc. For Class A operation of a BJT amplifier, V_{CE} should be close to (*V_{CC}, 1/2 V_{CC}, zero volts, −0.7 V*) ______. So the transistor is (*conducting, not conducting*) ______ while no signal is applied at V_{in}. Confirm whether the transistor is conducting or not conducting by determining the voltage across the collector resistor (R_4) and dividing by the resistor value to determine the amount of collector current: I_C = ______ mA.
3. Connect the function generator (sine-wave mode) or audio oscillator, and use the oscilloscope to set the input signal to 1 kHz at 0.5 V peak-to-peak. If the signal you find at V_{out} is clipped on either or both alternations, **reduce the input signal level until the clipping action is no longer observed.**	—	—

PROJECT **78** CONTINUED

BJT AMPLIFIER CLASSES OF OPERATION

BJT Class A Amplifier *(Continued)*

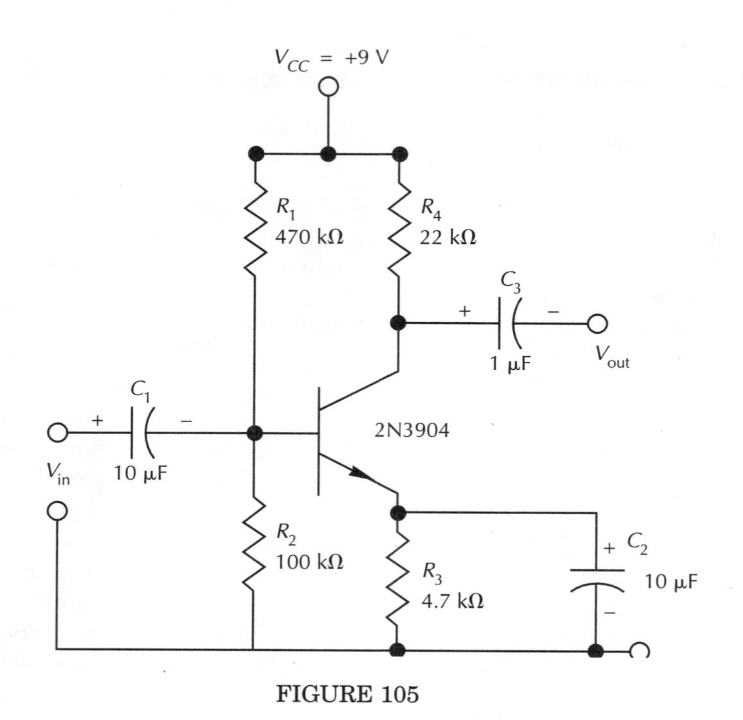

FIGURE 105

PROJECT PURPOSE

The purpose of this project is to demonstrate the dc biasing and ac amplification of a Class A, common-emitter BJT amplifier circuit.

PARTS NEEDED

- ☐ DMM/VOM
- ☐ VVPS (dc)
- ☐ CIS
- ☐ Dual-trace oscilloscope
- ☐ Function generator or audio oscillator
- ☐ NPN silicon transistor: 2N3904 (or equivalent)
- ☐ Resistors:
  - 4.7 kΩ
  - 22 kΩ
  - 100 kΩ
  - 470 kΩ
- ☐ Capacitors:
  - 1 μF
  - 10 μF (2)

ACTIVITY	OBSERVATION	CONCLUSION
4. Leave one channel of the oscilloscope connected to V_{in}, and connect the second channel of the oscilloscope to V_{out}. If the signal you find at V_{out} is not a clean sinusoidal waveform, reduce the V_{in} signal level until the waveform at the collector is a clean sine wave. Sketch the waveforms and determine the readings specified in the Observation column.	V_{in} waveform: 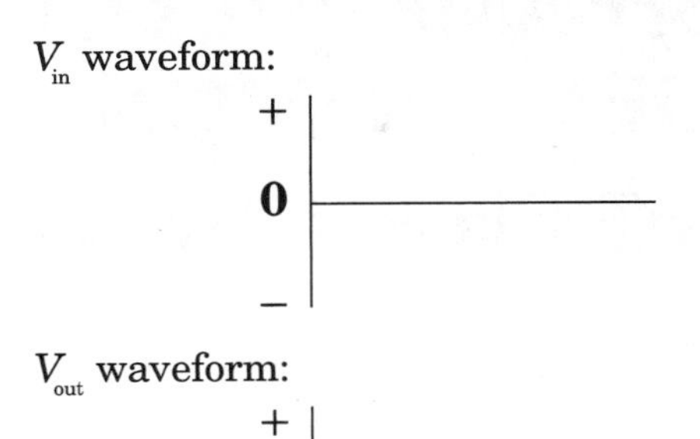 V_{out} waveform: 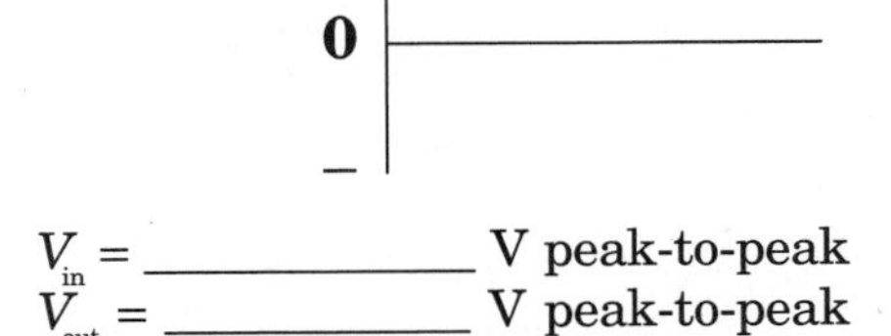 V_{in} = ______ V peak-to-peak V_{out} = ______ V peak-to-peak	Comparing the phases of the waveforms at V_{in} and V_{out}, you can say they are (*in phase, 180° out of phase*) ______. What is the voltage gain of this amplifier? ______
5. Slowly decrease the voltage level of the signal to one-half the amount used in step 4. Sketch the waveforms and determine the readings specified in the Observation column.	V_{in} waveform: 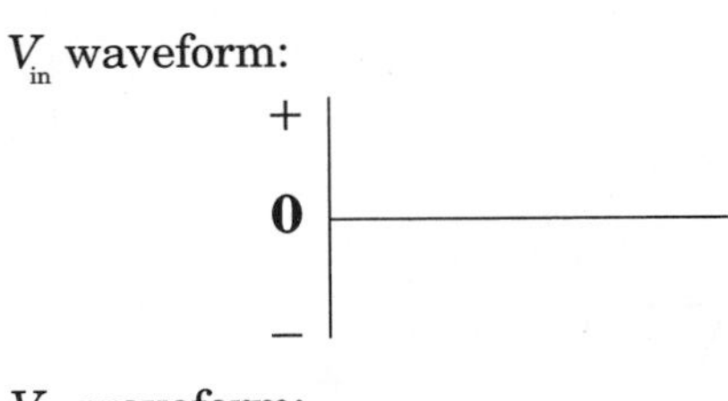 V_{out} waveform: 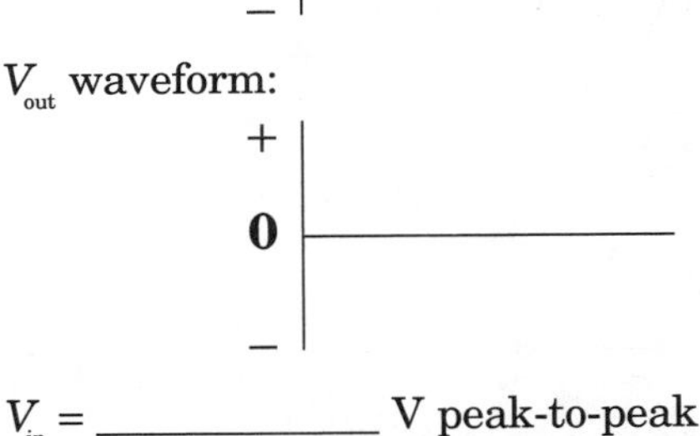 V_{in} = ______ V peak-to-peak V_{out} = ______ V peak-to-peak	The distortion in the output waveform of this amplifier is a sign that the input signal is too large for this circuit design. (*True, False*) ______

PROJECT 79

BJT AMPLIFIER CLASSES OF OPERATION

BJT Class B Amplifier

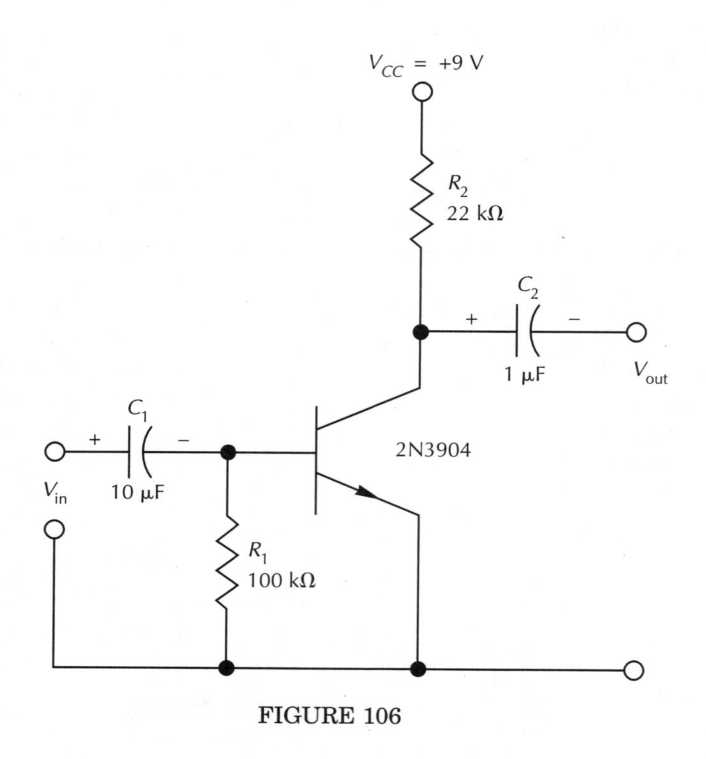

FIGURE 106

PROJECT PURPOSE

The purpose of this project is to demonstrate the dc biasing and ac amplification of a Class B, common-emitter BJT amplifier circuit.

PARTS NEEDED

- ☐ DMM/VOM
- ☐ VVPS (dc)
- ☐ CIS
- ☐ Dual-trace oscilloscope
- ☐ Function generator or audio oscillator
- ☐ NPN silicon transistor: 2N3904 (or equivalent)
- ☐ Resistors:
  - 22 kΩ
  - 100 kΩ
  - 470 kΩ
- ☐ Capacitors:
  - 1 μF
  - 10 μF

ACTIVITY	OBSERVATION	CONCLUSION

1. Connect the circuit as shown in Figure 106.

—

—

2. Set the dc source for a V_{CC} of +9 V. Connect the signal source (function generator in the sine-wave mode or an audio oscillator) to V_{in}, but make sure its output is at zero volts. Consider the negative side of the source as common, and measure the dc voltages cited in the Observation column.

V_E (voltage from emitter to common) = ______ Vdc

V_C (voltage from collector to common) = ______ Vdc

V_B (voltage from base to common) = ______ Vdc

For Class B operation of a BJT amplifier, V_{CE} should be close to (V_{CC}, *1/2* V_{CC}, *zero volts*, *−0.7 V*) ______. This means the transistor is (*conducting*, *not conducting*) ______ while no signal is applied at V_{in}. Confirm whether the transistor is conducting or not conducting by determining the voltage across the collector resistor (R_2) and dividing by the resistor value to determine the amount of collector current: I_C = ______ mA.

3. Connect one channel of the oscilloscope to V_{in}, and connect the second channel of the oscilloscope to V_{out}.

—

—

4. Adjust the signal source for an input signal of 1 kHz at 0.5 V peak-to-peak. If the signal you find at V_{out} is clipped ("flattened") on **both** half cycles, **reduce the input signal level until the clipping occurs on just one of the half cycles**. Sketch the waveforms and determine the readings specified in the Observation column.

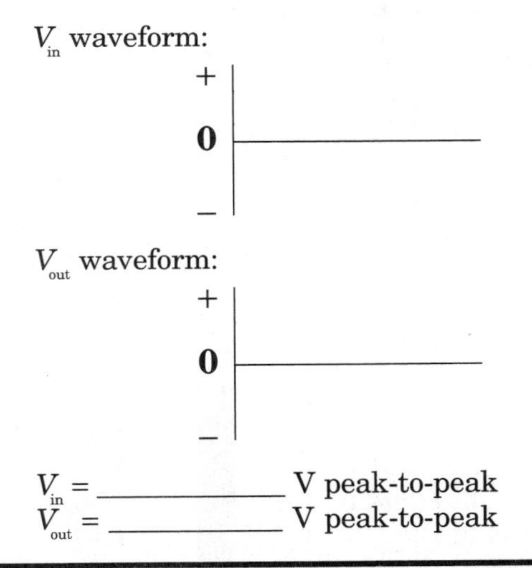

V_{in} = ______ V peak-to-peak

V_{out} = ______ V peak-to-peak

Comparing the phases of the waveforms at V_{in} and V_{out}, you can say they are (*in phase*, *180° out of phase*) ______.

The distortion noted on one alternation of the output waveform of this amplifier is a sign that something is wrong. (*True*, *False*) ______

5. Connect a 470-kΩ resistor between the base of the transistor and V_{CC}. Sketch the waveform for V_{out}.

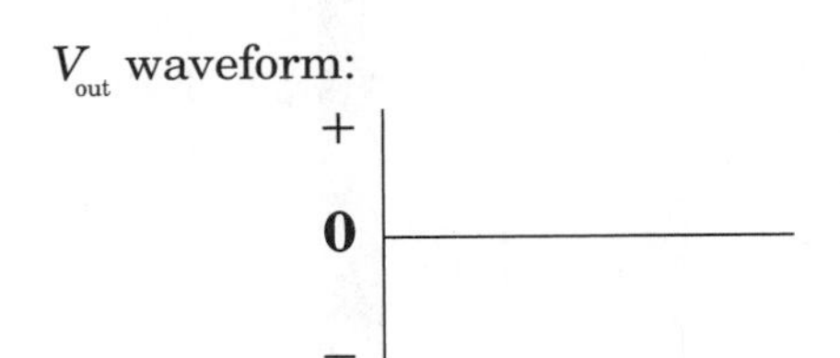

Adding a resistor between the base and V_{CC} gave this amplifier some (*forward*, *reverse*) ______ emitter-base bias that it didn't have before. This means the amplifier is no longer operating as a Class B amplifier. (*True*, *False*) ______

PROJECT

80

BJT AMPLIFIER CLASSES OF OPERATION

BJT Class C Amplifier

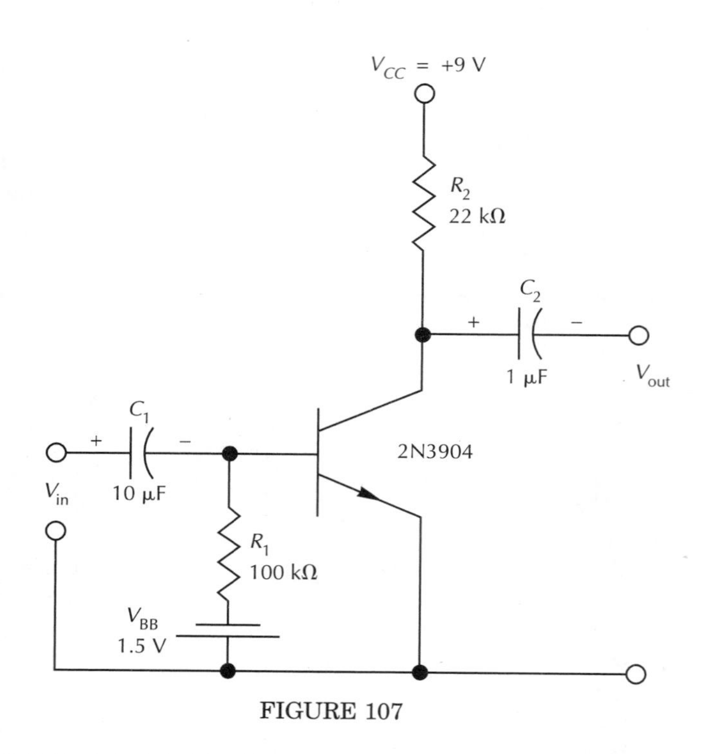

FIGURE 107

PROJECT PURPOSE

The purpose of this project is to demonstrate the dc biasing and ac amplification of a Class C, common-emitter BJT amplifier circuit.

PARTS NEEDED

- ☐ DMM/VOM
- ☐ VVPS (dc)
- ☐ CIS
- ☐ Dual-trace oscilloscope
- ☐ Function generator or audio oscillator
- ☐ NPN silicon transistor: 2N3904 (or equivalent)
- ☐ 1.5-volt cell
- ☐ Resistors:
  - 22 kΩ
  - 100 kΩ (2)
  - 470 kΩ
- ☐ Capacitors:
  - 1 μF
  - 10 μF

ACTIVITY	OBSERVATION	CONCLUSION
1. Connect the circuit as shown in Figure 107.	—	—
2. Set the dc source for a V_{CC} of +9 V. Connect the signal source (function generator in its sine-wave mode or the audio oscillator) to V_{in}, but make sure its output is at zero volts. Consider the negative side of the source as common, and measure the dc voltages required in the Observation column.	V_E (voltage from emitter to common) = ________ Vdc V_C (voltage from collector to common) = ________ Vdc V_B (voltage from base to common) = ________ Vdc	For Class C operation of a BJT amplifier, V_{CE} for an NPN transistor should be (*equal to V_{CC}, 1/2 V_{CC}, zero volts, less than zero volts*) ________. So the transistor is (*conducting, not conducting*) ________ while no signal is applied at V_{in}. Use the formula $I_C = (V_{CC} - V_C)/R_C$ to determine the actual amount of collector current when no signal is applied: I_C = ______ mA. When there is no signal applied to this amplifier, it can be said that the emitter-base junction is (*forward, reverse*) ________ biased.
3. Connect one channel of the oscilloscope to V_{in}, and connect the second channel of the oscilloscope to V_{out}.	—	—
4. Adjust the signal source for an input signal of 1 kHz at 0.5 V peak-to-peak. If the signal you find at V_{out} is clipped on **both** alternations, **reduce the input signal level until the clipping occurs on just one of the alternations**. Sketch the waveforms and determine the readings specified in the Observation column.	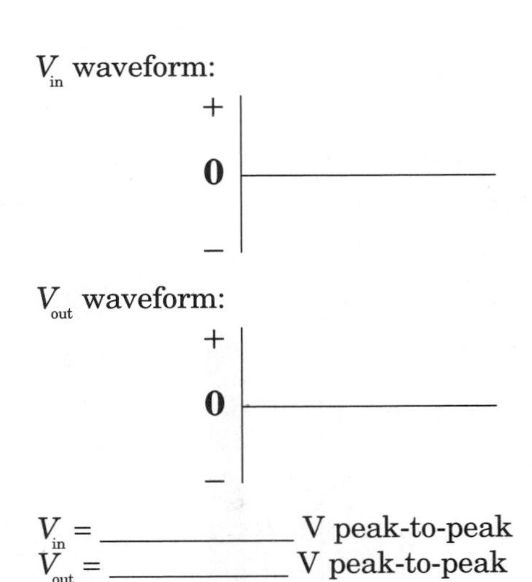 V_{in} = ________ V peak-to-peak V_{out} = ________ V peak-to-peak	Comparing the phases of the waveforms at V_{in} and V_{out}, you can say they are (*in phase, 180° out of phase*) ________. The distortion noted on one alternation of the output waveform of this amplifier indicates conduction during (*more than, equal to, less than*) ________ 180° of the input waveform. This type of distortion for a Class C amplifier is a sure sign that something is wrong. (*True, False*) ________

PROJECT
80
CONTINUED

BJT AMPLIFIER CLASSES OF OPERATION
BJT Class C Amplifier *(Continued)*

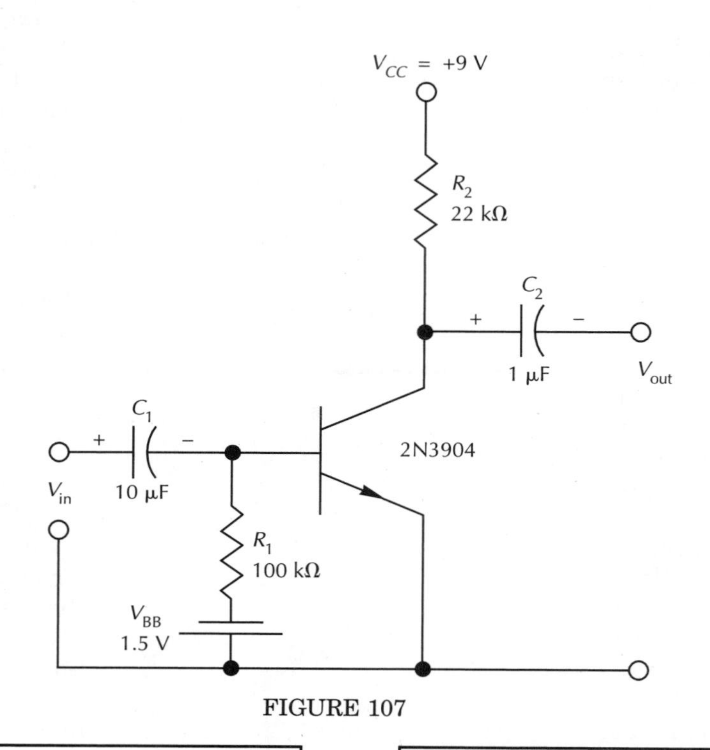

FIGURE 107

PROJECT PURPOSE

The purpose of this project is to demonstrate the dc biasing and ac amplification of a Class C, common-emitter BJT amplifier circuit.

PARTS NEEDED

- ☐ DMM/VOM
- ☐ VVPS (dc)
- ☐ CIS
- ☐ Dual-trace oscilloscope
- ☐ Function generator or audio oscillator
- ☐ NPN silicon transistor: 2N3904 (or equivalent)
- ☐ 1.5-volt cell
- ☐ Resistors:
  - 22 kΩ
  - 100 kΩ (2)
  - 470 kΩ
- ☐ Capacitors:
  - 1 μF
  - 10 μF

ACTIVITY | OBSERVATION | CONCLUSION

5. Connect a 470-kΩ resistor between the base of the transistor and V_{CC}. Sketch the waveform for V_{out}.

V_{out} waveform:

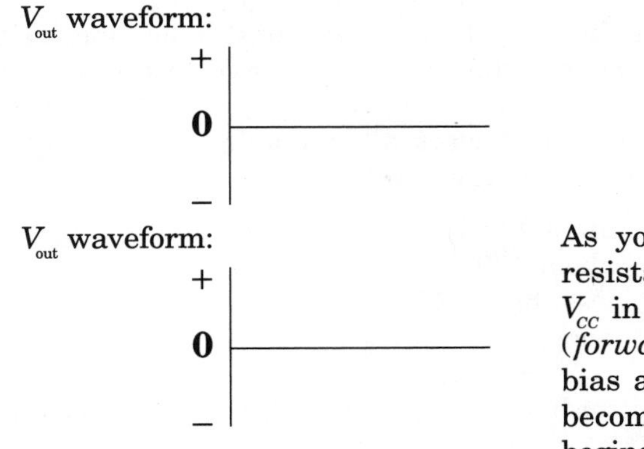

—

6. Remove the 470-kΩ resistor you added in Activity step 5 and replace it with a 100-kΩ resistor (between the base of the transistor and V_{CC}). Sketch the waveform for V_{out}.

V_{out} waveform:

As you decrease the amount of resistance between the base and V_{CC} in this circuit, the amount of (*forward*, *reverse*) ____________ bias at the emitter-base junction becomes less, and the transistor begins conducting over a (*greater*, *lesser*) ______________ portion of the input ac waveform.

BJT AMPLIFIER CLASSES OF OPERATION

Complete the following review questions, indicating the appropriate response by placing a check in the box next to the correct answer.

1. A Class A amplifier actually amplifies ______________ of an ac sine-wave that is applied to the input.

   ☐ all 360°
   ☐ about 180°
   ☐ less than 180°
   ☐ none

2. The emitter-base junction of a Class A common-emitter BJT amplifier is ____________ biased.

   ☐ forward
   ☐ reverse
   ☐ zero

3. The output waveform of a Class A common-emitter BJT amplifier is ________ compared to its input waveform.

   ☐ distorted
   ☐ shifted 90°
   ☐ shifted 180°

4. A Class A amplifier is _______________ when there is no signal applied at the input.

   ☐ conducting
   ☐ nonconducting

5. The emitter-base junction of a Class B common-emitter BJT amplifier is _____________ biased.

   ☐ forward
   ☐ reverse
   ☐ zero

6. A Class B amplifier actually amplifies ______________ of an ac sine-wave that is applied to the input.

   ☐ all 360°
   ☐ about 180°
   ☐ much less than 180°
   ☐ none

7. A Class B amplifier is _______________ when there is no signal applied at the input.

   ☐ conducting
   ☐ nonconducting

BJT AMPLIFIER CLASSES OF OPERATION

8. A Class C amplifier actually amplifies ____________ of an ac sine-wave that is applied to the input.

    - ☐ all 360°
    - ☐ about 180°
    - ☐ much less than 180°
    - ☐ none

9. The emitter-base junction of a Class C common-emitter BJT amplifier is ____________ biased.

    - ☐ forward
    - ☐ reverse
    - ☐ zero

10. Of all the classes of amplifier operation, the Class C amplifier is noted for

    - ☐ the least amount of signal distortion, and the highest amount of efficiency
    - ☐ the least amount of signal distortion, but the least amount of efficiency
    - ☐ the greatest amount of signal distortion, but the highest amount of efficiency
    - ☐ the greatest amount of signal distortion, and the least amount of efficiency

JFET CHARACTERISTICS AND AMPLIFIERS

Objectives

You will connect a JFET with proper biasing, and measure the current voltage levels required for plotting a pair of drain characteristic curves. You will also connect and observe the operation of a JFET amplifier circuit.

In completing these projects, you will connect circuits, make measurements, perform calculations, draw conclusions, and be able to answer questions about the following items related to JFET biasing and amplifier characteristics.

- Effects of bias-voltage levels and polarities upon the amount of drain current
- Effects of bias-voltage levels and polarities on the class of amplifier operation
- Differences between N-channel and P-channel JFET amplifier operation
- Features that distinguish classes of JFET amplifiers

PROJECT/TOPIC CORRELATION INFORMATION

PROJECT		TEXT CHAPTER	SECTION	RELATED TEXT TOPIC(S)
81	JFET Biasing and Drain Characteristic Curves	28	28-1	JFET Characteristic Curves and Ratings
82	Common-Source JFET Amplifier	28	28-1	Biasing JFET Amplifiers JFET Amplifier Circuits

PROJECT 81

JFET CHARACTERISTICS AND AMPLIFIERS

JFET Biasing and Drain Characteristic Curves

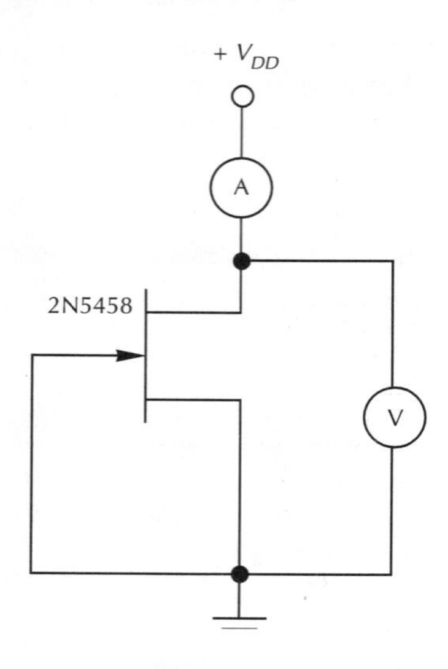

FIGURE 108

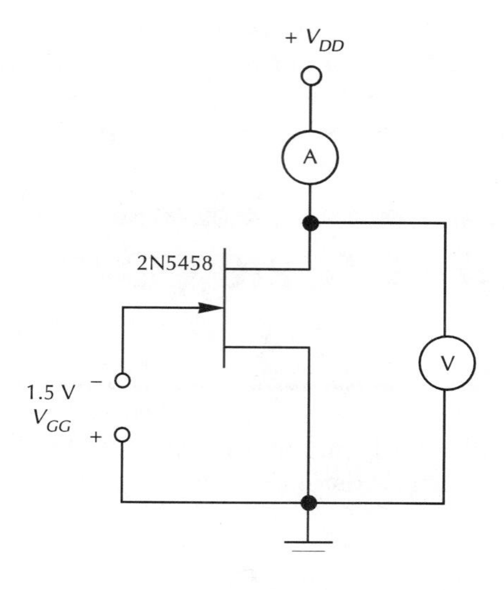

FIGURE 109

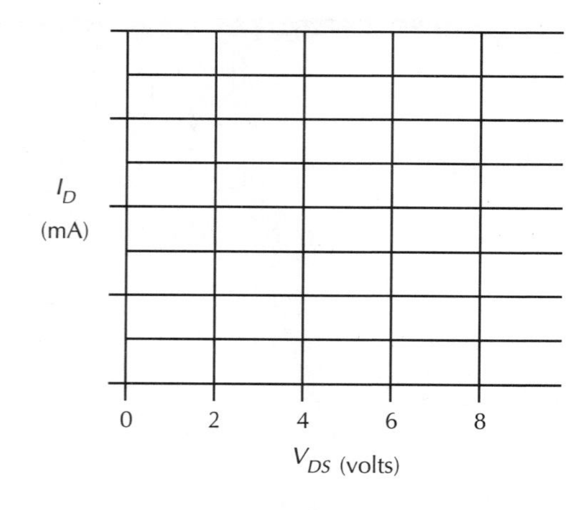

FIGURE 110

PROJECT PURPOSE

The purpose of this project is to demonstrate how the amounts of drain voltage and gate voltage affect the flow of drain current. You will develop a set of two drain characteristic curves for the JFET.

PARTS NEEDED

- ☐ DMM/VOM (2)
- ☐ VVPS (dc)
- ☐ CIS
- ☐ 1.5-volt cell
- ☐ N-channel JFET: 2N5458 (or equivalent)

Ideally, you will use two meters for this experiment: one for monitoring V_{DS} and a second for monitoring I_D.

ACTIVITY	OBSERVATION	CONCLUSION
1. Connect the circuit exactly as shown in Figure 108.	—	—
2. Set the VVPS so that the source-to-drain voltage (V_{DS}) is zero volts, then measure and record the amount of drain current (I_D) in the Observation column. Adjust the VVPS so that V_{DS} is at 2 V and record the amount of drain current. Continue to increase the value of V_{DS} according to the data in the Observation column and record the corresponding amount of drain current.	V_{GG} = 0 V V_{DS} = 0 V; I_D = ______ mA V_{DS} = 2 V; I_D = ______ mA V_{DS} = 4 V; I_D = ______ mA V_{DS} = 6 V; I_D = ______ mA V_{DS} = 8 V; I_D = ______ mA V_{DS} = 10 V; I_D = ______ mA	The maximum amount of drain current flows through the channel region of a JFET when V_{GS} is at zero volts. (*True, False*) ______. The maximum amount of drain current is called (*pinch-off current, leakage current,* I_{DSS}) ______. The source-gate junction in Figure 108 (*has no bias, is reverse biased, is forward biased*) ______. Plot the data you found for V_{GS} = 0 on the graph in Figure 110.
3. Turn off the VVPS and insert the 1.5-volt cell into the gate-source circuit of the JFET as shown in Figure 109.	—	—
4. Turn on the VVPS and set it for a V_{DS} of zero volts. Repeat the procedure in Activity step 2, recording the drain currents in the Observation column.	V_{GG} = −1.5 V V_{DS} = 0 V; I_D = ______ mA V_{DS} = 2 V; I_D = ______ mA V_{DS} = 4 V; I_D = ______ mA V_{DS} = 6 V; I_D = ______ mA V_{DS} = 8 V; I_D = ______ mA V_{DS} = 10 V; I_D = ______ mA	An N-channel JFET is (*an enhancement-mode, a depletion-mode*) ______ device. As such, the amount of drain current (*increases, decreases*) ______ as the V_{GS} increases in the negative direction. The source-gate junction in Figure 109 (*has no bias, is reverse biased, is forward biased*) ______. Plot the data you found for V_{GS} = −1.5 V on the graph in Figure 110.

PROJECT

82

JFET CHARACTERISTICS AND AMPLIFIERS

Common-Source JFET Amplifier

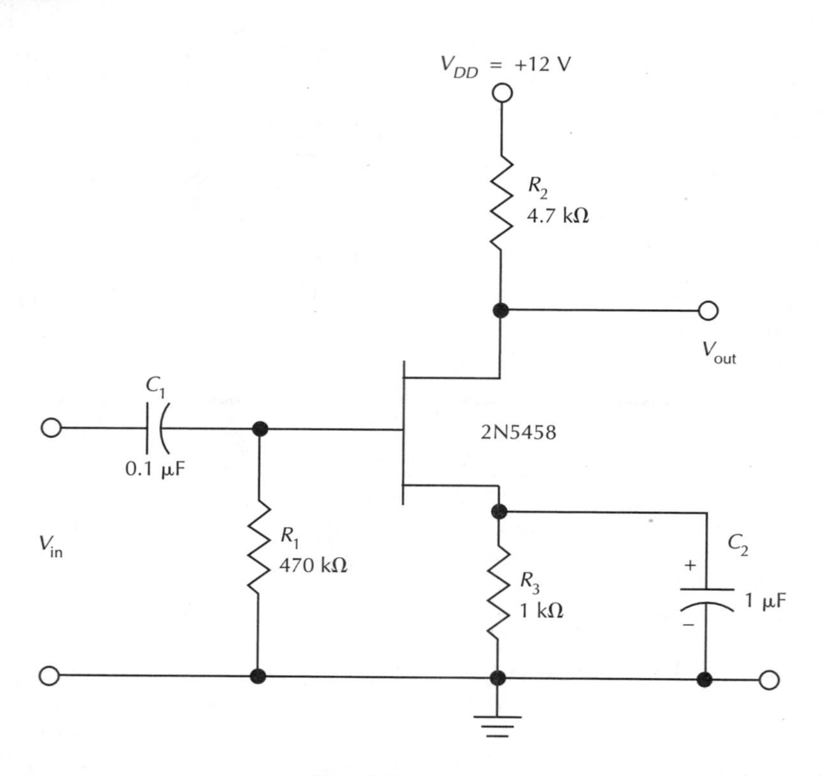

FIGURE 111

PROJECT PURPOSE

For this project you will construct a common-source JFET amplifier circuit, and you will use the DMM/VOM and dual-trace oscilloscope to determine the circuit's main operating features.

PARTS NEEDED

- ☐ DMM/VOM
- ☐ VVPS (dc)
- ☐ Dual-trace oscilloscope
- ☐ Function generator or audio oscillator
- ☐ CIS
- ☐ N-channel JFET: 2N5458 (or equivalent)
- ☐ Resistors:
  470 kΩ
  1 kΩ
  4.7 kΩ
- ☐ Capacitors:
  0.1 μF
  1 μF

ACTIVITY / OBSERVATION / CONCLUSION

1. Connect the circuit as shown in Figure 111. Set the dc source for a V_{DD} of +12 V. Connect the signal source (audio oscillator or function generator in the sine-wave mode) to V_{in}, but make sure its output is set for zero volts output.

—

—

2. Measure the dc voltages with respect to circuit common as required in the Observation column.

V_S (voltage from source to common) = ______________________ Vdc
V_G (voltage from gate to common) = ______________________ Vdc
V_D (voltage from drain to common) = ______________________ Vdc

Use the formula, $V_{DS} = V_D - V_S$, to determine the voltage drop between the source and drain. V_{DS} = ______. Based on the voltages you observed in this step, do you have good reason to suppose this is a Class A amplifier? ________ Explain your response. ________

---

3. Connect one channel of the oscilloscope to V_{in}, and connect the second channel of the oscilloscope to V_{out}. Adjust the signal source at V_{in} for a signal of 1 kHz at 0.5 V peak-to-peak. If the signal you find at V_{out} is distorted on either or both alternations, reduce the input signal level until the distortion is no longer observed. Sketch the waveforms and determine the readings specified in the Observation column.

V_{in} waveform:

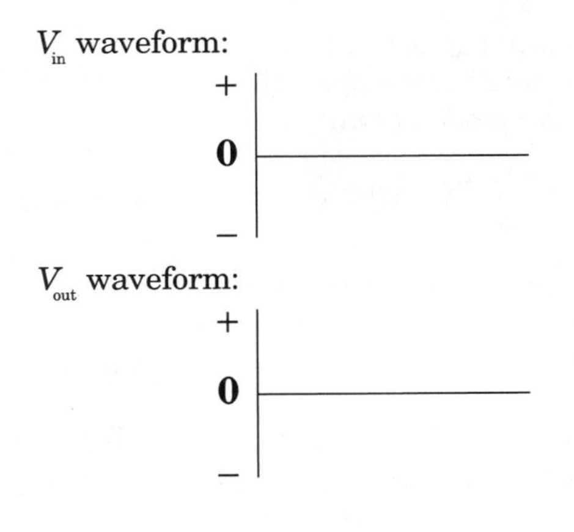

V_{out} waveform:

V_{in} = ____________ V peak-to-peak
V_{out} = ____________ V peak-to-peak

The oscilloscope traces indicate that the output voltage from a common-source amplifier is (*in phase, 180° out of phase*) with its input waveform. Use the readings you found for V_{in} and V_{out} to determine the voltage gain (A_V) of this circuit. A_V = ____________. (Recall that $A_V = V_{out}/V_{in}$.) You have seen in this circuit that the voltage gain of a common-source amplifier can be greater than 1. (*True*, *False*) ____________

---

4. Decrease the setting of the dc source for a V_{DD} of +9 V. If either peak of the waveform at V_{out} is clipped, reduce the input signal level until the clipping effect is no longer apparent. Gather the information requested in the Observation column.

V_{in} waveform:

+
0
–

V_{out} waveform:

+
0
–

V_{in} = ____________ V peak-to-peak
V_{out} = ____________ V peak-to-peak

Use the readings for V_{in} and V_{out} to determine the voltage gain of this circuit. A_V = __________. Decreasing the value of V_{DD} increases the voltage gain (*True*, *False*) ______.

JFET CHARACTERISTICS AND AMPLIFIERS

Complete the following review questions, indicating the appropriate response by placing a check in the box next to the correct answer.

1. When a JFET is properly used, the drain current is proportional to the gate current.

   ☐ True
   ☐ False

2. The source-gate junction of a JFET requires little bias current and signal current because it is

   ☐ normally reverse biased
   ☐ normally forward biased
   ☐ normally not used
   ☐ made of a nonconductive metal oxide film

3. A JFET should always be biased for

   ☐ depletion mode operation
   ☐ enhancement mode operation
   ☐ sine-wave mode operation

4. For an N-channel JFET, making the gate more negative with respect to the source

   ☐ causes drain current to increase
   ☐ causes drain current to decrease
   ☐ has no significant effect on drain current

5. For a common-source amplifier that uses an N-channel JFET, making the gate more negative with respect to the source causes the drain voltage to increase in the __________ direction.

   ☐ positive
   ☐ negative

6. The main difference between an N-channel JFET amplifier and a P-channel JFET amplifier is

   ☐ the N-channel version inverts the incoming signal, whereas the corresponding P-channel version does not
   ☐ the polarity of the bias voltages
   ☐ the N-channel version operates in the depletion mode, whereas the corresponding P-channel version operates in the enhancement mode

7. A common-drain JFET amplifier

   ☐ is impractical
   ☐ has a very low input impedance
   ☐ is sometimes called a voltage follower
   ☐ inverts the input signal

8. When an N-channel JFET is operating as a Class B amplifier, you would expect to find

   ☐ a relatively large negative voltage on the gate
   ☐ a relatively small negative voltage on the gate
   ☐ a relatively large positive voltage on the gate

9. You can distinguish a common-gate JFET amplifier by noting that

   ☐ the input is applied at the gate and the output is taken from the gate
   ☐ the input is applied to the drain and the output is taken from the source
   ☐ the input is applied to the gate and the output is taken from the source
   ☐ none of these

10. Inserting a resistor in series with the source terminal of an N-channel JFET can make the source more positive than the gate. This principle is the basis for

   ☐ preventing the JFET from overheating
   ☐ increasing the current through the drain terminal
   ☐ establishing self-bias
   ☐ establishing Class C operation

OPERATIONAL AMPLIFIERS

Objectives

You will connect basic inverting and noninverting op-amp circuits, and then you will compare their input and output signal levels and phases. You will also study the operation of a Schmitt trigger circuit built with op-amp circuitry.

In completing these projects, you will connect circuits, observe waveforms with an oscilloscope, draw conclusions, and answer questions about the following items related to op-amp circuits.

- Explain how the location of input and feedback resistors determines whether the circuit operates as an inverting or noninverting amplifier
- Explain how the ratio of feedback to input resistance affects the voltage gain of op-amp circuits
- Describe the operation of a simple op-amp Schmitt trigger circuit

PROJECT/TOPIC CORRELATION INFORMATION

PROJECT		TEXT CHAPTER	SECTION	RELATED TEXT TOPIC(S)
83	Inverting Op-Amp Circuit	29	29-2	An Inverting Amplifier
84	Noninverting Op-Amp Circuit	29	29-3	A Noninverting Amplifier
85	Op-Amp Schmitt Trigger Circuit	29	29-5	An Op-Amp Schmitt Trigger Circuit

PROJECT

83

OPERATIONAL AMPLIFIERS
Inverting Op-Amp Circuit

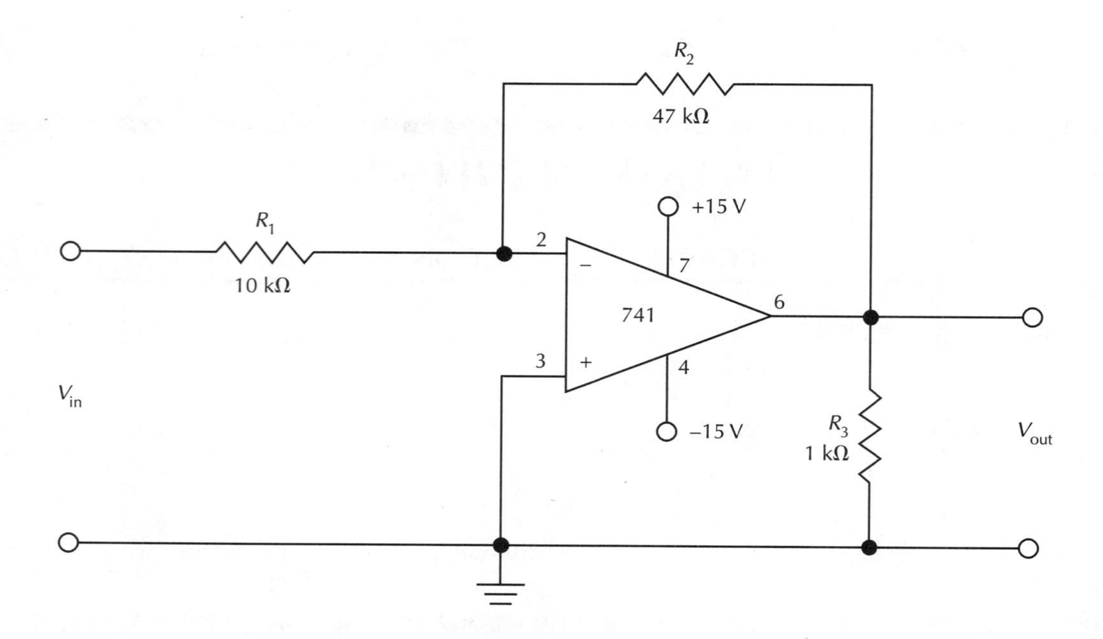

FIGURE 112

PROJECT PURPOSE

This project demonstrates the operation of a basic inverting op-amp.

PARTS NEEDED

- ☐ DMM/VOM
- ☐ Dual-trace oscilloscope
- ☐ ±15-Vdc power supply
- ☐ Function generator or audio oscillator
- ☐ CIS
- ☐ Operational amplifier: 741
- ☐ Resistors:
  - 1 kΩ
  - 10 kΩ
  - 47 kΩ
  - 100 kΩ

You will need the following formula to complete the work.

$$A_V = -(R_2/R_1)$$

where:

A_V is the voltage gain
R_2 is the value of the feedback resistor
R_1 is the value of the input resistor

ACTIVITY	OBSERVATION	CONCLUSION
1. Connect the circuit exactly as shown in Figure 112. Make sure the positive terminal of the ±15-Vdc supply is connected to pin 7 of the 741 IC, the negative terminal is connected to pin 4, and the COMMon terminal is connected to the common line as shown in the figure. Be sure you take all meter and oscilloscope readings with respect to the common line of the circuit (and not the −15-V connection of the power source). Measure and record the dc voltages requested in the Observation column.	Dc voltage from pin 7 to common = ______ V Dc voltage from pin 4 to common = ______ V	—
2. Connect the function generator (sine-wave mode) or audio oscillator to V_{in}. Connect one channel of the oscilloscope to V_{in}, and connect the second channel of the oscilloscope to V_{out}. Adjust the signal source for an input signal of 1 kHz at 1 V peak-to-peak. Sketch the waveforms and determine the readings specified in the Observation column.	Peak-to-peak voltage from V_{in} to common = ______ V Peak-to-peak voltage from V_{out} to common = ______ V V_{in} waveform: + 0 – V_{out} waveform: + 0 –	What is the calculated voltage gain of the circuit as shown in Figure 112? ______ What is the actual voltage gain of the circuit as determined by the measured values of V_{in} and V_{out}? ______ The output waveform is shifted (*0°, 90°, 180°*) ______ relative to the input waveform.

PROJECT
83
CONTINUED

OPERATIONAL AMPLIFIERS
Inverting Op-Amp Circuit *(Continued)*

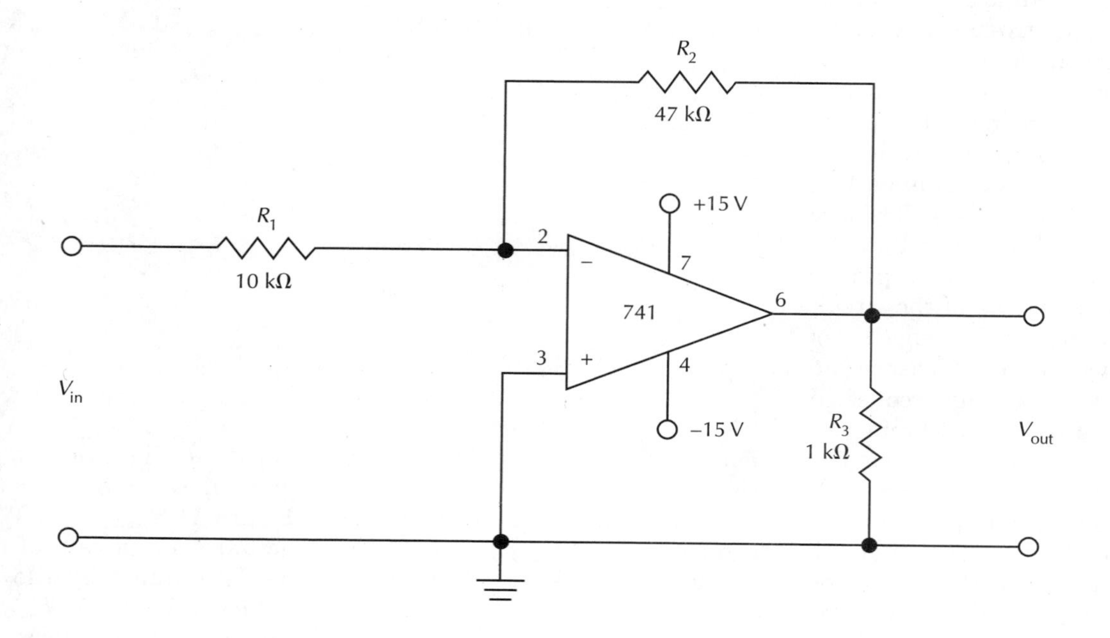

FIGURE 112

PROJECT PURPOSE

This project demonstrates the operation of a basic inverting op-amp.

PARTS NEEDED

- ☐ DMM/VOM
- ☐ Dual-trace oscilloscope
- ☐ ±15-Vdc power supply
- ☐ Function generator or audio oscillator
- ☐ CIS
- ☐ Operational amplifier: 741
- ☐ Resistors:
  - 1 kΩ
  - 10 kΩ
  - 47 kΩ
  - 100 kΩ

ACTIVITY	OBSERVATION	CONCLUSION
3. Increase the level of V_{in} to 2 V peak-to-peak; measure and record the values of V_{in} and V_{out}.	Peak-to-peak voltage from V_{in} to common = ______ V Peak-to-peak voltage from V_{out} to common = ______ V	What is the actual voltage gain of the circuit as determined by the measured values of V_{in} and V_{out} in Activity step 3? ______ Does increasing the value of V_{in} have any significant effect upon the value of V_{out}? ______. Does increasing the value of V_{in} have any significant effect upon the voltage gain of the circuit? ______.
4. Replace R_2 with a resistor having a value of 100 kΩ.	—	—
5. Adjust the signal source for 1 V peak-to-peak. Sketch the waveforms and determine the readings specified in the Observation column.	Peak-to-peak voltage from V_{in} to common = ______ V Peak-to-peak voltage from V_{out} to common = ______ V	What is the calculated voltage gain of the circuit as when R_2 = 100 kΩ? ______. What is the actual voltage gain of the circuit as determined by the measured values of V_{in} and V_{out}? ______ Does increasing the value of R_2 have any significant effect upon the voltage gain of the circuit? ______.

PROJECT 84

OPERATIONAL AMPLIFIERS
Noninverting Op-Amp Circuit

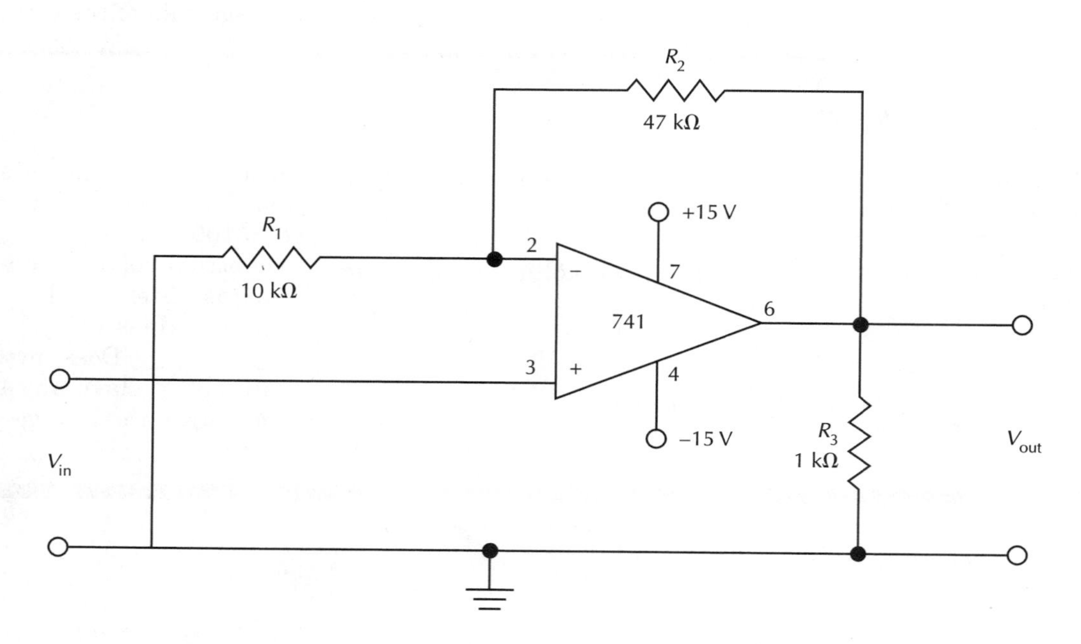

FIGURE 113

PROJECT PURPOSE

In this project you will observe the operation of a noninverting op-amp circuit, including a voltage-follower version. You will also see the effects of overdriving an op-amp circuit that has excessive voltage gain.

PARTS NEEDED

- ☐ DMM/VOM
- ☐ Dual-trace oscilloscope
- ☐ ±15-Vdc power supply
- ☐ Function generator or audio oscillator
- ☐ CIS
- ☐ Operational amplifier: 741
- ☐ Resistors:
  - 1 kΩ
  - 10 kΩ
  - 47 kΩ
  - 100 kΩ
  - 1 MΩ

The following formula can be helpful.

$$A_V = (R_2/R_1) + 1$$

where:

A_V is the voltage gain of a noninverting op-amp
R_2 is the value of the feedback resistor
R_1 is the value of the input resistor

ACTIVITY	OBSERVATION	CONCLUSION
1. Connect the circuit exactly as shown in Figure 113. Make sure the positive terminal of the ±15-Vdc supply is connected to pin 7 of the 741 IC, the negative terminal is connected to pin 4, and the COMMon terminal is connected to the common line as shown in the figure. Be sure you take all meter and oscilloscope readings with respect to the common line of the circuit (and not the −15-V connection of the power source). Measure and record the dc voltages requested in the Observation column.	Dc voltage from pin 7 to common = ______ V Dc voltage from pin 4 to common = ______ V	—
2. Connect the function generator (sine-wave mode) or audio oscillator to V_{in}. Connect one channel of the oscilloscope to V_{in}, and connect the second channel of the oscilloscope to V_{out}. Adjust the signal source for an input signal of 1 kHz at 1 V peak-to-peak. Sketch the waveforms and determine the readings specified in the Observation column.	Peak-to-peak voltage from V_{in} to common = ______ V Peak-to-peak voltage from V_{out} to common = ______ V V_{in} waveform: + 0 − V_{out} waveform: + 0 −	What is the calculated voltage gain of the circuit as shown in Figure 113? ______. What is the actual voltage gain of the circuit as determined by the measured values of V_{in} and V_{out}? ______. The output waveform is shifted (*0°, 90°, 180°*) ______ relative to the input waveform.

PROJECT
84
CONTINUED

OPERATIONAL AMPLIFIERS
Noninverting Op-Amp Circuit
(Continued)

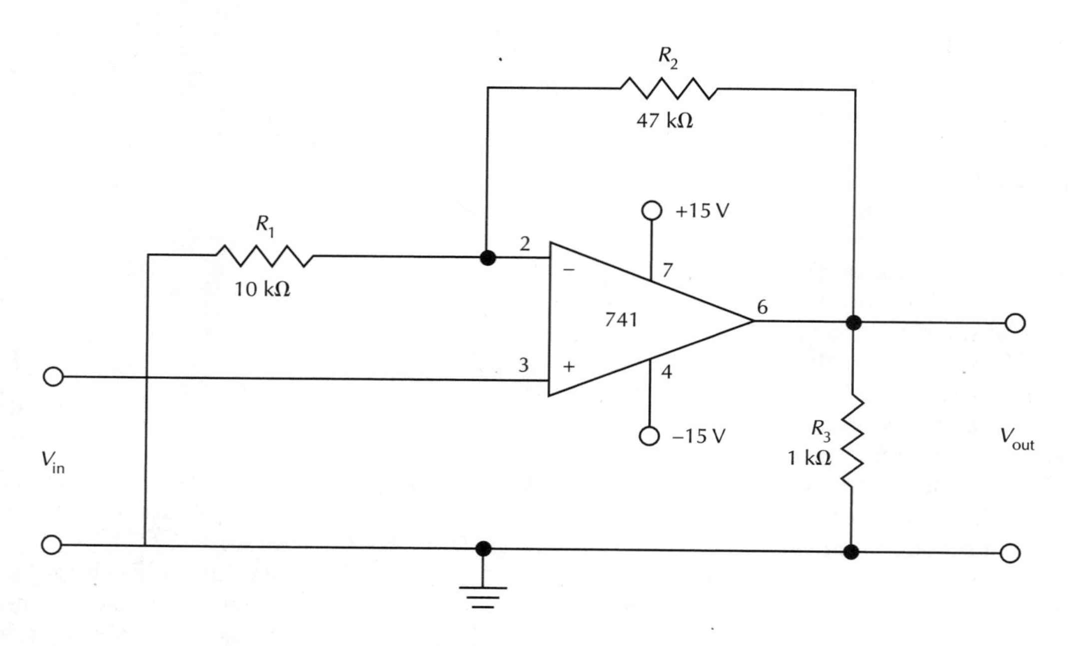

FIGURE 113

PROJECT PURPOSE

In this project you will observe the operation of a noninverting op-amp circuit, including a voltage-follower version. You will also see the effects of overdriving an op-amp circuit that has excessive voltage gain.

PARTS NEEDED

- ☐ DMM/VOM
- ☐ Dual-trace oscilloscope
- ☐ ±15-Vdc power supply
- ☐ Function generator or audio oscillator
- ☐ CIS
- ☐ Operational amplifier: 741
- ☐ Resistors:
  - 1 kΩ
  - 10 kΩ
  - 47 kΩ
  - 100 kΩ
  - 1 MΩ

ACTIVITY	OBSERVATION	CONCLUSION
3. Replace R_2 with a resistor having a value of 10 kΩ.	—	—
4. Leave the signal source at 1 V peak-to-peak. Measure and record the data specified in the Observation column.	Peak-to-peak voltage from V_{in} to common = ________ V Peak-to-peak voltage from V_{out} to common = ________ V	What is the calculated voltage gain of the circuit when R_2 = 10 kΩ? ________. What is the actual voltage gain of the circuit as determined by the measured values of V_{in} and V_{out}? ________. Explain why this particular circuit might be called a voltage follower. ________ ________
5. Replace R_2 with a resistor having a value of 1 MΩ.	—	—
6. Leave the signal source at 1 V peak-to-peak. Sketch the waveforms and determine the readings specified in the Observation column.	Peak-to-peak voltage from V_{in} to common = ________ V Peak-to-peak voltage from V_{out} to common = ________ V	What is the calculated voltage gain of the circuit when R_2 = 1 MΩ? ________. Explain why the output waveform is distorted. ________

V_{in} waveform:

+

0

–

V_{out} waveform:

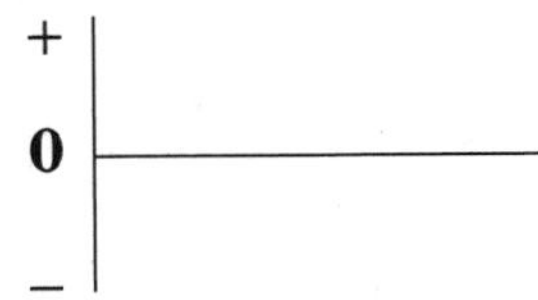

PROJECT
85

OPERATIONAL AMPLIFIERS
Op-Amp Schmitt Trigger Circuit

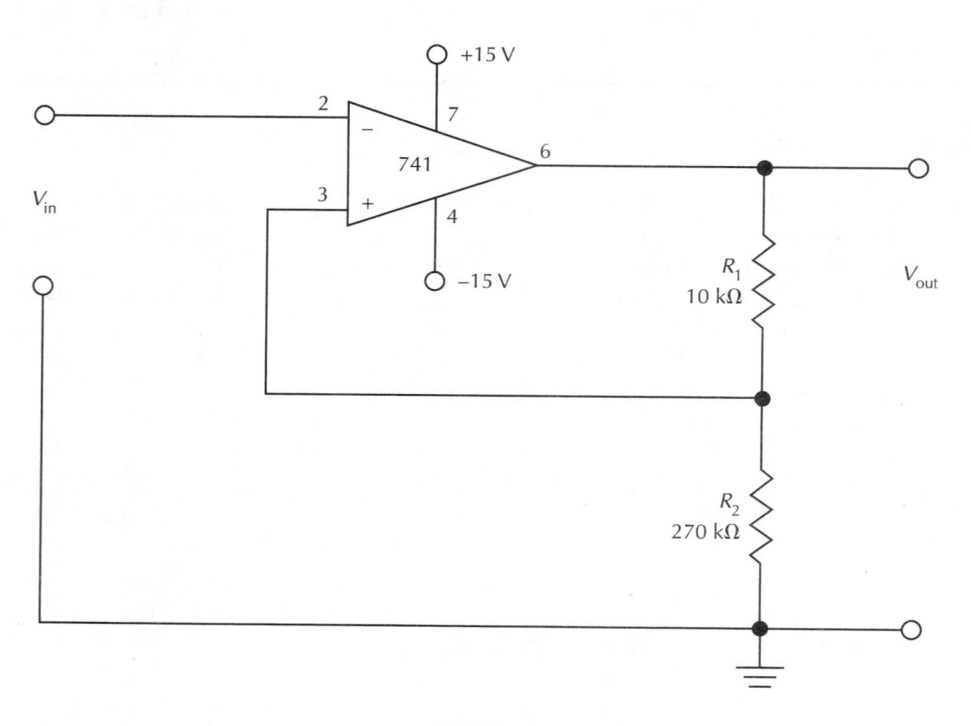

FIGURE 114

PROJECT PURPOSE

In this project you will observe the application of an op-amp as Schmitt trigger circuit.

PARTS NEEDED

- ☐ DMM/VOM
- ☐ Dual-trace oscilloscope
- ☐ ±15-Vdc power supply
- ☐ Function generator or audio oscillator
- ☐ CIS
- ☐ Operational amplifier: 741
- ☐ Resistors:
  - 10 kΩ
  - 270 kΩ

ACTIVITY	OBSERVATION	CONCLUSION
1. Connect the circuit exactly as shown in Figure 114. Make sure the positive terminal of the ±15-Vdc supply is connected to pin 7 of the 741 IC, the negative terminal is connected to pin 4, and the common terminal is connected to the common line as shown in the figure. (Be sure you take all meter and oscilloscope readings with respect to the common line of the circuit.)	—	—
2. Measure and record the dc voltages requested in the Observation column.	Dc voltage from pin 7 to common = ______ V Dc voltage from pin 4 to common = ______ V	Explain why having an ac input waveform makes it necessary to have both positive and negative supply voltages for an op-amp circuit. ______
3. Connect the function generator (sine-wave mode) or audio oscillator to V_{in}. Connect one channel of the oscilloscope to V_{in}, and connect the second channel of the oscilloscope to V_{out}. Adjust the signal source for an input signal of 1 kHz at 1 V peak-to-peak.	—	—
4. Sketch the waveforms and determine the readings specified in the Observation column.	Peak-to-peak voltage from V_{in} to common = ______ V Peak-to-peak voltage from V_{out} to common = ______ V V_{in} waveform: 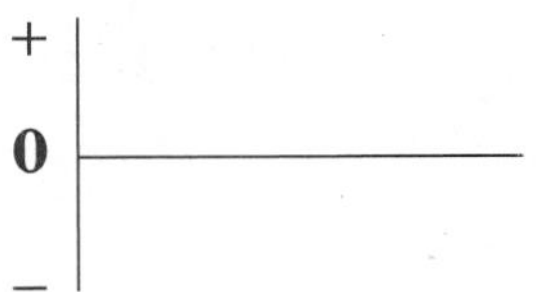 V_{out} waveform: 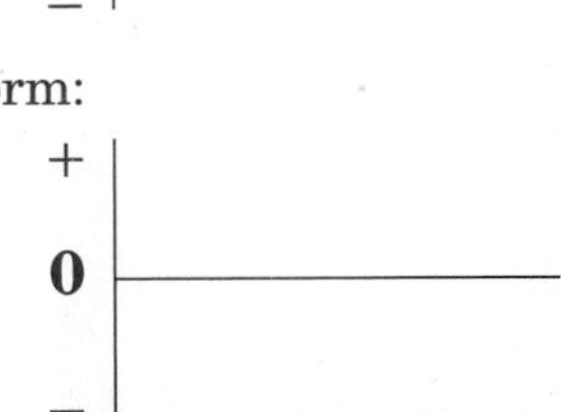	Calculate the following values for the Schmitt trigger circuit in Figure 114: *UTP* (calculated) = ____ V. *LTP* (calculated) = ____ V. Hysteresis = ______ V. If you superimpose oscilloscope waveforms for V_{in} and V_{out}, you should be able to determine the actual values of: *UTP* (measured) = ______ V. *LTP* (measured) = ______ V. Hysteresis = ______ V. The polarity of the output waveform is shifted (*0°, 90°, 180°*) ______ relative to the input waveform. This is an example of a nonlinear amplifier circuit. (*True, False*) ______.

PROJECT 85 CONTINUED

OPERATIONAL AMPLIFIERS
Op-Amp Schmitt Trigger Circuit
(Continued)

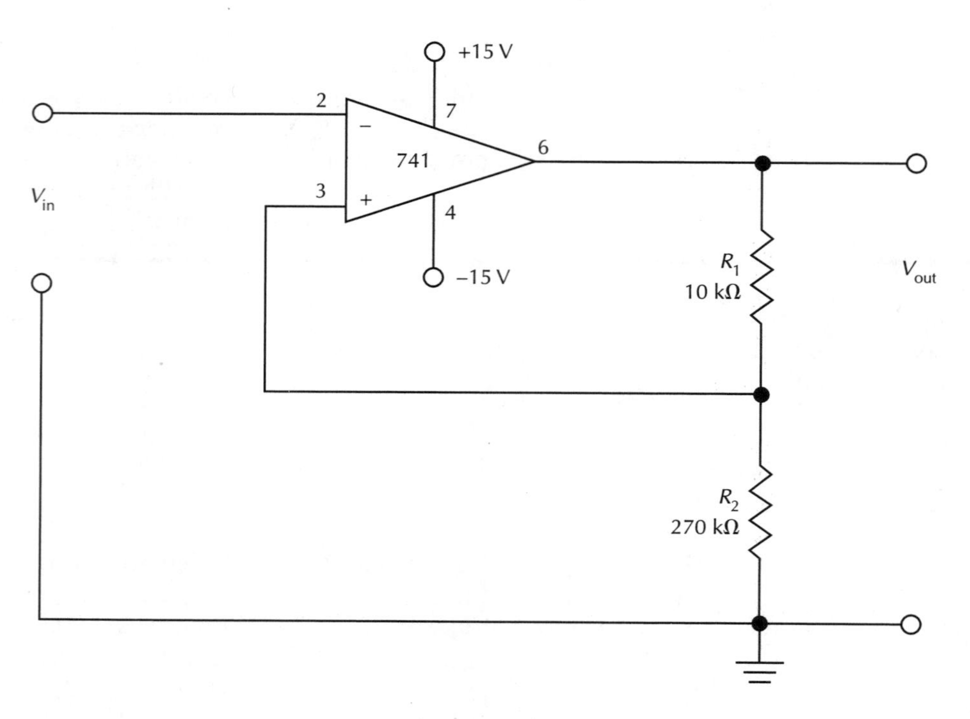

FIGURE 114

PROJECT PURPOSE

In this project you will observe the application of an op-amp as Schmitt trigger circuit.

PARTS NEEDED

- ☐ DMM/VOM
- ☐ Dual-trace oscilloscope
- ☐ ±15-Vdc power supply
- ☐ Function generator or audio oscillator
- ☐ CIS
- ☐ Operational amplifier: 741
- ☐ Resistors:
  - 10 kΩ
  - 270 kΩ

ACTIVITY	OBSERVATION	CONCLUSION
5. Adjust the signal source for an input signal 0.1 V peak-to-peak. Sketch the waveforms and determine the readings specified in the Observation column.	Peak-to-peak voltage from V_{in} to common = ______ V Peak-to-peak voltage from V_{out} to common = ______ V V_{in} waveform: + 0 – V_{out} waveform: + 0 –	Describe how the waveforms are different from those in Activity step 4. ______ How do you account for the differences? ______

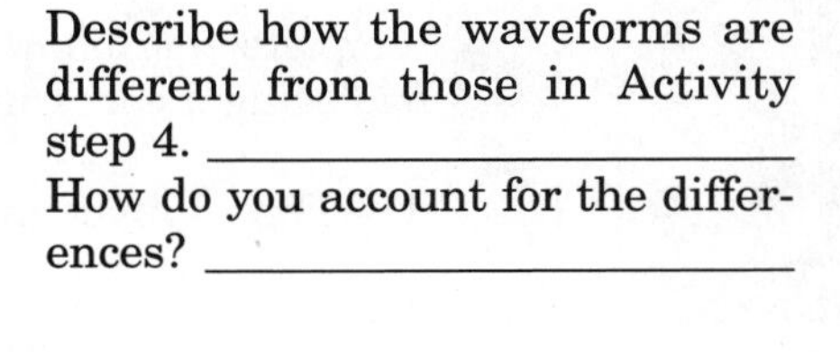

OPERATIONAL AMPLIFIERS

Complete the following review questions, indicating the appropriate response by placing a check in the box next to the correct answer.

1. For an inverting op-amp

   - ☐ the input and feedback resistors are both connected to the non-inverting input
   - ☐ the input resistor is connected to the inverting input and the feedback resistor is connected to the noninverting input
   - ☐ the input resistor is connected to the noninverting input and the feedback resistor is connected to the inverting input
   - ☐ the input and feedback resistors are both connected to the inverting input

2. The voltage gain of an inverting op-amp is determined by

   - ☐ dividing the amount of feedback resistance by the amount of input resistance
   - ☐ dividing the amount of input resistance by the amount of feed-back resistance
   - ☐ dividing the amount of input voltage by the value of the input resistance
   - ☐ multiplying the amount of input voltage by the amount of output voltage

3. An inverting op-amp circuit can never be used as a voltage follower because

   - ☐ the output voltage is always larger than the input voltage
   - ☐ the output waveform is always out of phase with the input wave-form
   - ☐ the output is nonlinear

4. What is the voltage gain of an inverting op-amp circuit when the input resistance equals the feedback resistance?

   - ☐ Zero
   - ☐ One
   - ☐ Cannot be determined without knowing the values

5. For a noninverting op-amp

   - ☐ the input is applied to the noninverting input and the feedback resistor is connected to the noninverting input
   - ☐ the input is applied to the inverting input and the feedback resistor is connected to the noninverting input
   - ☐ the input is applied to the noninverting input and the feedback resistor is connected to the inverting input
   - ☐ the input is applied to the inverting input and the feedback resistor is connected to the inverting input

6. The voltage gain of an noninverting op-amp

   - ☐ can be determined by dividing the amount of feedback resistance by the amount of input resistance
   - ☐ can be determined by dividing the amount of output voltage by the amount of input voltage
   - ☐ can be determined by dividing the amount of input voltage by the value of the feedback resistance
   - ☐ is always 1

7. What is the voltage gain of a noninverting amplifier when the feedback resistance is zero ohms?

   - ☐ Zero
   - ☐ One
   - ☐ Two
   - ☐ Cannot be determined without knowing the resistor values

8. As long as the output of an op-amp remains undistorted, the amount of input voltage has little to do with the amount of voltage gain for the circuit.

   - ☐ True
   - ☐ False

9. One of the main purposes of a Schmitt trigger circuit is to transform a sine waveform into a rectangular waveform.

   - ☐ True
   - ☐ False

10. The fact that the input signal for an op-amp Schmitt trigger circuit goes to the inverting input accounts for

    - ☐ the flattening of the output waveform
    - ☐ the very high voltage gain of the circuit
    - ☐ the 180° phase shift of the waveform
    - ☐ none of these

OSCILLATORS AND MULTIVIBRATORS

Objectives

In completing these projects, you will connect circuits, make oscilloscope measurements, perform calculations, draw conclusions, and answer questions about the following items related to Wien-bridge oscillators and 555-type astable multivibrators.

- Point out which components in the circuits determine their operating frequencies
- Describe how each frequency-determining component affects the operating frequency
- Explain the basic principles of lead-lag phase shift in a Wien-bridge oscillator
- Explain the basic principles of capacitor charge and discharge in a 555-type astable multivibrator

PROJECT/TOPIC CORRELATION INFORMATION

PROJECT	TEXT CHAPTER	SECTION	RELATED TEXT TOPIC(S)
86 Wien-Bridge Oscillator Circuit	30	30-4	Wien-Bridge Oscillator
87 555 Astable Multivibrator	30	30-5	Astable Multivibrators

PROJECT
86

OSCILLATORS AND MULTIVIBRATORS
Wien-Bridge Oscillator Circuit

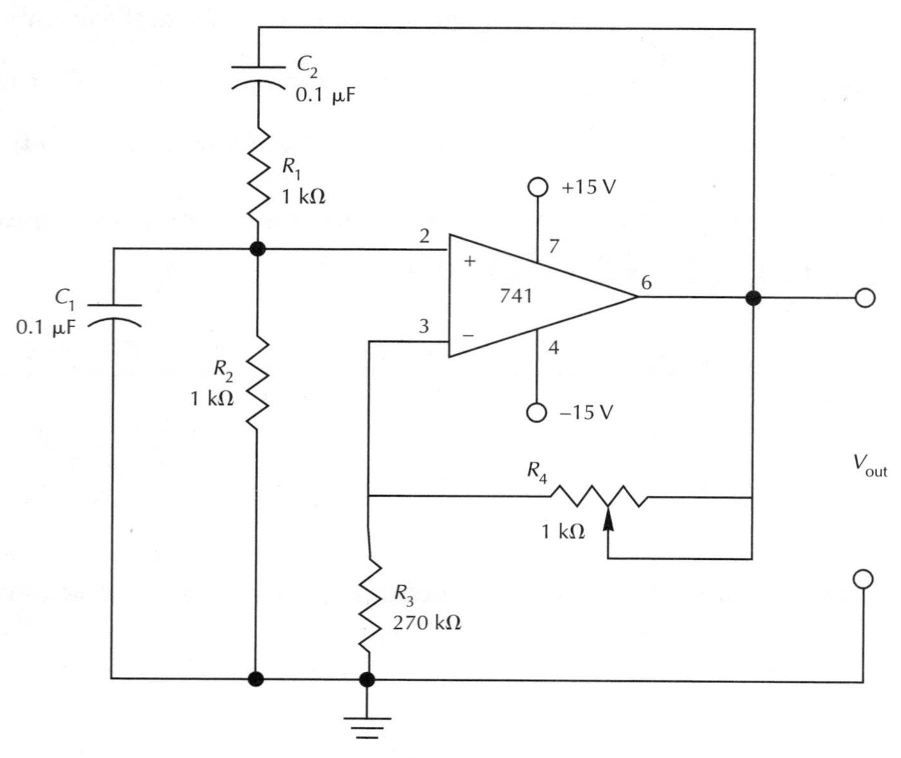

FIGURE 115

PROJECT PURPOSE

This project gives you a chance to check out the operation of a Wien-bridge oscillator circuit.

PARTS NEEDED

- ☐ Oscilloscope
- ☐ ±15-Vdc power supply
- ☐ CIS
- ☐ Operational amplifier: 741
- ☐ Resistors:
  270 Ω,
  1 kΩ (2)
  2.7 kΩ (2)
  27 kΩ
  1-kΩ linear potentiometer
- ☐ Capacitors:
  0.1 μF (2)

You will need the following formula for calculating the operating frequency of a Wien-bridge oscillator.

$$f_r = 1/2\pi RC$$

where:

f_r is the operating frequency of the circuit
R is the value assigned to both resistors in the lead-lag network
C is the value assigned to both capacitors in the lead-lag network

ACTIVITY	OBSERVATION	CONCLUSION
1. Connect the circuit shown in Figure 115.	—	List the components that make up the lead-lag network. ___________. This op-amp is connected as (*an inverting, a noninverting*) ______ ____________________ amplifier.
2. Connect the oscilloscope to V_{out} and adjust potentiometer R_3 until you see a maximum level of stable oscillation.	—	—
3. Sketch two complete cycles of the output waveform in the space provided in the Observation column.	Output waveform: + 0 –	Describe the form of the output waveform as sinusoidal, rectangular, sawtooth, or triangular: ____________________.
4. Measure the peak-to-peak amplitude of the waveform at V_{out}.	V_{out} (peak-to-peak) = __________ V	The amplitude of the output waveform is mainly determined by (*the gain of the amplifier, the values of the components in the lead-lag circuit*) ____________________.
5. Measure the period of the waveform at V_{out}.	*Period* = ________________ ms	Calculate the operating frequency of this circuit according to the formula for the operating frequency of a Wien-bridge oscillator. _____ kHz. Use the formula, $f = 1/period$, to determine the actual operating frequency. ______ kHz. Account for the difference, if any, between the calculated and measured frequency. ____________________

PROJECT
86
CONTINUED

OSCILLATORS AND MULTIVIBRATORS
Wien-Bridge Oscillator Circuit
(Continued)

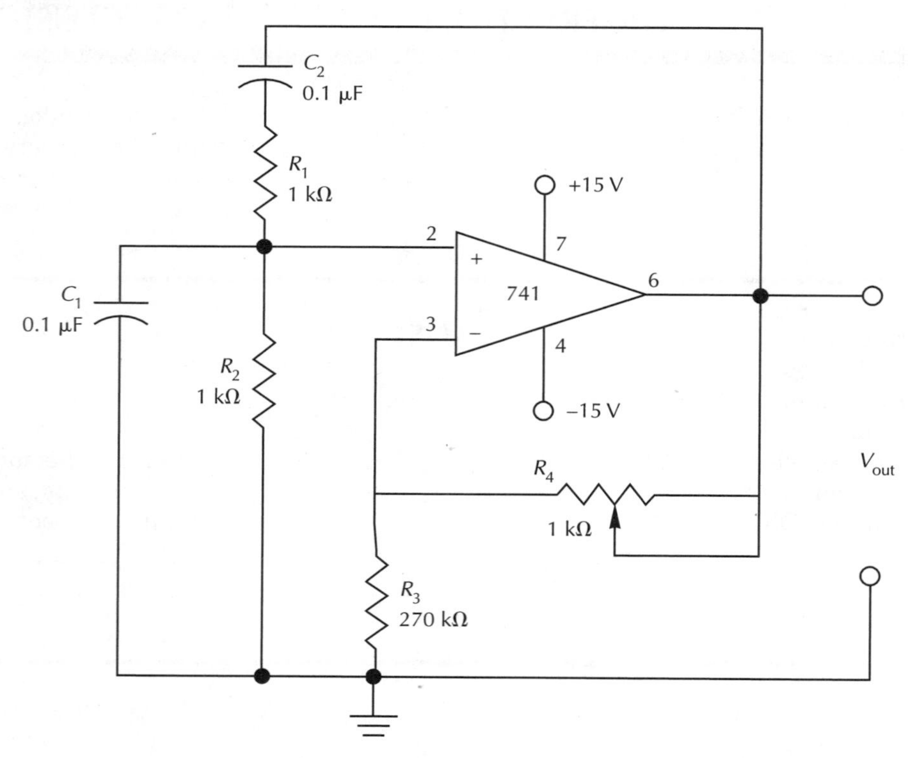

FIGURE 115

PROJECT PURPOSE

This project gives you a chance to check out the operation of a Wien-bridge oscillator circuit.

PARTS NEEDED

- ☐ Oscilloscope
- ☐ ±15-Vdc power supply
- ☐ CIS
- ☐ Operational amplifier: 741
- ☐ Resistors:
  270 Ω,
  1 kΩ (2)
  2.7 kΩ (2)
  27 kΩ
  1-kΩ linear potentiometer
- ☐ Capacitors:
  0.1 μF (2)

ACTIVITY	OBSERVATION	CONCLUSION
6. Replace resistors R_1 and R_2 in the circuit with 2.7-kΩ resistors. Adjust potentiometer R_3, if necessary, to get a maximum level of stable oscillation.	—	According to the formula for the operating frequency of a Wien-bridge oscillator, increasing the value of the lead-lag resistors should (*increase, decrease, have no effect on*) ______ the operating frequency.
7. Measure the period of the waveform at the output.	*Period* = ______ ms	The actual operating frequency of the circuit is now ______ Hz.

PROJECT 87

OSCILLATORS AND MULTIVIBRATORS

555 Astable Multivibrator

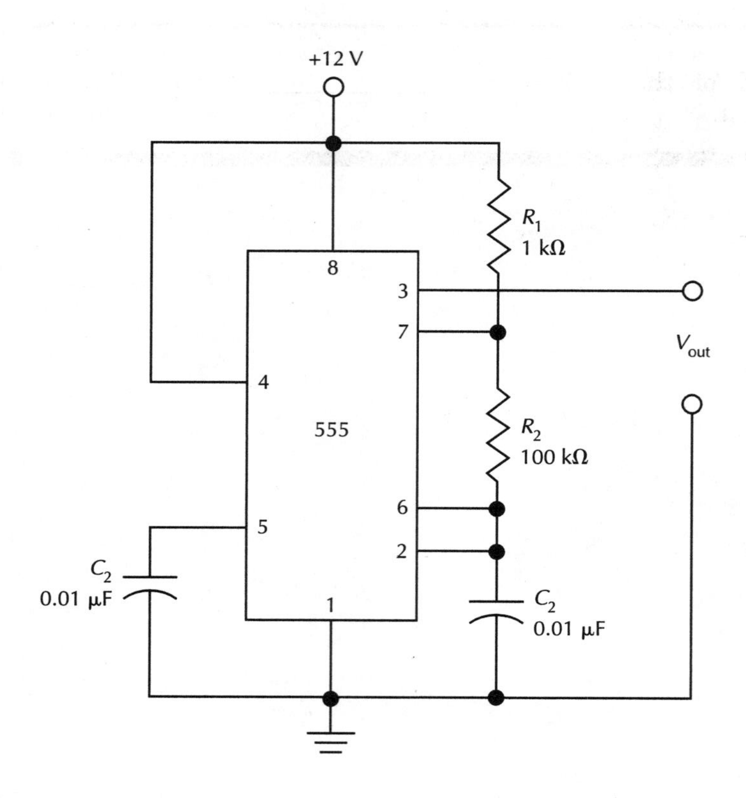

FIGURE 116

PROJECT PURPOSE

In this project you will use a 555 timer IC to construct and observe waveforms for an astable multivibrator.

PARTS NEEDED

- ☐ DMM/VOM
- ☐ Oscilloscope
- ☐ VVPS (dc)
- ☐ CIS
- ☐ 555 timer IC
- ☐ Resistors:
  - 1 kΩ
  - 10 kΩ
  - 100 kΩ
- ☐ Capacitors:
  - 0.01 μF (2)

You will need the following formula for calculating the operating frequency of this astable multivibrator.

$$f = 1/0.69C(R_A + 2R_B)$$

where:
f is the output frequency
C is the value of the timing capacitor
R_A and R_B are the values of the timing resistors

ACTIVITY	OBSERVATION	CONCLUSION
1. Connect the circuit exactly as shown in Figure 116.	—	Which resistor in the formula for operating frequency corresponds to resistor R_1 in this circuit? ________ Which corresponds to resistor R_2? ________
2. Connect the oscilloscope to V_{out} and sketch two complete cycles of the output waveform in the space provided in the Observation column.	Output waveform: + 0 −	Describe the form of the output waveform as sinusoidal, rectangular, sawtooth, or triangular: ________.
3. Measure the peak-to-peak amplitude and period of the waveform at V_{out}.	V_{out} (peak-to-peak) = ________ V *Period* = ________ ms	Calculate the operating frequency of this circuit according to the formula for the operating frequency of 555-type astable multivibrators. ________ kHz. Use the formula, $f = 1/period$, to determine the actual operating frequency. ________ kHz. Account for the difference, if any, between the calculated and measured frequency. ________
4. Replace resistor R_2 with a 10-kΩ resistor.	—	According to the formula for the operating frequency of a 555-type astable multivibrator, decreasing the value of either resistor should (*increase, decrease, have no effect on*) ________ the operating frequency.
5. Measure the peak-to-peak amplitude and period of the waveform at V_{out}.	V_{out} (peak-to-peak) = ________ V *Period* = ________ ms	The actual operating frequency of the circuit is now ________ Hz.

OSCILLATORS AND MULTIVIBRATORS

Complete the following review questions, indicating the appropriate response by placing a check in the box next to the correct answer.

1. At the frequency of oscillation, the lead-lag *RC* network in a Wien-bridge oscillator produces an overall phase shift of ____________ degrees.

   ☐ 0
   ☐ 90
   ☐ 180
   ☐ 270

2. In an op-amp version of a Wien-bridge oscillator, the lead-lag network is connected to the __________________ input of the op-amp.

   ☐ inverting
   ☐ noninverting
   ☐ voltage offset

3. In a Wien-bridge oscillator, decreasing the value of either or both resistors in the lead-lag network

   ☐ increases the operating frequency
   ☐ decreases the operating frequency
   ☐ has no effect on the operating frequency

4. In a Wien-bridge oscillator, increasing the value of either or both capacitors in the lead-lag network

   ☐ increases the operating frequency
   ☐ decreases the operating frequency
   ☐ has no effect on the operating frequency

5. In a 555 astable multivibrator, the charge path for the timing capacitor is through

   ☐ just one timing resistor
   ☐ both timing resistors
   ☐ neither timing resistor

6. In a 555 astable multivibrator, the discharge path for the timing capacitor is through

   ☐ just one timing resistor
   ☐ both timing resistors
   ☐ neither timing resistor

7. In a 555 astable multivibrator, decreasing the value of either or both of the timing resistors

   - ☐ increases the operating frequency
   - ☐ decreases the operating frequency
   - ☐ has no effect on the operating frequency

8. For a 555-type astable multivibrator, the discharge time (time the waveform is near zero volts) is always greater than the charge time (time the waveform is near the value of the positive supply voltage).

   - ☐ True
   - ☐ False

9. In a 555 astable multivibrator, increasing the value of the timing capacitor

   - ☐ increases the operating frequency
   - ☐ decreases the operating frequency
   - ☐ has no effect on the operating frequency

10. According to the formulas for the operating frequency of Wien-bridge oscillators and 555-type astable multivibrators, increasing the amount of supply voltage should

    - ☐ increase the operating frequency
    - ☐ decrease the operating frequency
    - ☐ have no effect on the operating frequency

SCR OPERATION

Objectives

You will connect basic SCR circuits and note how they can be used to control dc and ac power.

In completing these projects, you will connect and operate circuits, measure voltage levels, observe waveforms with an oscilloscope, draw conclusions, and answer questions about the following items related to SCRs.

- Conditions for starting, sustaining, and stopping the conduction of an SCR in dc and ac circuits
- Distribution of voltages and currents in SCR control circuits
- Reasons for differing levels of load power in SCR control circuits

PROJECT/TOPIC CORRELATION INFORMATION

PROJECT	TEXT CHAPTER	SECTION	RELATED TEXT TOPIC(S)
88 SCRs in DC Circuits	31	31-1	Theory of SCR Operation SCR Circuits
89 SCRs in AC Circuits	31	31-1	SCR Circuits

PROJECT 88

SCR OPERATION
SCRs in DC Circuits

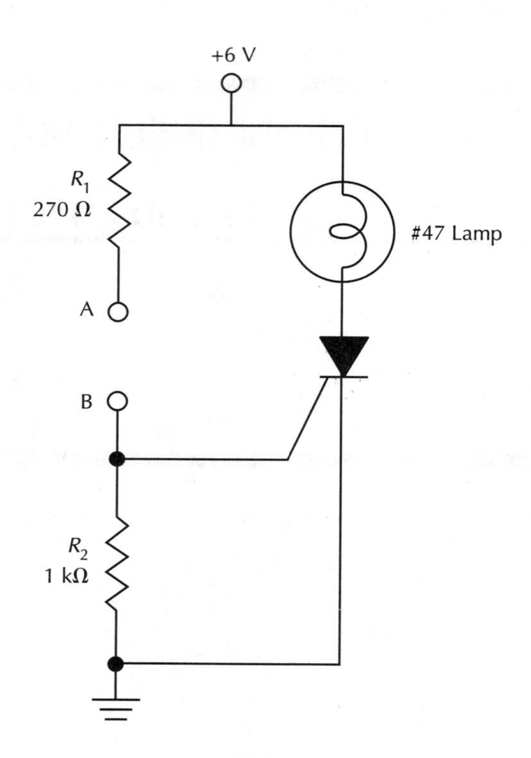

FIGURE 117

PROJECT PURPOSE

The purpose of this project is to demonstrate how an SCR in a dc circuit cannot conduct until it is gated on; and once gated on, remains on until the power source is interrupted.

PARTS NEEDED

- ☐ DMM/VOM
- ☐ Oscilloscope
- ☐ VVPS (dc)
- ☐ CIS
- ☐ 6-V incandescent lamp: #47 (or equivalent)
- ☐ SCR: 2N5060 (or equivalent)
- ☐ Resistors:
  - 270 Ω
  - 1 kΩ

NOTE: Make all meter readings with respect to the common connections for the circuit.

ACTIVITY	OBSERVATION	CONCLUSION
1. Connect the circuit exactly as shown in Figure 117. Make sure the 6-Vdc power is turned **off**. Make sure there is **no connection** between points A and B of the circuit.	—	—
2. Turn on the dc power supply and make sure it is set for 6 volts. **NOTE:** The lamp should not be on at this time. Measure the voltages at the anode and at the gate of the SCR. Record the data in the Observation column.	Anode voltage = ____________ V Gate voltage = ____________ V The lamp is (*on*, *off*) ____________.	The cathode-anode terminals of the SCR in this circuit are (*forward*, *reverse*, *not*) ____________ biased. The cathode-gate terminals are (*forward*, *reverse*, *not*) ____________ biased. Calculate the voltage across the lamp by subtracting the anode voltage (recorded in the Observation column) from the dc supply voltage: Lamp (load) voltage = ______ V. Under the conditions observed here, the SCR must be in its (*on*, *off*) __________ state. This means there is (*current*, *no current*) __________ flowing through the SCR and through the load.
3. With 6 Vdc still applied to the circuit, make a jumper-wire connection between points A and B. Provide the data required in the Observation column.	The lamp is (*on*, *off*) ____________. Anode voltage = ____________ V Gate voltage = ____________ V	The cathode-anode terminals of the SCR in this circuit are (*forward*, *reverse*, *not*) ____________ biased. The cathode-gate terminals are (*forward*, *reverse*, *not*) ____________ biased. Calculate the voltage across the lamp: Lamp (load) voltage = ________ V. Under the conditions observed here, the SCR is in its (*on*, *off*) ______ state. This means there is (*current*, *no current*) __________ flowing through the SCR and through the load.
4. With 6 Vdc still applied to the circuit, remove your jumper-wire connection between points A and B. Provide the data required in the Observation column.	The lamp is (*on*, *off*) ____________. Anode voltage = ____________ V Gate voltage = ____________ V	The cathode-anode terminals of the SCR in this circuit are (*forward*, *reverse*, *not*) ____________ biased. The cathode-gate terminals are (*forward*, *reverse*, *not*) __________ biased.

PROJECT 88 CONTINUED

SCR OPERATION

SCRs in DC Circuits *(Continued)*

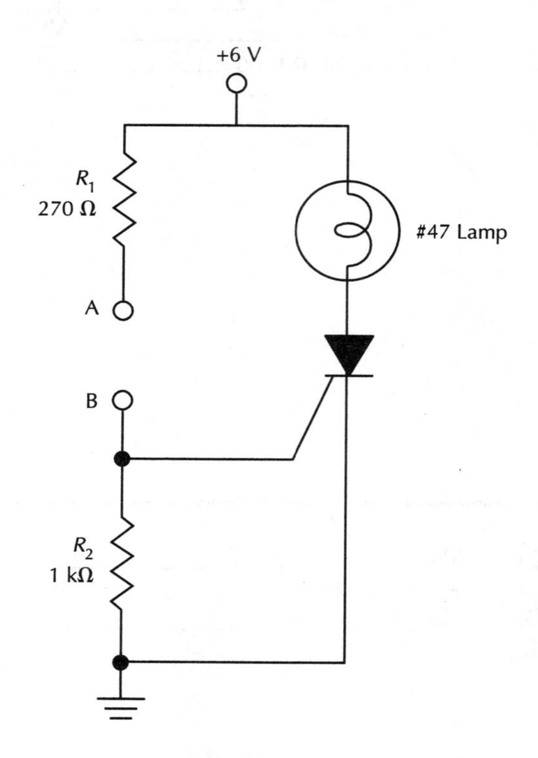

FIGURE 117

PROJECT PURPOSE

The purpose of this project is to demonstrate how an SCR in a dc circuit cannot conduct until it is gated on; and once gated on, remains on until the power source is interrupted.

PARTS NEEDED

- ☐ DMM/VOM
- ☐ Oscilloscope
- ☐ VVPS (dc)
- ☐ CIS
- ☐ 6-V incandescent lamp: #47 (or equivalent)
- ☐ SCR: 2N5060 (or equivalent)
- ☐ Resistors:
  - 270 Ω
  - 1 kΩ

ACTIVITY	OBSERVATION	CONCLUSION
5. Turn off the dc power supply. Monitor the dc power supply output voltage with your voltmeter, and wait for the voltage to drop all the way to zero.	—	—
6. Make sure there is **no connection** between points A and B of the circuit, and turn on the dc power supply. Provide the data required in the Observation column.	The lamp is (*on*, *off*) ____________.	Explain what is necessary to get the SCR conducting once you have restored dc power to the circuit. ________________ ________________

PROJECT 89

SCR OPERATION
SCRs in AC Circuits

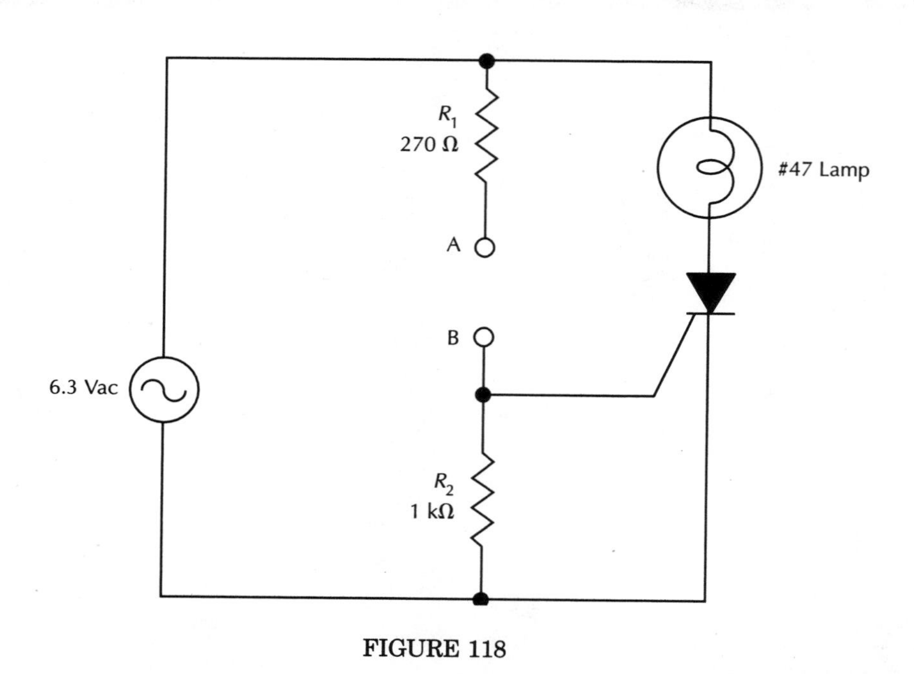

FIGURE 118

PROJECT PURPOSE

The purpose of this project is to demonstrate how an SCR in an ac circuit cannot conduct until it is forward biased between the anode and cathode, and is gated on. You will also see that the SCR is switched off when the applied ac voltage reverses polarity.

PARTS NEEDED

- ☐ DMM/VOM
- ☐ Dual-trace oscilloscope
- ☐ VVPS (dc)
- ☐ 6.3-Vac source
- ☐ CIS
- ☐ 6-V incandescent lamp: #47 (or equivalent)
- ☐ SCR: 2N5060 (or equivalent)
- ☐ Resistors:
  - 270 Ω
  - 1 kΩ

NOTE: Make all oscilloscope readings with respect to the common connections for the circuit.

ACTIVITY	OBSERVATION	CONCLUSION
1. Connect the circuit exactly as shown in Figure 118. Connect one channel of the oscilloscope to the 6.3-Vac source, and connect the second channel of the oscilloscope to the anode of the SCR. Make sure there is **no connection** between points A and B of the circuit.	—	—
2. With ac power applied to the circuit, note the state of the lamp and the waveforms. Record your findings in the Observation column.	The lamp is (*on, off*) ______. V_{ac} source waveform: 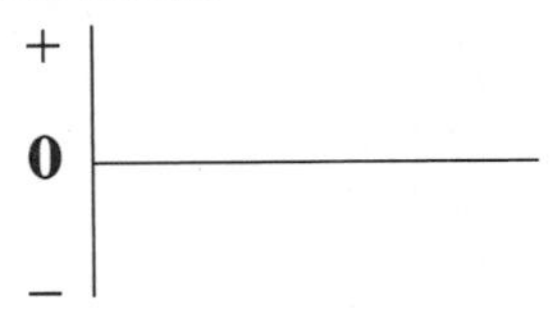 SCR anode waveform: + 0 – V_{ac} voltage = _____V peak-to-peak SCR anode voltage = _________ V peak-to-peak	The SCR is evidently (*conducting on both half cycles, conducting on every other half cycle, not conducting at all*) ______. Explain your answer. ______ In this particular operating state, the SCR is acting as (*an open, a closed*) ______ switch.
3. Connect a jumper wire between points A and B in the circuit. With ac power still applied to the circuit, note the state of the lamp and the waveforms appearing on the oscilloscope. Record your findings in the Observation column.	The lamp is (*on, off*) ______. V_{ac} source waveform: 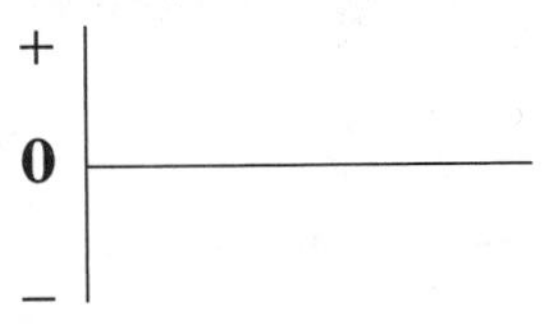 SCR anode waveform: + 0 – V_{ac} voltage = _____V peak-to-peak SCR anode voltage = _________ V peak-to-peak	The SCR is evidently (*conducting on both half cycles, conducting on every other half cycle, not conducting at all*) ______. Explain your answer. ______ In this particular operating state, the SCR is acting as a (*full-wave, half-wave*) ______ rectifier.

PROJECT
89
CONTINUED

SCR OPERATION
SCRs in AC Circuits *(Continued)*

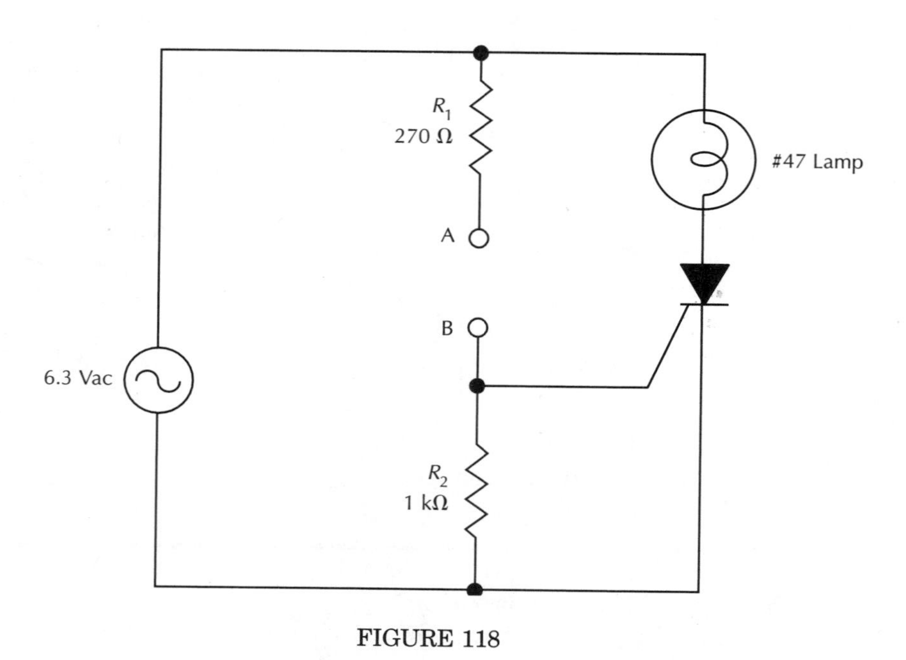

FIGURE 118

PROJECT PURPOSE

The purpose of this project is to demonstrate how an SCR in an ac circuit cannot conduct until it is forward biased between the anode and cathode, and is gated on. You will also see that the SCR is switched off when the applied ac voltage reverses polarity.

PARTS NEEDED

- ☐ DMM/VOM
- ☐ Dual-trace oscilloscope
- ☐ VVPS (dc)
- ☐ 6.3-Vac source
- ☐ CIS
- ☐ 6-V incandescent lamp: #47 (or equivalent)
- ☐ SCR: 2N5060 (or equivalent)
- ☐ Resistors:
  - 270 Ω
  - 1 kΩ

ACTIVITY	OBSERVATION	CONCLUSION

4. Without interrupting the source of ac power, remove the jumper wire you connected between points A and B in the circuit. Note the circuit's response and record your findings in the Observation column.

The lamp is (*on, off*) ____________.

V_{ac} source waveform:

+

0

–

SCR anode waveform:

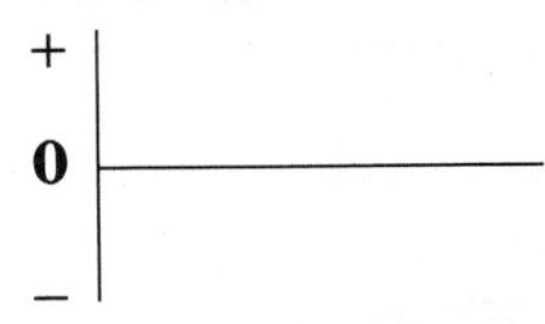

V_{ac} voltage = ______V peak-to-peak

SCR anode voltage = __________ V peak-to-peak

The SCR is evidently (*conducting on both half cycles, conducting on every other half cycle, not conducting at all*) ____________________.

Explain your answer. __________

SCR OPERATION

Complete the following review questions, indicating the appropriate response by placing a check in the box next to the correct answer.

1. In order to begin conduction, an SCR

   - ☐ must be forward biased at the anode, but the gate bias makes no difference
   - ☐ must be forward biased at the anode and gate at the same time
   - ☐ must be forward biased at the gate, but the anode-cathode bias makes no difference

2. In order to sustain conduction of an SCR,

   - ☐ it must remain forward biased at the anode, but the gate bias makes no difference
   - ☐ it must remain forward biased at both the anode and the gate
   - ☐ it must be forward biased at the gate; however, the anode-cathode bias makes no difference

3. In order to stop conduction of an SCR,

   - ☐ the gate must be at zero volts or reverse biased; the anode bias makes no difference
   - ☐ the anode must be at zero volts or reverse biased; the gate bias makes no difference
   - ☐ the gate and the anode must be reverse biased at the same time

4. In an ac circuit, conduction of the SCR automatically stops when the ac waveform reverses the anode voltage from positive to negative.

   - ☐ True
   - ☐ False

5. For an SCR operating in a dc circuit

   - ☐ the load current equals the source current minus the SCR current, and the load voltage equals the voltage drop across the SCR
   - ☐ the load current equals the SCR current, and the load voltage equals the source voltage minus the SCR voltage
   - ☐ the load current equals about one-half the SCR current, and the load voltage about one-half the source voltage

6. For an SCR that is switched on and forward biased in an ac circuit, the load voltage is equal to the instantaneous values of the source voltage minus the forward voltage drop across the SCR.

   - ☐ True
   - ☐ False

7. If an SCR operating in an ac circuit could be gated on 90° into its forward-biasing half cycle (instead of 0°),

   - ☐ the load would have more power applied to it
   - ☐ the load would have less power applied to it
   - ☐ there would be no effect on load power

8. At best, an SCR operating from an ac source can only conduct on alternate half cycles (180° of the full ac waveform).

   - ☐ True
   - ☐ False

9. An SCR that controls an incandescent lamp in a dc circuit allows the lamp to burn at full brightness because

   - ☐ nearly all dc power is being applied to the lamp
   - ☐ the SCR is acting as a closed switch
   - ☐ there is only a small forward voltage drop across the SCR
   - ☐ all of the above

10. An SCR that controls an incandescent lamp in an ac circuit does not allow the lamp to burn at full brightness because

    - ☐ dc power is more energetic than ac power
    - ☐ the SCR cannot allow power more than 180° of the ac waveform to be applied
    - ☐ the forward voltage drop of an SCR is relatively large
    - ☐ all of the above